Introductory

NUCLEAR REACTOR DYNAMICS

Introductory

NUCLEAR REACTOR DYNAMICS

Karl O. Ott
Robert J. Neuhold

AMERICAN NUCLEAR SOCIETY
La Grange Park, Illinois, USA

To our daughters,

Martina and Monika

and

Mary, Lisa, and Monica.

Library of Congress Cataloging in Publication Data

Ott, Karl O. (Karl Otto), 1925–
Introductory nuclear reactor dynamics.

Includes bibliographies and index.
1. Nuclear reactors. 2. Nuclear reactor kinetics.
I. Neuhold, Robert J. II. Title. III. Title: Nuclear reactor dynamics.
QC786.5.086 1985 621.48′3 85-22903
ISBN 0-89448-029-4

ISBN: 0-89448-029-4
Library of Congress Catalog Card Number: 85-22903
ANS Order No. 350011

555 North Kensington Avenue
La Grange Park, Illinois 60525

Printed in the United States of America

CONTENTS

Four: STATIC PERTURBATION THEORY

Five: THE POINT KINETICS EQUATIONS

Six: SOLUTION OF BASIC KINETICS PROBLEMS

Seven: MICROKINETICS

Eight: APPROXIMATE POINT KINETICS

Nine: MEASUREMENT OF REACTIVITIES

PREFACE

Time dependencies of neutron flux that comprise the area of reactor kinetics or dynamics have been important since the field of nuclear engineering came into being. In early years, the time variation of the neutron population as a whole was considered. With the deployment of large power reactors, the space and energy dependencies of these time variations affected the transients and had to be evaluated.

The presentation of the material follows this historical pattern. The "point kinetics" is developed first in three inductive steps that readily lead to the "space and energy dependent kinetics" as a natural extension of point kinetics. The largest part of the material is concerned with evaluation and discussion of the various solutions to kinetics problems. Reactivity measurements and the associated analysis problems are discussed extensively. The treatments are generally restricted to the time domain. Stability theory, Laplace transformations, and the then-needed complex analysis are not covered in this text.

Nearly all of the illustrations have been specifically devised for this text to facilitate insight into various aspects of kinetics and to develop a feel for time-dependent phenomena.

The material of this text has been taught by one of the authors (K.O.O.) for a number of years as a three-hour dual-level introductory course. Parts of the material, however, such as the more sophisticated reactivity measurement analysis and some of the space-energy dependent dynamics, should be deleted from a one-semester introductory course.

Material related to reactor statics is based on *Introductory Nuclear Reactor Statics* by K. O. Ott and W. A. Bezella. The static perturbation theory presented in this volume is an extension of the statics text. Therefore, these texts are companion volumes; they complement each other in the coverage of the neutronics of nuclear reactors.

This course has also been taught by Dr. Robert C. Borg and more recently, in 1981, by Dr. Donald J. Malloy (ANL), who included the analysis of special transients of the Purdue reactor, PUR-1, as assignments—a practice that subsequently became a permanent feature of the course. We are also grateful for his many suggestions for improving this text.

The course was originally conceived as an advanced graduate-level course. Through consideration of the comments and reactions of many students and, in particular, through the close cooperation of the authors, it developed into the present form that emphasizes an introductory presentation. Specific contributions by students are the calculations required for the numerous illustrations of this text; gratefully acknowledged are the contributions by Bob Borg, Bob Burns, Nelson Hanan, Ken Koch, Larry Luck, Don Malloy, Paul Maudlin, and Sai Chi Mo as well as the Computer Science Center of Purdue University. Special thanks are due to Ken Koch for his thorough corrections of the manuscript, to Jerry Andrews for preparing most of the index, and to Sai Chi Mo for the QX1 calculations depicted on the cover.

The authors thank very much Profs. W. E. Kastenberg (UCLA) and D. A. Meneley (University of New Brunswick, Canada) for their reviews and helpful suggestions. Additional thanks go to Dr. W. K. Terry for his careful review of the galley proofs. We are especially grateful for the detailed review, cogent comments, and the large number of suggested improvements by Prof. R. L. Murray; they contributed significantly to the refinement of the entire manuscript.

We also thank Lorretta Palagi of the American Nuclear Society for her skillfull editing and for her continued patience and care in the publishing of this manuscript. In addition, the authors have incurred indebtedness to the secretaries Mmes. Georgia Ehrman for patiently typing the various versions of the manuscript and Lee Harmon for her help in the final stages with the galley and page proofs and the cover.

Special thanks are due the authors' wives, Gunhild and Colleen, for their encouragement, support, and patience throughout the many years of manuscript preparation.

Karl O. Ott
Purdue University

Robert J. Neuhold
U.S. Department of Energy

December 1985

One

BASIC TOPICS AND NOMENCLATURE

1-1 Basic Time-Dependent Phenomena in Nuclear Reactors

Time-dependent phenomena in nuclear reactors may be subdivided into three distinctively different classes. The time constants of the individual phenomena in the three classes differ by orders of magnitude. In addition, different physical phenomena are treated in each class; it is not just a case of the same phenomenon occurring at different speeds:

1. *short time phenomena,* which typically occur in time intervals of milliseconds to seconds; in special cases, the time intervals may extend to many minutes
2. *medium time phenomena,* which occur over hours or days corresponding to the mean buildup and decay times of certain fission products that strongly affect the reactivity
3. *long time phenomena,* with variations developing over several months or years.

These time-dependent phenomena basically include changes in the neutron flux as well as causally related changes in the reactor system, i.e., composition or temperature. The causal relationship between the neutron flux and the physical reactor system may occur in either direction; that is, changes in the composition or temperature of the system may cause a change in the flux, or changes in the flux may alter the composition or temperature and thus the density and absorption characteristics of the system. Changes in the system can also be externally induced, for example, by the motion of an independent neutron source, or of control or shutdown rods, resulting in neutron flux changes. If the flux changes cause changes in the reactor and these changes subsequently "act back" on the flux, the phenomenon is termed "feedback" (see Chapters 10 and 11).

The "short time phenomena" include more or less rapid changes in the neutron flux due to intended or accidental changes in the system. The latter changes may influence the flux through feedback. Short time phenomena include flux transients important for:

1. accident analysis and safety
2. experiments with time-dependent neutron fluxes
3. reactor operation, such as startup, load change, and shutdown (even though some startup procedures may take hours)
4. analysis of stability with respect to neutron flux changes.

"Medium time phenomena" are generally associated with the buildup, burnup, and beta decay of two fission products (^{135}Xe and ^{149}Sm) in thermal reactors. These two fission products have very high thermal neutron capture cross sections and thus require special attention in thermal reactors. Since the treatment of medium time phenomena is methodologically different from kinetics, it is addressed in an appendix and is not included in the main text (see App. A).

"Long time phenomena" include particularly the burnup and buildup of fissionable isotopes, as well as the buildup, beta decay, and burnup of most of the fission products. In the fast neutron energy range, the cross sections of all fission products are so small that they do not affect the flux and the reactivity as strongly as in thermal reactors.

Other long time phenomena occurring in reactors that have only a minimal effect on the neutron flux include swelling of the structural material, changes in the fuel pellets due to burnup, etc.

Since short, medium, and long time phenomena are physically different phenomena resulting in different sets of equations, different concepts and solution approaches are utilized. These are the strongest reasons for separating these time phenomena into three different categories with different names.

1-2 Kinetics Versus Dynamics

The nomenclature used in textbooks and publications for the different categories of time-dependent phenomena in nuclear reactors is not unique. The two basic names in use are *kinetics* and *dynamics*. A few authors (e.g., Ref. 1) subsume *all* time-dependent phenomena under "dynamics," including burnup and buildup of isotopes. Most authors, however, consider long time phenomena to represent a separate category, namely "fuel cycle problems." The latter widely used practice is followed in this text.

Essentially three names are in use for the class of *short* time phenomena:

1. kinetics, for the entire class of short time phenomena (e.g., Ref. 2)
2. dynamics, also for the entire class of short time phenomena (e.g., Ref. 3)
3. dynamics, as a general heading for the entire class of short time phenomena, with two subheadings: (a) kinetics, for short time phenomena without feedback and (b) dynamics, in the narrower sense, for short time phenomena with feedback (e.g., Ref. 4).

The latter nomenclature is used in this book since it is probably in more widespread use and the structure of the problem seems to suggest such a nomenclature. It is convenient to have a special name for the range of problems (kinetics problems, kinetics equations) in which only the time behavior of neutrons need be considered. If feedback is important, the system of kinetics equations must be completed by another, often larger, set of equations describing the various feedback effects. It is convenient to have a different name for the completed set of equations (dynamics equations, dynamics problems). Since the completed set of equations describes the general problem, dynamics is also used as a general heading.

This text is concerned with the short time variations of the neutron flux as a function of time, i.e., with the typical topics of kinetics and dynamics. Stability analysis, even though it is also applied to short term variations, is not covered here since a number of good texts exist in this area; e.g., Refs. 3, 5, and 6.

Review Questions

1. Describe briefly the three categories of time dependencies occurring in nuclear reactors.
2. State three areas of kinetics or dynamics applications.
3. Considering the nomenclature, what do various authors consider to be the subject of "dynamics" or "kinetics"?
4. What is the main difference in the balance equations for the neutron flux in reactor dynamics and fuel cycle analysis?

REFERENCES

1. H. S. Isbin, *Introductory Nuclear Reactor Theory,* Reinhold Publishing Corp., New York (1963).
2. Milton Ash, *Nuclear Reactor Kinetics,* McGraw-Hill Book Co., New York (1965).

3. D. L. Hetrick, *Dynamics of Nuclear Reactors,* The University of Chicago Press, Chicago (1971).
4. G. I. Bell and Samuel Glasstone, *Nuclear Reactor Theory,* Van Nostrand Reinhold Co., New York (1970).
5. Z. Akcasu, G. S. Lellouche, and L. M. Shotkin, *Mathematical Methods in Nuclear Reactor Dynamics,* Academic Press, New York (1971).
6. L. E. Weaver, *Reactor Dynamics and Control,* American Elsevier Publishing Company, Inc., New York (1968).

Two

DELAYED NEUTRONS

In static reactor problems, the prompt and the delayed fission neutrons always appear together as the total number of fission neutrons. The fact that some of the fission neutrons are emitted as delayed neutrons has no consequence for static problems. The time dependence of the neutron flux, however, may be strongly influenced by the small fraction of the fission neutrons that is produced after time delays of about a second to several minutes. Even though this fraction is small, it may play a dominant role in many kinetics phenomena. Therefore, the production rates of delayed neutrons must be considered in detail.

In $^{235}U/^{238}U$-fueled thermal reactors at higher burnup, as well as in plutonium-fueled fast reactors, several isotopes contribute comparably or significantly to the production of delayed neutrons. Thus, the delayed neutron data of several fissionable isotopes must be considered. A further complication appears in fast reactors where fissions are induced by neutrons in a wide energy range. Therefore, the isotope as well as the energy dependencies of the delayed neutron production need to be considered and are discussed in this chapter. Particular emphasis is placed on features of the delayed neutron data that have an impact on the theoretical formulation of the kinetics equations. The discussion of the physics of delayed neutron production is limited to a survey of the material necessary to understand the presented delayed neutron data. For a detailed discussion of the relevant physics, see Refs. 1 through 4.

2-1 Production of Prompt and Delayed Neutrons Through Nuclear Fission

Nearly all of the neutrons produced as a result of a fission process are emitted "promptly," i.e., without noticeable delay. The prompt neutrons are emitted by the "direct" fission products immediately after the fission process since the excitation energy of the fission product nuclei

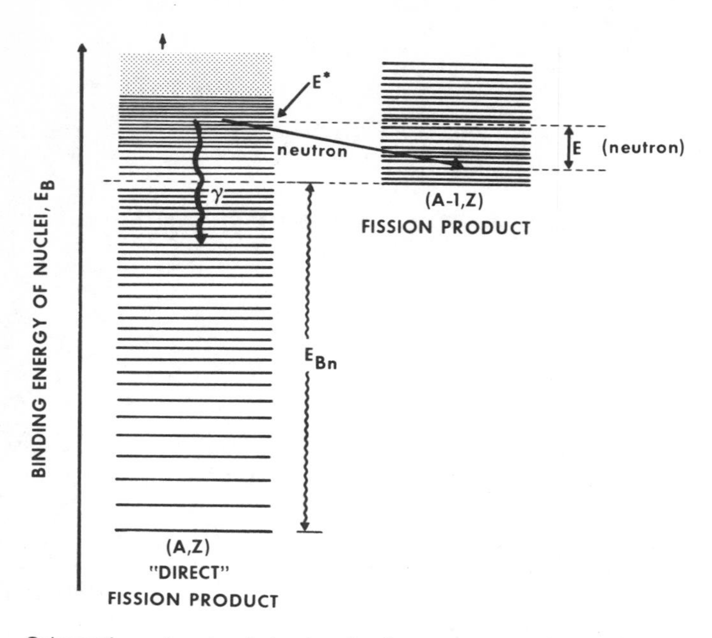

Fig. 2-1. Schematic nuclear level structure leading to prompt neutron emission.

is generally much larger than the separation energy for a neutron (i.e., neutron binding energy E_{Bn}). Figure 2-1 shows schematically the nuclear levels of the two nuclei involved in the production of prompt neutrons. The difference in the total binding energies of the nuclides (A,Z) and $(A - 1,Z)$ is merely the neutron binding energy E_{Bn} (note that $A = Z+N$). The excitation energy E^* in Fig. 2-1 is $>E_{Bn}$. The typical decay time of such excited states may be 10^{-15} s or less, which is completely negligible.

Since E^* is larger than the neutron binding energy, there is competition between the neutron and gamma-ray emissions that leads to a strong straggling in the actual number of neutrons emitted from a fission event. Some fission product pairs may—with a certain probability—lose their excitation energy only by emission of gamma rays. In other cases, as many as six prompt neutrons may be emitted after a fission event.[1] The *average* number of prompt fission neutrons is denoted by ν_p.

After the fission process and the production of prompt neutrons is completed, the reaction products consist of

1. two radioactive nuclei
2. several prompt neutrons
3. several gamma rays.

None of the resultant fission product nuclei can directly emit an additional neutron. However, some of the fission product nuclei may decay into daughter nuclei for which the excitation energy is larger than the neutron binding energy; i.e., after a beta decay, the typical level scheme situation depicted in Fig. 2-1 may in some cases occur again. Such nuclei may then immediately emit a neutron, which has been delayed by the comparatively long time it took such a nucleus to undergo a beta decay. Practically, the entire delay comes from the beta decay and only a negligibly small amount comes from the actual neutron emission. This interpretation of the mechanism of the production of delayed neutrons was already given in 1939 by Bohr and Wheeler.[5]

The terms "precursor" nuclei and "emitter" nuclei are commonly used to denote the specific parent and daughter nuclei that produce delayed neutrons. Figure 2-2 shows schematically the comparative nuclear level structure of the three nuclei involved in the production of delayed neutrons. The exceptional feature that allows the emission of a delayed neutron is that E_β^{max}, the maximum electron energy in the beta decay, must be larger than the energy E_{Bn} required to separate a neutron, i.e.,

$$E_\beta^{max} > E_{Bn} \quad . \tag{2.1}$$

Some beta decays can then lead to excited states in the emitter nucleus above E_{Bn}. From such an excited state, the nucleus has a chance to emit

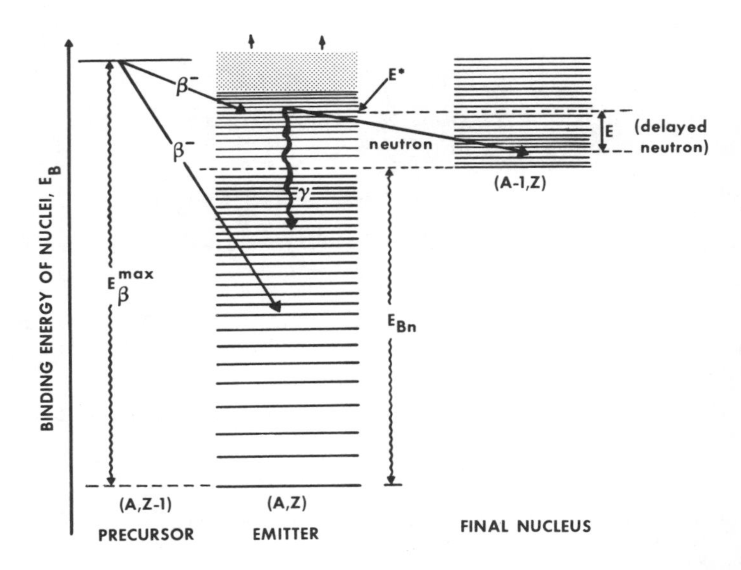

Fig. 2-2. Schematic nuclear level structure leading to the emission of delayed neutrons.

a neutron, or alternatively emit a gamma quantum and assume temporarily an excited state below E_{Bn}. Consequently, not all of the possible "precursor" nuclei decays will lead to delayed neutron production. Since only those "parent nuclei" decays that really yield delayed neutrons are relevant for dynamics, "precursors" are defined as only that fraction of the parent nuclei that yields delayed neutrons.

It is thus obvious that the production of delayed neutrons after the precursor nucleus has decayed into the emitter nucleus is basically the same physical process as the production of prompt neutrons. However, the difference in the excitation energy and the neutron binding energy in delayed neutron emitter nuclei is normally much smaller than in the highly excited "direct" fission products. Therefore, delayed neutrons are emitted on the average with considerably smaller energies than prompt neutrons.

2-2 Total Delayed Neutron Yields

The average total number of neutrons, ν, which is composed of prompt and delayed neutrons, is expressed in terms of the relevant yields (neutrons per fission):

$$\nu = \nu_p + \nu_d \quad . \tag{2.2}$$

The production of delayed neutrons was often described in the older literature by the delayed neutron "fraction,"[a]

$$\beta^{ph} = \frac{\nu_d}{\nu} \quad . \tag{2.3}$$

The delayed neutron fraction, ν_d, appears then as a derived quantity, i.e., as the product of ν and β^{ph}:

$$\nu_d = \nu\beta^{ph} \quad , \tag{2.4}$$

rather than directly as in Eq. (2.2). There are, however, two good reasons to favor the description in Eq. (2.2):

1. The yield ν_d is the more basic quantity and the use of the more basic quantity is preferred in general, unless there is an advantage to replacing it by a less basic or artificially defined quantity.
2. Theoretical arguments[6] as well as newer measurements[7–9] show

---

[a]The somewhat clumsy notation β^{ph} = β-physical is used to denote the ratio of Eq. (2.3); the short notation β is reserved for the frequently occurring β-effective, often denoted by β_{eff}.

that ν_d in the range in which nearly all delayed neutrons are produced ($E \lesssim 4$ MeV) is practically independent of the energy of the fission-inducing neutron, i.e.,

$$\nu_d(E) = \nu_d \simeq \text{constant for } 0 \lesssim E \lesssim 4 \text{ MeV} \quad .$$

The prompt yield ν_p (and thus ν, because of the small contribution of ν_d) shows, however, a significant energy variation in this range (an increase of ~20%).[10] Therefore, by using Eq. (2.3), an artificial energy dependence would be introduced into the description of delayed neutrons.

Figure 2-3 shows[7,8] schematically the energy dependence of ν_d for ^{235}U and ^{239}Pu. The lack of energy dependence below ~4 MeV is obvious. At higher energies, $\nu_d(E)$ decreases in general agreement with theoretical expectations[11] and recent measurements.[7,8] This decrease is in strong disagreement with earlier measurements, which showed an increase by a factor of ~2 for all isotopes (see, for example, the compilation in Ref. 12). Since the fraction of delayed neutrons produced above 4 MeV is very small, ν_d is assumed independent of the incident neutron energy in the entire range of neutronics calculations.

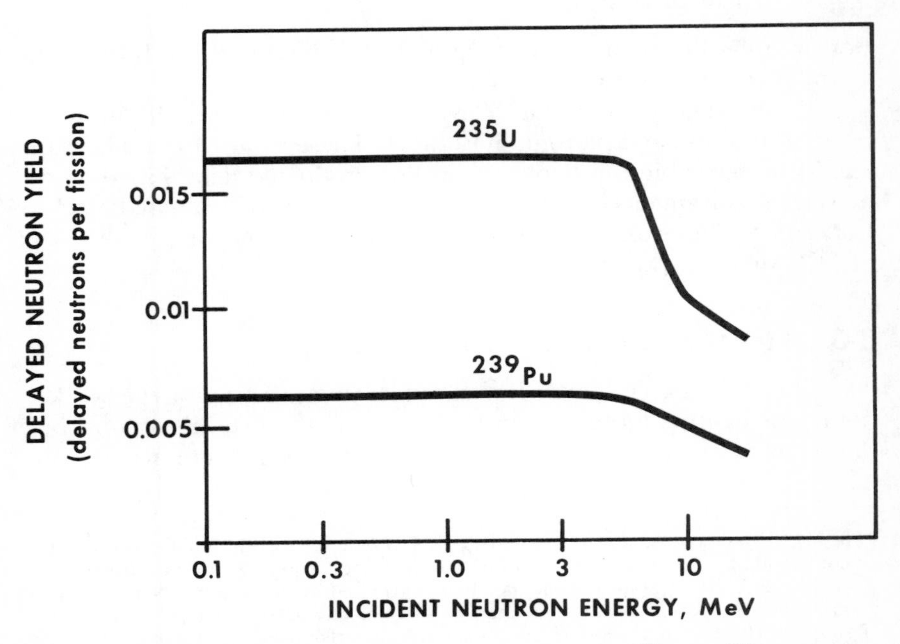

Fig. 2-3. Energy dependence of the delayed neutron yields of ^{235}U and ^{239}Pu (Ref. 7).

TABLE 2-I
Total Delayed Neutron Yields
(Delayed neutrons per fission)

Fission Nuclide	ν_d	Reference
^{233}U	0.0070 ±0.0004	14
^{235}U	0.0165 ±0.0005	14
^{238}U	0.0412 ±0.0017	14
^{239}Pu	0.0063 ±0.0003	14
^{240}Pu	0.0088 ±0.0006	14
^{241}Pu	0.0154 ±0.0015	15
^{242}Pu	0.016 ±0.005	8

The total delayed neutron yields, ν_{di}, are significantly different for different fissioned isotopes. There seem, however, to be two regularities[13] (compare also Table 2-I):

1. The total delayed neutron yield increases with increasing atomic weight for a given element.
2. The total delayed neutron yield decreases with increasing number of protons.

Both regularities seem to be suggested by corresponding shifts in the abundances of fission products.

Table 2-I gives the total delayed neutron yields for the most important uranium and plutonium isotopes. These values are used for the results presented in this book. Since 1969, evaluated data[9] based on the newest measurements have been compiled in the U.S. Evaluated Nuclear Data File.[b] The essential change is the 10 to 15% increase in the yield of ^{238}U that was reported earlier.[7]

2-3 Yields of Delayed Neutron Groups

The "normal" situation is that the binding energy of the weakest bound neutron is larger, often much larger, than the maximum beta decay energy of the neighboring isotope. However, there are ~40 out

---

[b]The U.S. Evaluated Nuclear Data File-B (ENDF/B) is a library of evaluated nuclear data recommended by a Cross-Section Evaluation Working Group for use in nuclear calculations. The ENDF/B library is compiled and distributed by the National Neutron Cross-Section Center located at the Brookhaven National Laboratory in Upton, New York. Release of the fourth version of the library called ENDF/B-IV was completed early in 1975. The ENDF/B-V library became available in 1983 for limited use.

of ~500 different fission product nuclides that have the special property required for being a delayed neutron emitter [Eq. (2.1)]. All 40 precursors have different lifetimes and therefore the corresponding neutrons will appear at different delay times. The consequence of the different precursor lifetimes is that the corresponding delayed neutrons will have a different effect on the time dependence of the neutron flux. Therefore, the different precursor lifetimes have to be accounted for in some way.

The most direct way to consider the effect of the differences in the precursor lifetimes is to take all of them into account individually. This approach has three serious drawbacks:

1. The lifetimes, and the abundancies in particular, of many precursors are not known accurately enough.
2. Even if the lifetimes and abundancies of all precursors were known accurately, their inclusion into the theoretical formulation of the kinetics problem would lead to an impractically lengthy set of differential equations.
3. Several precursors are themselves products of beta decays. The individual description of these precursors requires the inclusion of double or even multiple beta decay schemes. It has not been shown that this added complexity has any practical significance.

Since the use of the individual precursors in kinetics appears too complicated and still too inaccurate, an approach in which the delayed neutron data of all precursors are properly condensed needed to be found that allows a simple and yet sufficiently accurate theoretical description of reactor kinetics.

A suitable approach for obtaining condensed delayed neutron data was suggested historically by the fact that experimental knowledge on delayed neutron production was available primarily in the form of average source curves, $S_d(t)$. These source curves are obtained by exposing a sample of fissionable material to a very short neutron pulse, which instantaneously produces—through fission—a large number of precursors. The decay of these precursors results in the source of delayed neutrons $S_d(t)$ (Ref. 14). If n_f is the number of fissions that occurred in the sample during the flux pulse, then,

$$\nu_d n_f = \text{total number of precursors} \quad .$$

The term $S_d(t)$ describes the decay rate of these precursors and thus the production rate of delayed neutrons. Figure 2-4 shows a typical example of such a source curve, based on the data given in Ref. 14. The decay rate of the single isotope ^{87}Br with its mean decay time of ~80 s is also shown. The experimental points in the actual experiment are average

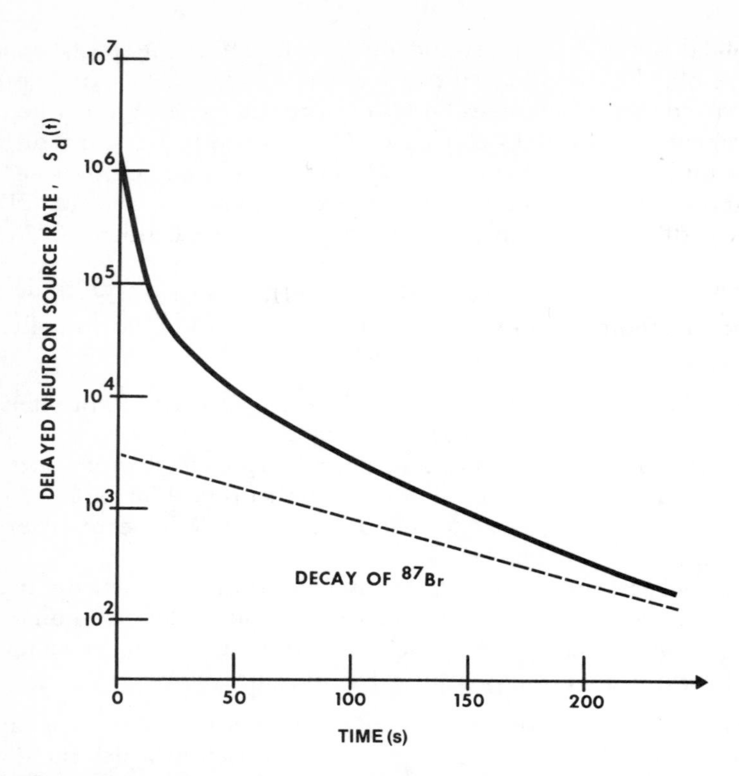

Fig. 2-4. Delayed neutron source following a fission pulse (based on data from Ref. 14).

values of 80 measured decay curves. Physically, the decay curve $S_d(t)$ is a superposition of contributions from all precursors including those that are themselves beta decay products. In Ref. 14, it was shown that this complicated superposition $S_d(t)$ can be fairly accurately represented by just six exponential functions without including a buildup term, representing production of precursors through beta decay:

$$S_d(t) = n_f \sum_{k=1}^{6} \nu_{dk} \lambda_k \exp(-\lambda_k t) \quad . \tag{2.5}$$

Equation (2.5) represents a delayed neutron source that results from the decay of six "average" groups[c] of precursors, all of them produced at

[c]The notation "group" may often lead to confusion, especially when different energy groups are used in the same equation or sentence. Therefore, we use consistently the notation "delay group" and reserve the notation "group" for energy group. To avoid this kind of confusion, some authors use the shorter term "family" for "delay group" introduced in Ref. 16.

$t = 0$. The "delayed group" yields ν_{dki} and the average "delayed group" decay constants λ_{ki} are obtained for each isotope i by a least-squares fit of the right side of Eq. (2.5) to an experimentally determined left side:

$$S_{di}(t) = n_{fi} \sum_{k=1}^{6} \nu_{dki} \lambda_{ki} \exp(-\lambda_{ki} t) \quad . \tag{2.6}$$

This six delay group structure is in general use in reactor kinetics. For a complete presentation of the data as well as comparisons with the results of other authors, see Ref. 1.

As long as ^{235}U is the dominant fuel in a thermal reactor, practically all of the delayed neutrons are produced from fission products of ^{235}U. Then the delayed neutron sources $S_d(t)$ consist of six terms as in Eq. (2.6). In high-burnup light water reactor fuel and in a fast breeder reactor (FBR), however, there are several nuclides that contribute to the delayed neutron source. The ^{238}U contribution in an FBR is comparable to the ^{239}Pu contribution because its delayed neutron yield is about seven times larger than that for ^{239}Pu, thereby largely compensating for its lower fission rate. Other contributions such as those of ^{240}Pu and ^{241}Pu may also be of practical significance. The contribution of ^{242}Pu is generally small. In any case, if more than one isotope is of importance, the simplicity of the condensed representation, Eq. (2.5), is lost. The total delayed neutron source has to be found as a sum of all isotopic contributions; e.g., in the case of a neutron pulse:

$$S_d(t) = \sum_i n_{fi} \sum_k \nu_{dki} \lambda_{ki} \exp(-\lambda_{ki} t) \quad . \tag{2.7}$$

Fortunately, the isotope dependence of the decay constants is not very pronounced. Most of them differ only within their statistical errors, which become apparent from Table 2-II in which the delay group decay constants of ^{239}Pu are compared with those of ^{235}U, ^{238}U, and ^{240}Pu. Therefore, the use of a single set of isotope-independent decay constants, λ_k, can be expected to fit the experimental delayed neutron sources with only an insignificant loss of accuracy compared to Eq. (2.6), i.e.,

$$S_{di}(t) = n_{fi} \sum_k \nu_{dki} \lambda_k \exp(-\lambda_k t) \quad . \tag{2.8}$$

The results of such a fit[17] are given in Table 2-III. The delay group yields in Table 2-III are used for the calculations presented in this book. The great advantage of having a single set of decay constants is that the precursor concentrations and the delayed neutron sources can be readily summed up for all isotopes, and macroscopic cross sections can be introduced in the same way as that for prompt fission neutrons (see Sec. 3-2A). Although the yields of Table 2-III have been suggested for use

TABLE 2-II
Comparison of Precursor Decay Constants in Six Delay Groups for Various Isotopes*
(s^{-1})

Delay Group	^{239}Pu	^{240}Pu
1	0.0129 ±0.0002	0.0129 ±0.0004
2	0.0311 ±0.0005	0.0313 ±0.0005
3	0.134 ±0.003	0.135 ±0.011
4	0.331 ±0.012	0.333 ±0.031
5	1.26 ±0.12	1.36 ±0.21
6	3.21 ±0.26	4.04 ±0.78
	^{238}U	^{235}U
1	0.0132 ±0.0003	0.0127 ±0.0002
2	0.0321 ±0.0006	0.0317 ±0.0008
3	0.139 ±0.005	0.115 ±0.003
4	0.358 ±0.014	0.311 ±0.008
5	1.41 ±0.07	1.40 ±0.081
6	4.02 ±0.21	3.87 ±0.37

*From Ref. 14.

in fast reactors, they are also applicable to thermal reactor problems. For thermal reactor problems, one would prefer the use of the λ_k values of ^{235}U instead of ^{239}Pu, as applied in Ref. 17. However, the differences in the λ's are so small that they cannot affect the accuracy of the resulting fit.

Instead of the consistent fit represented by Eq. (2.8), other more or less inconsistent eliminations of the isotopic dependence of the decay constants are in widespread use.

The longer living precursors that contribute significantly to delayed neutron production are probably all known (delay groups 1 to 3). Delay group yields may therefore also be defined by the composition of individual precursor yields within a proper lifetime bracket.[4] Some neutron emission probabilities and precursor production abundancies are known experimentally. Most of them, however, have to be estimated theoretically so that composite yields can be calculated. In Ref. 4, results for delay groups 1 to 4 of the isotopes ^{235}U, ^{238}U, and ^{239}Pu are presented. The values for groups 1 and 2 agree for all isotopes with the values of Table 2-III within their error limits (delay group 1 is assumed to consist of ^{87}Br only, and for delay group 2 there are only three major contributors). For delay groups 3 and 4, the agreement is still remarkably good even though most of the values are outside of the error limits of

TABLE 2-III
Delayed Neutron Group Yields Consistently Fit to a Single Set of Decay Constants*

Delay Group	ν_{dki}, $\frac{\text{neutrons}}{\text{fission}} \times 100$	
	^{233}U	^{235}U
1	0.053 ±0.003	0.060 ±0.005
2	0.197 ±0.012	0.364 ±0.013
3	0.175 ±0.025	0.349 ±0.024
4	0.212 ±0.013	0.628 ±0.015
5	0.047 ±0.014	0.179 ±0.014
6	0.016 ±0.006	0.070 ±0.005
	^{238}U	^{232}Th
1	0.049 ±0.005	0.143 ±0.012
2	0.540 ±0.027	0.776 ±0.046
3	0.681 ±0.092	0.843 ±0.099
4	1.526 ±0.096	2.156 ±0.103
5	0.836 ±0.033	0.838 ±0.053
6	0.488 ±0.036	0.204 ±0.034
	^{239}Pu	^{240}Pu
1	0.024 ±0.002	0.028 ±0.003
2	0.176 ±0.009	0.237 ±0.016
3	0.136 ±0.014	0.162 ±0.046
4	0.207 ±0.012	0.314 ±0.038
5	0.065 ±0.007	0.106 ±0.026
6	0.022 ±0.003	0.039 ±0.008
	^{241}Pu	^{242}Pu
1	0.019 ±0.004	0.036 ±0.013
2	0.369 ±0.010	0.263 ±0.086
3	0.276 ±0.045	0.270 ±0.075
4	0.534 ±0.089	0.607 ±0.177
5	0.310 ±0.043	0.279 ±0.055
6	0.032 ±0.007	0.145 ±0.059

*From Ref. 17. The new delayed neutron yields are to be used with the following set of isotope-independent decay constants (s^{-1}):

$\lambda_1 = 0.0129 \pm 0.0002$ $\quad \lambda_2 = 0.0311 \pm 0.0005$
$\lambda_3 = 0.134 \pm 0.003$ $\quad \lambda_4 = 0.331 \pm 0.012$
$\lambda_5 = 1.26 \pm 0.12$ $\quad \lambda_6 = 3.21 \pm 0.26$.

Table 2-III. Full agreement cannot be expected because of the inherent difference in the definitions of the composite yields[4] and the yields resulting from a least-squares fit of a source curve.[14]

2-4 Emission Spectra of Delayed Neutrons

The excitation energies of delayed neutron emitters are much smaller than the excitation energies of prompt neutron emitters (see Sec. 2-1). Whereas the prompt fission neutrons have an average energy near 2 MeV and their spectrum extends even beyond 10 MeV, the delayed neutrons are emitted with a much smaller average energy and thus with a softer spectrum than prompt neutrons.

Since different delay groups are composed of different precursors with different energy ranges ($E^* - E_{Bn}$; see Fig. 2-2), the emission spectra depend on the delay group; they are denoted by $\chi_{dk}(E)$.

If the energy dependence of the delayed neutron emission is included in Eq. (2.5), the following energy-dependent delayed neutron source results from a fission pulse for a single isotope:

$$S_d(E,t) = n_f \sum_k \chi_{dk}(E) \nu_{dk} \lambda_k \exp(-\lambda_k t) \quad . \tag{2.9}$$

Let us consider three extreme cases:

1. the "initial" emission spectrum:

$$S_d(E,0) \propto \sum_k \chi_{dk}(E) \nu_{dk} \lambda_k \tag{2.10}$$

2. the stationary emission spectrum, which results from integrating Eq. (2.9) over all times:

$$S_d^{stat}(E) \propto \sum_k \chi_{dk}(E) \nu_{dk} \tag{2.11}$$

3. the asymptotic emission spectrum, i.e., the case in which the longest living group provides the overriding contribution:

$$S_d^{as}(E) \propto \chi_{dk}(E) \text{ with } k = 1 \quad . \tag{2.12}$$

The comparison of these extreme cases shows that the composition of the total delayed neutron emission spectrum may vary strongly with time and the static spectrum may have a different composition compared to any time-dependent one.

For many years, the only emission spectra for delayed neutron groups were those measured by Batchelor and McK. Hyder[18] for the first four delay groups of ^{235}U fissioned by thermal neutrons. More recently, Fieg[19] measured delayed neutron emission spectra for ^{238}U, ^{239}Pu, and also ^{235}U, with fission initiated by 14-MeV neutrons. The yields at 14 MeV are significantly lower than those in the dominating fission range (see Fig. 2-3). This suggests possible delayed neutron energy spectrum differences between the fission range and 14 MeV. Fortunately, the dependencies on isotopes and on the initiating neutron energy appear to

TABLE 2-IV
Average Energies for Various Delay Groups, Isotopes, and Fissioning Energies

Isotope	Average Energies in Delay Groups (keV) 1	2	3	4	5	Reference
^{235}U (thermal)	250 ±20	460 ±10	405 ±20	450 ±20	—	18
^{235}U (thermal)	277 ±28	484 ±48	447 ±45	432 ±43	—	19
^{235}U (14 MeV)	286 ±29	458 ±46	432 ±43	480 ±48	—	19
^{238}U (14 MeV)	278 ±28	468 ±47	443 ±44	425 ±43	382 ±40	19
^{239}Pu (14 MeV)	296 ±30	481 ±48	411 ±41	430 ±43	—	19

be relatively small, as indicated by the average energies presented in Table 2-IV, which shows that the average energies for each delay group are about the same for all three isotopes and for thermal and 14-MeV fission (i.e., all error bands overlap). This independence is not surprising—delay group 1 consists only of the prescursor ^{87}Br in all cases. All other groups consist of contributions of several precursors. A variation in the emission spectra and, thus, in the average energies of the other delay group neutrons is due to small variations in the relative abundance of different precursor isotopes. Also, the variation of the average energy with decreasing lifetime of the precursor groups appears to be very minimal, hardly outside of the experimental errors (10% for Ref. 19 and 2 to 8% for Ref. 18). Therefore, the common practice of using the emission spectrum measured for delay group 4 for delay groups 5 and 6 also seems to be reasonable. However, the emission spectra should be extended down to low energies,[20] i.e., below the experimental lower limit of ~100 keV in Ref. 18. A set of such χ_d values in a half lethargy group structure is presented in Table 2-V along with the total neutron emission spectrum.[21] Evaluations[9] for ENDF/B-IV recommend the use of the data of Ref. 19 for all delay groups except delay group 2. Recently reported measurements[22] are recommended for delay group 2.

Homework Problems

1a. Calculate the average energy, $\overline{E}_k$, of the delayed neutron groups 1 through 4, using the emission spectra $\chi_{dk}(E)$ given in Table 2-V.

b. Compare these values with $\overline{E}$ for the total $\chi(E)$ given in the same table and with the $\overline{E}_k$ of Table 2-IV.

TABLE 2-V
Delayed and Total Neutron Emission Spectra*

Group Number	u_g	E_g (eV)	χ_1	χ_2	χ_3	χ_4, χ_5, χ_6	χ
—	0	10×10^6	—	—	—	—	—
1	0.5	6.065×10^6	—	—	—	—	0.0325
2	1.0	3.679×10^6	—	—	—	—	0.1217
3	1.5	2.231×10^6	—	—	—	—	0.2109
4	2.0	1.353×10^6	—	—	—	0.009	0.2230
5	2.5	0.821×10^6	0.009	0.022	0.012	0.025	0.1728
6	3.0	0.498×10^6	0.021	0.066	0.070	0.062	0.1105
7	3.5	0.302×10^6	0.093	0.295	0.191	0.184	0.0628
8	4.0	0.183×10^6	0.088	0.110	0.094	0.128	0.0316
9	4.5	0.111×10^6	0.156	0.098	0.097	0.157	0.0168
10	5.0	0.674×10^5	0.174	0.108	0.156	0.109	0.0083
11	5.5	0.409×10^5	0.171	0.107	0.142	0.109	0.0040
12	6.0	0.248×10^5	0.131	0.088	0.118	0.099	0.0019
13	6.5	0.150×10^5	0.121	0.079	0.080	0.089	0.0009
14	7.0	0.912×10^4	0.020	0.013	0.015	0.015	0.0004
15	7.5	0.553×10^4	0.010	0.009	0.012	0.010	0.0002
16	8.0	0.335×10^4	0.005	0.004	0.005	0.004	0.0001
17	8.5	0.203×10^4	0.001	0.001	—	—	—
18	9.0	0.123×10^4	—	—	—	—	—
19	9.5	749	—	—	—	—	—
20	10.0	454	—	—	—	—	—
21	10.5	275	—	—	—	—	—
22	11.0	167	—	—	—	—	—
23	11.5	101	—	—	—	—	—
24	12.0	61.4	—	—	—	—	—
25	12.5	37.3	—	—	—	—	—
26	13.0	22.6	—	—	—	—	—

*From Ref. 21.

2. Estimate the fractions of prompt and delayed neutrons that can cause fast fission in ^{238}U, assuming a sharp threshold at 1 MeV. Use the $\chi(E)$ and $\chi_{dk}(E)$ as given in Table 2-V and apply linear interpolation in the group around 1 MeV.

3a. Give the general formula for the source density of the precursor production in delay group k for a mixture of isotopes using isotope-dependent ν_{dk}. Introduce (average) macroscopic $\nu_d \Sigma_f$ in analogy to $\nu \Sigma_f$ in reactor statics.

b. Complete the source term of problem 3a with a radioactive decay term and obtain the balance equation for the precursor density $C_k(\mathbf{r},t)$.

c. Discuss the problems that will arise if the precursor decay constants depend on the isotopes present in the mixture.

Review Questions

1. What are delayed neutron precursors?
2. Give the typical level scheme of precursor-emitter-daughter nuclei that leads to the production of delayed neutrons.
3. What is the known or estimated approximate number of possible precursors?
4. Name the quantities used to describe the physics of precursors and delayed neutrons.
5. Give approximately the total precursor yields of ^{235}U, ^{238}U, and ^{239}Pu.
6. Sketch the typical E dependence of $\nu_d(E)$ for ^{235}U and ^{239}Pu.
7. What is the typical difference in the emission spectra of prompt and delayed neutrons? Give typical values for the respective average emission energies.
8. How many delayed neutron groups (families) are generally used per fissioning isotope?
9. What is the disadvantage of using isotope-dependent decay constants?
10. Describe briefly the experiment for which delayed neutron data are derived.
11. Describe briefly the way in which delayed neutron data are obtained from the experimental results.
12. Give the approximate mean lifetime of the slowest and the fastest decaying precursor group.

REFERENCES

1. G. R. Keepin, *Physics of Nuclear Reactors,* Addison-Wesley Publishing Co., Reading, Massachusetts (1965).
2. *Proc. Panel Delayed Fission Neutrons,* International Atomic Energy Agency, Vienna (1968).
3. *Proc. 2nd IAEA Symp. Physics and Chemistry of Fission,* International Atomic Energy Agency, Vienna (1969).
4. L. Tomlinson, "Theory of Delayed Neutron Physics," *Nucl. Technol.,* **14,** 42 (1972).
5. N. Bohr and J. A. Wheeler, "The Mechanism of Nuclear Fission," *Phys. Rev.,* **56,** 426 (1939).
6. G. R. Keepin, "Interpretation of Delayed Neutron Phenomena," *J. Nucl. Energy,* **7,** 13 (1958).
7. C. F. Masters, M. M. Thorpe, and D. B. Smith, "The Measurement of Absolute Delayed Neutron Yields from 3.1- and 14.9-MeV Fission," *Nucl. Sci. Eng.,* **36,** 202 (1969).

8. M. S. Krick and A. E. Evans, "The Measurement of Total Delayed-Neutron Yields as a Function of Energy of the Neutron Inducing Fission," *Nucl. Sci. Eng.*, **47,** 311 (1972).
9. S. A. Cox, "Delayed Neutron Data—Review and Evaluation," ANL/NDM-5, Neutron Data and Measurement Series, Argonne National Laboratory (Apr. 1974).
10. G. R. Keepin, *Physics of Nuclear Reactors,* p. 54, Addison-Wesley Publishing Co., Reading, Massachusetts (1965).
11. R. J. Tuttle, "Delayed Neutron Data for Reactor Physics Analysis," AI-AEC-13044, Research Development Report, U.S. Atomic Energy Commission (Nov. 1972); see also G. R. Keepin, "Physics of Delayed Neutrons—Recent Experimental Results," *Nucl. Technol.*, **14,** 53 (1972).
12. G. R. Keepin, *Physics of Nuclear Reactors,* p. 98, Addison-Wesley Publishing Co., Reading, Massachusetts (1965).
13. G. R. Keepin, *Physics of Nuclear Reactors,* p. 99, Addison-Wesley Publishing Co., Reading, Massachusetts (1965).
14. G. R. Keepin, T. F. Wimmett, and R. K. Zeigler, "Delayed Neutrons from Fissionable Isotopes of Uranium, Plutonium, and Thorium," *Phys. Rev.*, **107,** 1044 (1957).
15. S. A. Cox, "Delayed-Neutron Studies from the Thermal-Neutron Induced Fission of Pu^{241}," *Phys. Rev.*, **123,** 1735 (1961).
16. D. A. Meneley, L. C. Kvitek, and D. M. O'Shea, "MACH 1, A One-Dimensional Diffusion Theory Package," ANL-7223, Argonne National Laboratory (1966).
17. J. E. Cahalan and K. O. Ott, "Delayed Neutron Data for Fast Reactor Analysis," *Nucl. Sci. Eng.*, **50,** 208 (1973).
18. R. Batchelor and H. R. McK. Hyder, "The Energy of Delayed Neutrons from Fission," *J. Nucl. Energy,* **3,** 7 (1956).
19. G. Fieg, "Measurement of Delayed Fission Neutron Spectra of U-235, U-238, and Pu-239 with Proton Recoil Proportional Counters," *Proc. Natl. Topl. Mtg. New Developments in Reactor Physics and Shielding,* Kiamesha Lake, New York, September 12–15, 1972, CONF-720901, p. 687, U.S. Atomic Energy Commission, Technical Information Center (1972); see also G. Fieg, "Measurements of Delayed Fission Neutron Spectra of ^{235}U, ^{238}U, and ^{239}Pu with Proton Recoil Proportional Counters," *J. Nucl. Energy,* **26,** 585 (1972).
20. S. Ramchandran and K. R. Birney, "Impact of the New LASL Delayed Neutron Data on FTR Design," *Proc. Natl. Topl. Mtg. New Developments in Reactor Physics and Shielding,* Kiamesha Lake, New York, September 12–15, 1972, CONF-720901, p. 660, U.S. Atomic Energy Commission, Technical Information Center (1972).
21. A 26-Group Constant Set for MACH 1 (based on ENDF/B-I data), Argonne National Laboratory (1970).
22. S. Shalev and F. M. Cuttler, "The Energy Distribution of Delayed Fission Neutrons," *Nucl. Sci. Eng.*, **51,** 52 (1973).

Three

PRELIMINARY FORMULATION OF THE POINT KINETICS EQUATIONS

The diffusion theory balance equation for the time-dependent neutron flux depends on space and energy. As in reactor statics, the complete neutron balance equation would also depend on angle. For many applications, especially in kinetics, it is neither necessary nor feasible to solve such a complicated set of equations including the space-dependent set of precursor equations. The equations can be condensed into purely time-dependent ones, the so-called "point kinetics" equations. The term "point kinetics" is widely used to indicate the formal independence of space and energy, which results from integration over the respective variables.

The point kinetics equations are derived in preliminary formulations in this chapter. These formulations allow meaningful insight into kinetics concepts and provide the basis for the "exact" treatment presented in Chapter 5. It should be clearly understood, however, that the preliminary formulations are too inaccurate for practical applications. This holds true particularly for the one-group formulations, which are merely included to illustrate the meaning of the otherwise more complicated formulas.

3-1 Intuitive Point Kinetics—The Basic Concepts

3-1A The Prompt Neutron Balance Equation

The simple case in which an off-critical reactor contains no independent source and all delayed neutrons are assumed to be promptly emitted is considered first. Off-criticality then means that the production and the loss of neutrons do not balance. Since the production and loss of prompt neutrons are proportional to the number of neutrons present, the difference in production and loss is also proportional to $n(t)$, the number of neutrons in the reactor. This off-balance thus leads to a time variation, which is proportional to $n(t)$:

$$\dot{n}(t) = \frac{dn(t)}{dt} \propto n(t) \quad , \tag{3.1}$$

or

$$\dot{n}(t) = \alpha n(t) \quad , \tag{3.2}$$

where α represents the "inverse period." If α is a constant, the solution of Eq. (3.2) is given by

$$n(t) = n_0 \exp(\alpha t) \quad . \tag{3.3}$$

The neutron population will exponentially increase or decrease depending on the sign of α. A stationary flux can be obtained only for $\alpha = 0$.

Since α results from the off-balance of neutron production and loss, it can be expressed readily as the difference between the corresponding reaction rates. In a reactor consisting of a single homogeneous composition where all neutrons (not just thermal neutrons) can be treated by a simple one-group theory with leakage described by a DB^2 term,[a] one obtains:

$$\frac{dn(t)}{dt} = \nu\Sigma_f\hat{\phi}(t) - (\Sigma_a + DB^2)\hat{\phi}(t) \quad . \tag{3.4}$$

All terms in Eq. (3.4) have dimensions of neutrons per second since $\hat{\phi}$(neutrons cm/s) represents a spatial integral of $\phi(\mathbf{r},t)$ over the entire homogeneous reactor composition.

In the one-group model, the total integrated flux and the number of neutrons are related by an average velocity $\bar{v}$:

$$\hat{\phi} = \bar{v}n \quad . \tag{3.5}$$

By inserting Eq. (3.5) into Eq. (3.4), either n or $\hat{\phi}$ may be eliminated. Retaining $\hat{\phi}$ yields:

$$\frac{1}{\bar{v}}\frac{d\hat{\phi}}{dt} = (\nu\Sigma_f - \Sigma_a - DB^2)\hat{\phi} \quad , \tag{3.6}$$

and dividing Eq. (3.6) by $\nu\Sigma_f$ yields the familiar multiplication constant

[a]This model is subsequently referred to as a "simple" one-group model. For a more complete discussion of such models, see, for example, Ref. 1.

k (k_{eff} is denoted here by k):

$$k = \frac{\nu\Sigma_f}{\Sigma_a + DB^2} \quad , \tag{3.7}$$

and from it, the reactivity:

$$\rho = \frac{k - 1}{k} \quad . \tag{3.8}$$

Equation (3.6) is then reduced to the following simple form:

$$\frac{1}{\bar{v}\nu\Sigma_f}\frac{d\hat{\phi}}{dt} = \rho\hat{\phi} \quad . \tag{3.9}$$

3-1B Average Neutron Generation Time and Lifetime

The coefficient on the left side of Eq. (3.9) is called the "average neutron generation time," Λ:

$$\Lambda = \frac{1}{\bar{v}\nu\Sigma_f} \quad . \tag{3.10}$$

The name "generation time" was chosen since Λ describes the average time between two birth events in successive generations: The quantity $1/\Sigma_f$ is the mean free path for fission, i.e., the average distance a neutron travels from its birth to a fission event. Then,

$$\frac{1}{\bar{v}}\left[\frac{\text{s}}{\text{cm}}\right] \cdot \frac{1}{\Sigma_f}\,[\text{cm}] = \Delta t_f\,[\text{s}] \tag{3.11}$$

is the average time between the birth of a neutron and a fission event it may cause. Since a fission reaction produces ν neutrons, the average time between birth and the birth of a single neutron in the next generation is obtained by dividing Eq. (3.11) by ν:

$$\frac{1}{\nu}\Delta t_f = \frac{1}{\bar{v}}\frac{1}{\nu\Sigma_f} = \Lambda \quad . \tag{3.12}$$

In a similar way, the quantity

$$\ell = \frac{1}{\bar{v}}\frac{1}{\Sigma_a + DB^2} \tag{3.13}$$

can be interpreted as the "average neutron lifetime" in the reactor. The average traveling distance between birth and absorption or leakage ("death") of a neutron is $1/(\Sigma_a + DB^2)$. Division by $\bar{v}$ gives the required time, i.e., the average lifetime ℓ.

The lifetime in an infinite system, i.e., in a system without leakage, is readily obtained from Eq. (3.13) by setting $B^2 = 0$:

$$\ell_\infty = \frac{1}{\bar{v}}\frac{1}{\Sigma_a} \quad . \tag{3.14}$$

The interpretations of lifetime and generation time are illustrated in Fig. 3-1.

The relation between the lifetime ℓ in a finite system and the lifetime in an infinite system ℓ_∞ is readily derived from Eqs. (3.13) and (3.14). Their relation is the same as the relation between k and k_∞ in the simple one-group model:

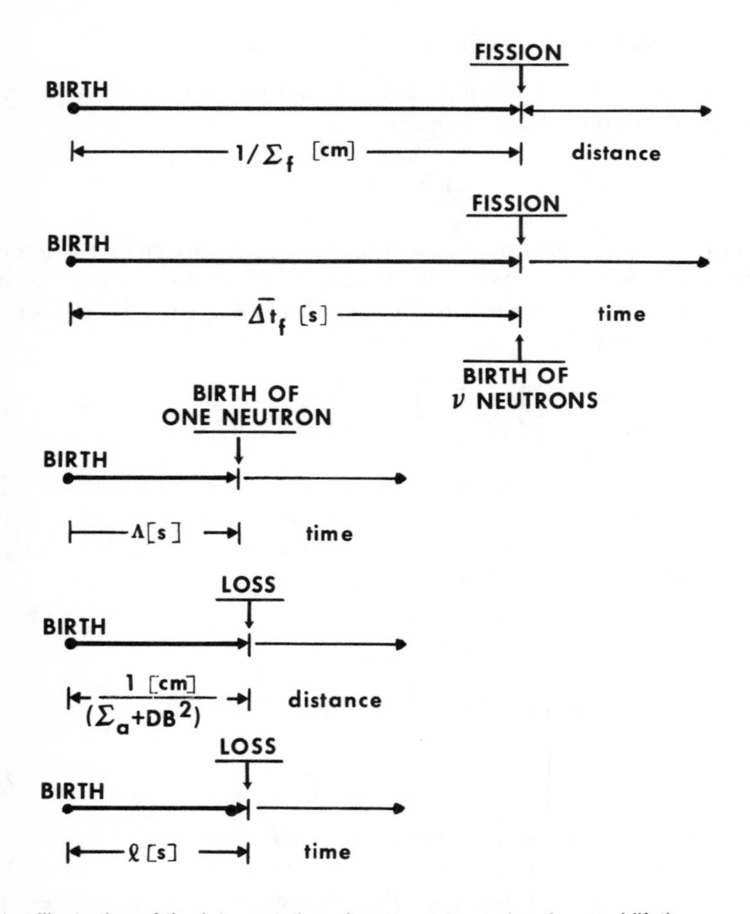

Fig. 3-1. Illustration of the interpretation of neutron generation time and lifetime.

$$k = \frac{\nu\Sigma_f}{\Sigma_a}\frac{1}{1 + L^2B^2} = \frac{k_\infty}{1 + L^2B^2} \tag{3.15a}$$

and

$$\ell = \frac{\ell_\infty}{1 + L^2B^2} \ , \tag{3.15b}$$

where L is the one-group diffusion length, $\sqrt{D/\Sigma_a}$.

The one-group generation time is not directly related to the size of the system. Only the one-group lifetime contains DB^2. However, lifetime and generation time are directly proportional to each other and the proportionality constant is just the multiplication constant k [Eq. (3.7)]:

$$\ell = k \cdot \Lambda \ . \tag{3.16}$$

Equation (3.16) shows that the average lifetime is smaller than the average neutron generation time if the reactor is subcritical, and ℓ is larger than Λ if the reactor is supercritical. This is easily understood from the following arguments, which pertain to prompt fission neutrons only. If the lifetime of a neutron is shorter than the time required to reproduce itself, $\ell < \Lambda$, the neutron population in a reactor will decrease, $k < 1$; if a neutron lives long enough to reproduce more than just one generation, $\ell > \Lambda$, the population will increase. For a critical reactor, $k = 1$, ℓ and Λ are the same.

By introducing either the generation time or the lifetime into the prompt neutron kinetics equation, Eq. (3.9), one obtains:

$$\frac{d\hat{\phi}}{dt} = \frac{\rho}{\Lambda}\hat{\phi} \tag{3.17}$$

or

$$\frac{d\hat{\phi}}{dt} = \frac{\Delta k}{\ell}\hat{\phi} \ . \tag{3.18}$$

The two equations are equivalent. In some of the older literature, a kinetics equation based on the "number" of neutrons is employed. However, the quantities appearing in Eq. (3.17), i.e., flux, reactivity, and generation time, are generally preferred in the modern literature. The neutron flux is used in all reactor physics areas such as statics, static off-criticality problems, and the fuel cycle. It is, therefore, consistent and convenient to also use the flux in dynamics. The reactivity has better additivity properties than Δk. This is important in dynamics because the reactivity is composed of combinations of various initiating and feedback effects. The use of ρ instead of Δk then enforces the use of the generation time rather than lifetime.

The quantities Λ and ℓ are often called "prompt" neutron generation or lifetime, respectively, since the delay of the precursor decay and, thus, the delay in generating the delayed neutrons, is conceptually disregarded. This does not affect the results, since Λ is introduced only as an abbreviation of the factor in front of the time derivative term. Its meaning as "generation time" is merely an interpretation of this factor; it is not used in calculations in any consequential manner.

Since prompt and delayed neutrons, once generated with a certain energy, have the same lifetime in a reactor, the use of the adjective "prompt" in front of the neutron lifetime could be misleading. It is therefore not used in this text. The same holds for the generation time.

3-1C The Effect of Delayed Neutrons—The Intuitive Point Kinetics Equations

Only prompt neutron production is directly associated with the fission process. The delayed neutrons appear after beta decay of precursors. To account for this split, the total ν in the first term of the right side of Eq. (3.6) is replaced by ν_p, the prompt yield, and the delayed neutrons are added in the form of the beta decay rate of their precursors:

$$\frac{1}{\bar{v}}\frac{d\hat{\phi}}{dt} = (\nu_p\Sigma_f - \Sigma_a - DB^2)\hat{\phi} + \sum_k \lambda_k \hat{C}_k \quad . \tag{3.19}$$

The delayed neutron precursor spatial distribution is integrated over the reactor and designated $\hat{C}_k$ in the same way as the other terms in Eq. (3.19). The reactivity is introduced by adding and subtracting $\nu_d\Sigma_f$ inside the parentheses on the right side of Eq. (3.19) and dividing by $\nu\Sigma_f$; one obtains:

$$\frac{1}{\bar{v}\nu\Sigma_f}\frac{d\hat{\phi}}{dt} = \left[\frac{\nu\Sigma_f - (\Sigma_a + DB^2)}{\nu\Sigma_f} - \frac{\nu_d\Sigma_f}{\nu\Sigma_f}\right]\hat{\phi} + \frac{1}{\nu\Sigma_f}\sum_k \lambda_k \hat{C}_k$$

and thus

$$\Lambda\frac{d\hat{\phi}}{dt} = (\rho - \beta)\hat{\phi} + \frac{1}{\nu\Sigma_f}\sum_k \lambda_k \hat{C}_k \quad , \tag{3.20}$$

with ρ given by its one-group approximation and β by

$$\beta = \frac{\nu_d\Sigma_f}{\nu\Sigma_f} = \frac{\sum_i (\nu_d\Sigma_f)_i}{\sum_i (\nu\Sigma_f)_i} \quad . \tag{3.21}$$

The "effective delayed neutron fraction," $\beta_{eff} = \beta$, can also be expressed in terms of contributions of the six delay groups:

$$\beta = \sum_{k=1}^{6} \beta_k \quad , \tag{3.22a}$$

with

$$\beta_k = \frac{\nu_{dk}\Sigma_f}{\nu\Sigma_f} \quad . \tag{3.22b}$$

The effective fraction of delayed neutrons, β, at this level of sophistication consists of the ratio of the delayed and total neutron production, expressed by the respective macroscopic cross sections.

Dividing Eq. (3.20) by Λ gives the typical form of the kinetics equation:

$$\frac{d\hat{\phi}}{dt} = \frac{\rho-\beta}{\Lambda}\hat{\phi} + \frac{1}{\Lambda\nu\Sigma_f}\sum_k \lambda_k \hat{C}_k \quad . \tag{3.23}$$

Equation (3.23) must be completed by equations describing the balance between production and decay of precursors:

$$\frac{d\hat{C}_k}{dt} = -\lambda_k\hat{C}_k + \nu_{dk}\Sigma_f\hat{\phi} \quad , \; k = 1, ..., 6 \quad . \tag{3.24}$$

The first term in Eq. (3.24) describes the decay; the second one the production of precursors. Note that ν_{dk} is used for two different meanings in Eqs. (3.22b) and 3.24. In Eq. (3.22b), ν_{dk} denotes the group yield of delayed neutrons and in Eq. (3.24), the group yield of precursors. Both yields are numerically the same due to the definition of precursors.

Equations (3.23) and (3.24) represent a set of seven differential equations for the seven unknown functions $\hat{\phi}(t)$ and $\hat{C}_k(t)$ with $k = 1, ..., 6$, and are termed the *intuitive point kinetics equations:*

$$\frac{d\hat{\phi}}{dt} = \frac{\rho-\beta}{\Lambda}\hat{\phi} + \frac{1}{\Lambda\nu\Sigma_f}\sum_k \lambda_k \hat{C}_k \tag{3.25a}$$

and

$$\frac{d\hat{C}_k}{dt} = -\lambda_k\hat{C}_k + \nu_{dk}\Sigma_f\hat{\phi} \quad . \tag{3.25b}$$

In the precursor balance equation, Eq. (3.25b), the precursor source, $\hat{S}_{pk}$, appears in its direct form, i.e., as

$$\hat{S}_{pk} = \nu_{dk}\Sigma_f\hat{\phi} \quad . \tag{3.26}$$

In some of the older literature, S_{dk} is rewritten as

$$\hat{S}_{pk} = \beta_k \nu \Sigma_f \bar{v} n \quad , \tag{3.27a}$$

or by using Eq. (3.12) as

$$\hat{S}_{pk} = \frac{\beta_k}{\Lambda} n \quad . \tag{3.27b}$$

This formulation seems to imply an inverse dependence of $\hat{S}_{pk}$ on Λ, which does not appear in the direct formulation, Eq. (3.26).

3-2 One-Group Point Kinetics

The intuitive point kinetics approach successfully provided the *basic form* of the equations as well as the fundamental *integral concepts,* which determine the time dependence of the neutron flux: ρ, β, and Λ. The definitions obtained for these integral parameters in the previous section are merely illustrations. To obtain better definitions of the integral parameters and to assess more clearly the applied assumptions, the derivation of the kinetics equations must start from an adequate neutronics equation as it is done in this section. However, important aspects, such as the adjoint flux weighting, are not yet included. So the equations derived in this section are not yet the ones that are practically applied; the complete derivation is deferred to Chapter 5, since the perturbation theory introduced in Chapter 4 is needed for the derivation of the practically applied equations.

3-2A The Diffusion Approximation as the Basis of Reactor Kinetics

The multigroup diffusion theory with appropriate cell-averaged group constants is a relatively good approximation for reactor statics problems. Deviations from the multigroup diffusion theory can be expected around the core reflector or core blanket interfaces and throughout fast reactor blankets. These spatial areas in which the diffusion theory is inaccurate have only a low "importance" for the reactivity. Since the reactivity essentially determines the time dependence of the flux, these inaccuracies of the diffusion theory are of lesser significance for dynamics than for statics problems. The derivation of the kinetics equation is thus based on the diffusion equations except for the treatment of control rods. The energy-dependent form of the diffusion equation rather than the multigroup version is employed to make the derivation more transparent. Operator notation is presented in App. B.

If the stationary neutron balance is perturbed at some area in space, the result will in general be a time dependence of the entire neutron flux. This is described by the time-dependent diffusion equation:

$$\frac{1}{v}\frac{\partial\phi(\mathbf{r},E,t)}{\partial t} = (\mathbf{F}_p - \mathbf{M})\Phi(\mathbf{r},E,t) + S_d(\mathbf{r},E,t) + S(\mathbf{r},E,t) \quad . \tag{3.28}$$

Equation (3.28) is the space-dependent analog to Eq. (3.25a); an independent source S has been included for completeness. The neutron velocity on the left side is proportional to the square root of E. The left side describes the change of the number of neutrons per $cm^3 \cdot s \cdot dE$. This change is composed of the corresponding changes of the three neutron sources and loss rates of the right side: $\mathbf{F}_p\Phi$, S_d, and S describe the sources of prompt, delayed, and independent source neutrons, respectively; $\mathbf{M}\Phi$ gives the rate of neutron loss due to absorption and leakage per $cm^3 \cdot s \cdot dE$.

The operator $\mathbf{F}_p$ describes the production of prompt neutrons only, whereas $\mathbf{M}$ is defined in the same way as it is for static problems:

$$\mathbf{F}_p\Phi = \sum_i \chi_{pi}(E)\int_{E'} \nu_{pi}(E')\Sigma_{fi}(\mathbf{r},E',t)\phi(\mathbf{r},E',t)\, dE' \quad ; \tag{3.29a}$$

$$\mathbf{M}\Phi = -\nabla\cdot D(\mathbf{r},E,t)\nabla\phi(\mathbf{r},E,t) + \Sigma_t(\mathbf{r},E,t)\phi(\mathbf{r},E,t) - \int_{E'} \Sigma_s(\mathbf{r},E'\rightarrow E,t)\phi(\mathbf{r},E',t)\, dE' \quad ; \tag{3.29b}$$

and

$$S_d(\mathbf{r},E,t) = \sum_k \lambda_k C_k(\mathbf{r},t)\chi_{dk}(E) \quad . \tag{3.29c}$$

The flux is dependent on position, energy, and time:

$$\Phi = \phi(\mathbf{r},E,t) \quad ; \tag{3.29d}$$

$C_k(\mathbf{r},t)$ denotes the concentration of the precursors resulting from fissions of all isotopes; λ_k and χ_{dk} are the corresponding decay constants and emission spectra of the delayed neutron source.

The isotope dependence of χ_{pi} is for the most part neglected: The prompt neutron emission spectra of ^{238}U and of the four plutonium isotopes that occur together in fast reactors are not significantly different; $\chi_{pi}(E)$ of ^{235}U differs somewhat from the emission spectra of the plutonium isotopes. For the prompt fission source, disregarding the isotope dependence of χ_p gives:

$$\mathbf{F}_p\Phi = \chi_p(E)\int_0^\infty \nu_p\Sigma_f(\mathbf{r},E',t)\phi(\mathbf{r},E',t)\, dE' \quad , \tag{3.30}$$

with the macroscopic cross section given by:

$$\nu_p\Sigma_f(\mathbf{r},E,t) = \sum_i \nu_{pi}(E)\Sigma_{fi}(\mathbf{r},E,t) \quad . \tag{3.31}$$

Equations (3.29) must be completed by the balance equations for the precursors. In most reactor dynamics problems, the physical transport of precursors can be neglected. The corresponding balance equations need not be complicated by a precursor transport term, i.e., the precursors can be assumed to decay at the location where they are produced. This approximation is often called the assumption of "stationary fuel." Then, the balance equation for the precursor concentrations produced by all isotopes is given in full analogy to Eq. (3.25b) by:

$$\frac{\partial C_k(\mathbf{r},t)}{\partial t} = -\lambda_k C_k(\mathbf{r},t) + \int_0^\infty \nu_{dk}\Sigma_f(\mathbf{r},E',t)\phi(\mathbf{r},E',t)\, dE' \quad , \tag{3.32}$$

with

$$\nu_{dk}\Sigma_f(\mathbf{r},E,t) = \sum_i \nu_{dki}\Sigma_{fi}(\mathbf{r},E,t) \quad , \tag{3.33}$$

in full analogy to Eq. (3.31).

Without the isotope independence of λ_k and $\chi_{dk}(E)$, a set of balance equations for each isotope (e.g., 36 balance equations for six fissionable isotopes) would need to be considered.

A typical dynamics problem in which the precursors may not decay at the point where they have been produced would involve melting or motion of fuel, or at least motion of gaseous fission products (^{87}Br). The description of this kind of precursor transport has been incorporated into some dynamics programs.[2]

The quantities of $\nu_{dk}\Sigma_f(E)$ show a very pronounced increase with increasing energy around 1 MeV in fast breeder reactors due to the large contribution of ^{238}U. This peculiar energy dependence is illustrated in Fig. 3-2 by plotting the contributions of ^{239}Pu and ^{238}U in a 1:8 mixture of these two isotopes:

$$\nu_{dk}\Sigma_f(E) = (\nu_{dki}N_i)_{239}\left[\sigma_{f239}(E) + \frac{(\nu_{dki}N_i\sigma_{fi})_{238}}{(\nu_{dki}N_i)_{239}}\right] \quad . \tag{3.34}$$

Figure 3-2 shows the two contributions to the bracket, the latter one for the six delay groups. The pair of curves, $8\sigma_{f238}(E)$ and $\sigma_{f239}(E)$, indicates the relative contributions to the total fission neutron source, which is less pronounced than the relative contributions to the delayed neutron source. The fairly regular increase with the delay group index (k) reflects the corresponding increase in the ν_{dk} ratios for ^{238}U and ^{239}Pu, as they can be determined from Table 2-III. The general implications of the

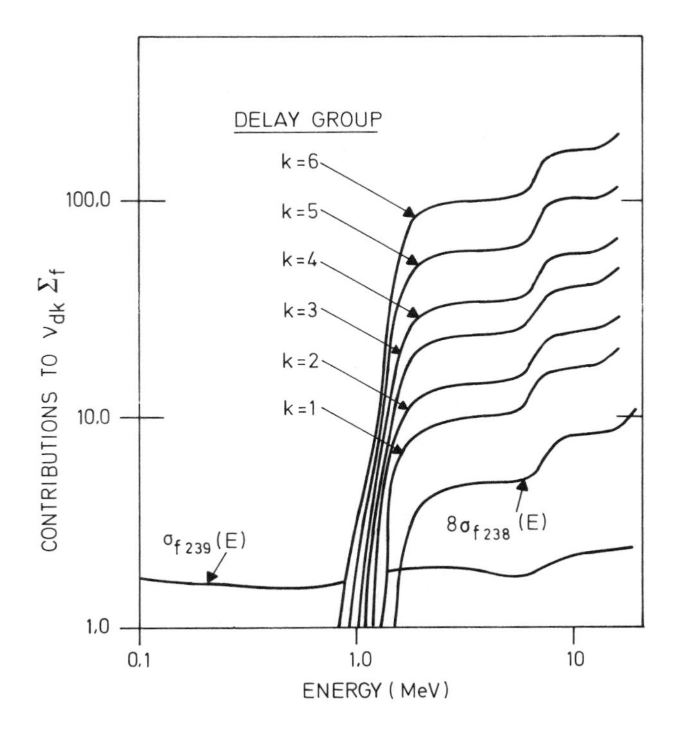

Fig. 3-2. Energy dependence of the relative contributions of ^{239}Pu and ^{238}U to $\nu_{dk}\Sigma_f(E)$; see Eq. (3.34).

increasing contribution of ^{238}U with the decreasing lifetime of the precursors is briefly discussed in Sec. 5-1E.

3-2B Derivation of the One-Group Point Kinetics Equations

To avoid complications that are unnecessary at this level of sophistication, three simplifications are introduced (the general derivation is presented in Chapter 5):

1. The time dependence of the flux $\phi(\mathbf{r},E,t)$ is assumed to be separable from its space and energy dependence:

$$\phi(\mathbf{r},E,t) = p(t)\psi(\mathbf{r},E) \quad . \tag{3.35}$$

2. The local leakage loss is formally described by

$$D(\mathbf{r},E)B^2(\mathbf{r},E)\phi(\mathbf{r},E,t) \quad , \tag{3.36}$$

where $B^2(\mathbf{r},E)$ is to be calculated from the initial flux distribution.
3. The term $\Sigma_f(\mathbf{r},E)$ is assumed independent of time.

In Eq. (3.35) the flux is separated into a purely time-dependent amplitude function, $p(t)$, and a time-independent shape function, $\psi(\mathbf{r},E)$. For convenience, this shape function is set equal to the initial flux assuming a stationary state for $t < 0$:

$$\psi(\mathbf{r},E) = \phi_0(\mathbf{r},E) \quad . \tag{3.37a}$$

The initial condition for $p(t)$ then becomes

$$p(0) = p_0 = 1 \quad . \tag{3.37b}$$

Using the separation (3.35) and Eq. (3.37a), the time dependence of the total reactor power is given by:

$$P(t) = p(t)C_{fsh}\int_V\int_0^\infty \Sigma_f(\mathbf{r},E)\phi_0(\mathbf{r},E)\,dE\,dV \quad , \tag{3.38}$$

where C_{fsh} denotes the conversion between fission and stationary heat production; $C_{fsh}^{-1} \simeq 3 \times 10^{10}$ fission/W·s. In most cases, the shape of the fission rate depends only weakly on time; thus, the flux amplitude function $p(t)$ can often be considered a good approximation to the *relative* power transient [a fact indicated by the notation $p(t)$ for the amplitude function]:

$$p(t) \simeq \frac{P(t)}{P_0} \quad . \tag{3.39}$$

The first step in the derivation of the one-group kinetics equations is to insert the three simplifications, Eqs. (3.35), (3.36), and (3.37), into Eq. (3.28) and integrate with respect to energy:

$$\int_0^\infty \frac{\psi(\mathbf{r},E)}{v(E)}\,dE\,\frac{dp}{dt} = \left\{\int_0^\infty \nu_p\Sigma_f(\mathbf{r},E)\psi(\mathbf{r},E)\,dE \right.$$
$$\left. - \int_0^\infty [\Sigma_a(\mathbf{r},E,t) + D(\mathbf{r},E)B^2(\mathbf{r},E)]\psi(\mathbf{r},E)dE\right\}p(t)$$
$$+ \sum_k \lambda_k C_k(\mathbf{r},t) + \int_0^\infty S(\mathbf{r},E,t)\,dE \quad . \tag{3.40}$$

Note that all normalized emission spectra $\chi_p(E)$ and $\chi_{dk}(E)$ disappear with the energy integration; the scattering term in $\Sigma_t\phi$ (i.e., $\Sigma_s\phi$) cancels with the scattering integral since all neutrons that "disappear" in scattering ($\Sigma_s\phi$) "reappear" again, through the scattering integral.

If Eq. (3.40) is integrated with respect to space, and divided by the integrated total shape function, $\hat{\psi}$,

$$\hat{\psi} = \int_V\int_0^\infty \psi(\mathbf{r},E)\,dE\,dV \quad , \tag{3.41}$$

the following one-group equation is obtained:

$$\overline{\left(\frac{1}{v}\right)}\frac{dp}{dt} = [\nu_p\Sigma_f - \Sigma_a - DB^2]p + \frac{1}{\hat{\psi}}\sum_k \lambda_k\hat{C}_k + \frac{1}{\hat{\psi}}\hat{S} \quad . \tag{3.42}$$

Equation (3.42) is very similar to the intuitive form, Eq. (3.19). However, the derivation provided the one-group definitions for all the quantities involved:

$$\overline{\left(\frac{1}{v}\right)} = \frac{1}{\hat{\psi}}\int_V\int_0^\infty \frac{\psi(\mathbf{r},E)}{v(E)}\, dE\, dV \quad , \tag{3.43a}$$

$$\nu_p\Sigma_f = \frac{1}{\hat{\psi}}\int_V\int_0^\infty \nu_p\Sigma_f(\mathbf{r},E)\psi(\mathbf{r},E)\, dE\, dV \quad , \tag{3.43b}$$

$$\Sigma_a(t) = \frac{1}{\hat{\psi}}\int_V\int_0^\infty \Sigma_a(\mathbf{r},E,t)\psi(\mathbf{r},E)\, dE\, dV \quad , \tag{3.43c}$$

$$DB^2 = \frac{1}{\hat{\psi}}\int_V\int_0^\infty D(\mathbf{r},E)B^2(\mathbf{r},E)\psi(\mathbf{r},E)\, dE\, dV \quad , \tag{3.43d}$$

$$\hat{C}_k(t) = \int_V C_k(\mathbf{r},t)\, dV \quad , \tag{3.43e}$$

and

$$\hat{S}(t) = \int_V\int_0^\infty S(\mathbf{r},E,t)\, dE\, dV \quad . \tag{3.43f}$$

As in the case of the intuitive point kinetics equations, the delayed neutron term $\nu_d\Sigma_f$ is added and subtracted in the brackets on the right side of Eq. (3.42). Dividing the result by $\nu\Sigma_f$ and introducing β yields:

$$\Lambda\frac{dp}{dt} = (\rho - \beta)p + \frac{1}{\hat{S}_{f0}}\sum_k \lambda_k\hat{C}_k + \frac{\hat{S}}{\hat{S}_{f0}} \quad . \tag{3.44}$$

Here, $\hat{S}_{f0}$ is the initial total source of fission neutrons if ψ is chosen to be equal to the initial flux shape, Eq. (3.37a):

$$\hat{S}_{f0} = \nu\Sigma_f\hat{\phi}_0 \quad . \tag{3.45}$$

The generation time Λ now depends on the average inverse velocity, rather than the inverse average value as in Eq. (3.12):

$$\Lambda = \frac{1}{\nu\Sigma_f}\overline{\left(\frac{1}{v}\right)} \quad . \tag{3.46}$$

The total sources of delayed and independent neutrons appear in Eq. (3.44) relative to the initial total source of fission neutrons. Note that part of the initial fission neutron source also appeared in Eq. (3.20) as a denominator, $\nu\Sigma_f$. To simplify the notation, the following relative sources are introduced:

$$\lambda_k \zeta_k(t) = \lambda_k \frac{\hat{C}_k(t)}{\hat{S}_{f0}} = \begin{array}{l}\text{relative neutron source of} \\ \text{delay group } k\end{array} , \quad (3.47)$$

with $\zeta_k(t)$ called the "reduced precursors" of delay group k, and

$$s(t) = \frac{\hat{S}(t)}{\hat{S}_{f0}} = \text{relative independent source} \quad . \quad (3.48)$$

After integrating the precursor balance equation, Eq. (3.32), with respect to space and dividing by $\hat{S}_{f0}$ one obtains a differential equation for the reduced precursors:

$$\frac{d\zeta_k(t)}{dt} = -\lambda_k \zeta_k(t) + \beta_k p(t) \quad . \quad (3.49)$$

This yields the improved definition of the effective β_k, i.e.,

$$\beta_k = \frac{\int_V \int_0^\infty \nu_{dk} \Sigma_f(\mathbf{r},E) \psi(\mathbf{r},E)\, dE\, dV}{\int_V \int_0^\infty \nu \Sigma_f(\mathbf{r},E) \psi(\mathbf{r},E)\, dE\, dV} \quad . \quad (3.50)$$

Equations (3.44) to (3.49) represent approximate *point kinetics equations*:

$$\frac{dp}{dt} = \frac{\rho - \beta}{\Lambda} p + \frac{1}{\Lambda} \sum_k \lambda_k \zeta_k + \frac{1}{\Lambda} s(t) \quad (3.51a)$$

and

$$\frac{d\zeta_k}{dt} = -\lambda_k \zeta_k + \beta_k p(t) \quad . \quad (3.51b)$$

The definitions of all quantities are expressed in terms of a one-group model. The general definitions actually applied are derived in Chapter 5.

Commonly, the $\frac{1}{\Lambda}$ factors in front of the sources are combined with these quantities:

$$c_k(t) = \frac{1}{\Lambda} \zeta_k(t) \quad (3.52a)$$

and

$$s_c(t) = \frac{1}{\Lambda}s(t) \quad . \tag{3.52b}$$

This results in the more familiar form of the point kinetics equations:

$$\frac{dp}{dt} = \frac{\rho - \beta}{\Lambda}p + \sum_k \lambda_k c_k + s_c \tag{3.53a}$$

and

$$\frac{dc_k}{dt} = -\lambda_k c_k + \frac{\beta_k}{\Lambda}p \quad . \tag{3.53b}$$

Although the form of Eqs. (3.53) of the kinetics equations is widely accepted, the form of Eqs. (3.51) is generally preferred in this text. There are essentially three reasons for departing from the common notation in this case:

1. The generation time appears explicitly only where it was formed in the derivation, i.e., in the equation for the flux and not in the balance equation for the precursors; compare Eqs. (3.51b) and (3.53b).
2. The kinetics approximation, which consists of forming the limit $\Lambda \to 0$, is of great importance in reactor dynamics. Only Eqs. (3.51) allow an elegant formulation of the theory in the zero-generation-time approximation (see Sec. 8-1).
3. The sources in Eqs. (3.51) have a simpler and more direct physical interpretation than those in Eqs. (3.53).

Homework Problems

1. Find $\overline{v}$ and the generation time Λ for thermal neutrons. Calculate $\overline{v}$ as the spectrum average for a normalized Maxwell spectrum:

$$\varphi_m(E)\, dE = \frac{E}{kT} \exp(-E/kT) \frac{dE}{kT} \quad .$$

   Use $T = 900$ K and $\nu\Sigma_f = 0.3$/cm.

2. Find $\overline{v}$ and $(\overline{1/v})$ for a two-group representation of a thermal reactor spectrum, composed for simplicity of a Maxwellian and a $1/E$ spectrum:

$$\varphi_1(E) = a/E \qquad \text{for } 0.2 \text{ eV} \le E \le 2 \text{ MeV}$$

$$\varphi_2(E) = \frac{bE}{(kT)^2} \exp(-E/kT) \quad \text{for } 0 \le E \le \infty \quad .$$

[The Maxwellian decreases fast enough so that no finite upper limit needs to be considered in $\varphi_2(E)$.]

a. Find a and b such that the two components of the normalized $\varphi(E)$ provide equal contributions to the energy integral.
b. Find the average velocities for both groups ($\overline{v}_1$ and $\overline{v}_2$).
c. Derive the two group definitions of $\overline{v}$ and $\overline{(1/v)}$.
d. Find the corresponding numerical values.

3a. Define a one-group $\nu\Sigma_f$ based on the two-group values $\nu\Sigma_{f1}$ = 0.015/cm and $\nu\Sigma_{f2}$ = 0.3/cm.
b. Calculate Λ with $\overline{v}$ and with $\overline{(1/v)}$.
c. Discuss the results.

Review Questions

1. Give the simplest form of the balance equation for the total number of neutrons in a reactor, $n(t)$. Use one-group constants and disregard delayed neutrons.
2. Bring the balance equation into a form with the flux and the number of neutrons as unknowns, respectively, and introduce the reactivity to describe the off-criticality.
3. Give the approximate one-group formulas and the physical interpretations of (a) the average generation time Λ and (b) the average lifetime ℓ.
4. Derive the relation between Λ and ℓ.
5. Derive the relation between the lifetimes in finite and infinite systems.
6. Give the point kinetics equation for all-prompt neutrons by using Λ and ℓ, respectively.

7a. Give the balance equation of the delayed neutron precursors in its most direct form.
b. In which way is the source term in this equation often rewritten?

8a. Give the formula for β-effective in the one-group approximation.
b. Which physical fact is described by β-effective in this approximation?

9. Derive the simplest forms of the point kinetics equations based on the concepts $\hat{\phi}$, ρ, β, and Λ, in the ζ notation.
10. Why is the diffusion equation a reasonable approximation for kinetics in large reactors?

11a. Give the space-, energy-, and time-dependent diffusion equation in operator form.

b. Define the operators.
c. Give the balance equations for the precursor densities for stationary fuel.

12a. What does the assumption of "stationary fuel" mean?
b. Give two examples where this assumption is violated.

13a. Separate the flux into amplitude and shape functions.
b. Give the time- and energy-dependent diffusion equation (use operators as far as reasonable).
c. Sketch the derivation of the one-group point kinetics equations and give the result.

14a. Give the balance equations for the precursor densities for stationary fuel.
b. Sketch the derivation of the precursor balance equations for point kinetics.
c. Give the results in the ζ and c formulations.

15. Give the definition for Λ in the one-group approximation.

16. In which form does the precursor production term appear in the ζ_k and c_k formulations, respectively?

REFERENCES

1. K. O. Ott and W. A. Bezella, *Introductory Nuclear Reactor Statics,* American Nuclear Society, La Grange Park, Illinois (1983).
2. D. A. Meneley, G. K. Leaf, A. J. Lindeman, T. A. Daly, and W. T. Shaw, "A Kinetics Model for Fast Reactor Analysis in Two Dimensions," p. 483 in *Dynamics of Nuclear Systems,* D. L. Hetrick, Ed., University of Arizona Press, Tucson (1972).

Four

STATIC PERTURBATION THEORY

Perturbation theories were developed primarily for special applications, e.g., Schrödinger's and Dirac's perturbation theories for quantum theoretical applications, but perturbation theory has become a well-developed discipline of mathematics. Most time-dependent phenomena of reactor dynamics are typical perturbation problems since nuclear reactors normally operate in a steady state. In most of the literature, the treatment of time-dependent neutron fluxes is based on equations that are formulated in analogy to static reactor perturbation problems. The same formulation is applied here. As a basis, static perturbation theory is briefly presented in this chapter. The derivations employ operator notation. The formal result can be converted readily to explicit formulas for the terms of the neutronics balance equation that are proportional to macroscopic cross sections. The leakage term, however, requires special considerations that are presented in Sec. B-5 of App. B.

In all perturbation problems discussed in this text, it is assumed that the perturbation does not change the spatial domain; i.e., boundary perturbations are not considered.

4-1 The Basic Reactor Eigenvalue Problem and the Perturbation Theory Approach

4-1A Motivations for a Perturbation Theory

Practical reactor configurations have quite complicated internal structures and boundaries. Therefore, complete and accurate theoretical modeling requires treatment of three-dimensional configurations with spatially varying heterogeneous material compositions. A correct treatment of such complicated configurations would be tedious and expensive. However, sufficient accuracy can normally be obtained from proper approximations. The complications that have been eliminated in these approximate solutions can then be treated as "perturbations" of simplified cases.

The neutron flux and the multiplication constant are of primary interest in static reactor problems. Most of the other needed quantities can be derived from the neutron flux and the multiplication constant.

The application of perturbation theory is very advantageous for the calculation of the perturbation of multiplication constants k ($= k_{eff}$) for complicated configurations. Static perturbation theory is therefore frequently and commonly applied to find perturbations in multiplication constants, especially when the calculation of the corresponding perturbed neutron flux is not required. The application of perturbation methods to find the flux perturbation does not simplify the treatment. Consequently, direct solutions are generally preferred.

4-1B Neutron Multiplication as an Eigenvalue Problem

For the quantitative description of neutron multiplication and the criticality of nuclear reactors, the concept of the multiplication constant has been introduced as an eigenvalue in a static neutron balance equation (see, for example, Ref. 1):

$$\mathbf{M}\Phi = \lambda\mathbf{F}\Phi \quad . \tag{4.1}$$

The migration and loss of neutrons, described by $\mathbf{M}\Phi$, is made equal to a modified source of fission neutrons, $\lambda\mathbf{F}\Phi$, by introducing the eigenvalue λ in front of the fission neutron source $\mathbf{F}\Phi$. The inclusion of an eigenvalue in Eq. (4.1) is required to obtain a nontrivial solution of the static neutron balance equation that is needed to describe the degree of off-criticality. The neutron fluxes obtained as a solution of Eq. (4.1) are λ mode fluxes. The multiplication constant k is then defined in terms of the eigenvalue λ as

$$k = \frac{1}{\lambda} \quad . \tag{4.2}$$

The migration and loss operator $\mathbf{M}$ and the fission source operator $\mathbf{F}$ for diffusion theory are given by the same expression as in the time-dependent case, Eqs. (3.29), except that the cross sections and the flux are time independent and the fission source includes prompt *and* delayed neutrons; Φ is equal to $\phi(\mathbf{r},E)$:

$$\mathbf{M}\Phi = -\nabla\cdot D(\mathbf{r},E)\nabla\phi(\mathbf{r},E) + \Sigma_t(\mathbf{r},E)\phi(\mathbf{r},E) - \int_{E'} \Sigma_s(\mathbf{r},E'\rightarrow E)\phi(\mathbf{r},E')\,dE' \quad , \tag{4.3a}$$

$$\mathbf{F}\Phi = \sum_i \chi_i(E) \int_0^\infty \nu_i(E')\Sigma_{fi}(\mathbf{r},E')\phi(\mathbf{r},E')\, dE' \quad . \tag{4.3b}$$

If the isotope dependence of $\chi_i(E)$ is omitted [see the discussion after Eqs. (3.29)], Eq. (4.3b) can be simplified by combining all isotopic contributions into a single quantity $\nu\Sigma_f$, i.e.,

$$\mathbf{F}\Phi = \chi(E) \int_0^\infty \nu\Sigma_f(\mathbf{r},E')\phi(\mathbf{r},E')\, dE' \tag{4.4a}$$

and

$$\nu\Sigma_f(\mathbf{r},E) = \sum_i \nu_i(E)\Sigma_{fi}(\mathbf{r},E) \quad . \tag{4.4b}$$

For a simple one-group model, with leakage described by a DB^2 term (see Sec. 3-1A), the operators **M** and **F** are given by

$$\mathbf{M} = DB^2 + \Sigma_a \tag{4.5a}$$

and

$$\mathbf{F} = \nu\Sigma_f \quad . \tag{4.5b}$$

The balance equation, Eq. (4.1), then reduces to

$$(DB^2 + \Sigma_a)\hat{\phi} = \lambda\nu\Sigma_f\hat{\phi} \quad . \tag{4.6}$$

The reactor is critical if $\lambda = 1$; i.e., if

$$\frac{1}{\lambda} = k = \frac{\nu\Sigma_f}{DB^2 + \Sigma_a} = 1 \quad . \tag{4.7}$$

The same simple neutronics model was applied in Sec. 3-1A, in the context of the discussion on intuitive kinetics equations. Equation (3.6) expresses approximately the time-dependent neutron balance in a reactor without an independent source ($S = 0$) and with all neutrons treated as prompt neutrons. If the reactor is not critical, i.e., if the term $\nu\Sigma_f - (\Sigma_a + DB^2)$ in Eqs. (3.6) or (4.6) is not zero, the flux must be either time dependent or equal to zero.

Thus, if a source-free reactor is off-critical and if a nontrivial static solution of the balance equation is desired, the balance equation must be artificially altered. This is conventionally achieved by multiplying the fission source with a factor $\lambda = 1/k$ [see Eqs. (4.1) and (4.6)]. The resulting balance equation is then mathematically an eigenvalue problem with λ as the eigenvalue and Φ as the eigenfunction, the so-called "λ mode," frequently designated by Φ_λ.

Equation (4.6) illustrates the role of the eigenvalue λ. Suppose

$$DB^2 + \Sigma_a \neq \nu\Sigma_f \quad . \tag{4.8}$$

The physical balance equation,

$$(DB^2 + \Sigma_a)\hat{\phi} = \nu\Sigma_f\hat{\phi} \quad , \tag{4.9}$$

then has only the trivial solution $\hat{\phi} \equiv 0$.

If the fission source is multiplied with a factor λ, such as in Eq. (4.6), a nontrivial solution[a] results, which provides a basis for measuring the degree of off-criticality in the form of an eigenvalue or its inverse:

$$\frac{1}{\lambda} = k = \frac{\nu\Sigma_f}{DB^2 + \Sigma_a} \neq 1 \quad . \tag{4.10}$$

The degree of off-criticality is defined as "reactivity," or more precisely, the "static reactivity":

$$1 - \lambda = 1 - \frac{1}{k} = \frac{\Delta k}{k} = \rho \quad . \tag{4.11}$$

The solution of the eigenvalue problem, Eq. (4.6), can be found by solving a simple algebraic equation. In general, the solution of eigenvalue problems, such as Eq. (4.1), is much more complicated. For a multigroup description of the energy dependence, Eq. (4.1) represents a set of partial differential equations. The solution of such an eigenvalue problem generally requires the application of direct inversion or iterative procedures implemented in computer codes.

4-1C The Basic Approach for the Calculation of an Eigenvalue Perturbation

A "clean" configuration, which is simpler than the actual system for which the eigenvalue $\lambda = 1/k$ is desired, is introduced as a reference system for the treatment of perturbations. The more complicated actual system is then called the "perturbed system." Let Eq. (4.1) describe the actual, complicated configuration; the clean or "unperturbed" balance equation is denoted by:

$$\mathbf{M}_0\Phi_0 = \lambda_0\mathbf{F}_0\Phi_0 \quad . \tag{4.12}$$

This unperturbed problem is chosen in such a way that its solution

[a]Flux normalization cannot be derived from the solution of the homogeneous eigenvalue problem and generally requires an additional equation to determine its value; for example, normalization of the power equal to the total reactor power.

can be readily determined. One then wants to find $\Delta\lambda = \lambda - \lambda_0$, utilizing the quantities λ_0 and Φ_0, which are known from the solution of the unperturbed problem; i.e., one wants to determine

$$\Delta\lambda = \lambda - \lambda_0 = -\frac{\Delta k}{k_0 k} = -\Delta\rho \quad , \tag{4.13}$$

where $\Delta\rho$ is the "reactivity increment," and

$$\Delta k = k - k_0 \quad . \tag{4.14}$$

Obviously, $\Delta\lambda$ is physically determined by the differences in the two systems, which are mathematically described by the differences in the corresponding operators:

$$\Delta\mathbf{M} = \mathbf{M} - \mathbf{M}_0 \tag{4.15a}$$

and

$$\Delta\mathbf{F} = \mathbf{F} - \mathbf{F}_0 \quad . \tag{4.15b}$$

In the simple one-group model, these differences are given by

$$\Delta\mathbf{M} = \Delta(DB^2) + \Delta\Sigma a \tag{4.16a}$$

and

$$\Delta\mathbf{F} = \Delta\nu\Sigma_f \quad . \tag{4.16b}$$

In general (e.g., for the energy-dependent diffusion equation), the difference between the leakage parts of two $\mathbf{M}$ operators (i.e., $-\nabla\cdot D\nabla + \nabla\cdot D_0\nabla$) cannot be formed explicitly. Then $\Delta\mathbf{M}$ is to be understood merely as an abbreviation for $\mathbf{M} - \mathbf{M}_0$; its subsequent treatment in scalar products requires special mathematical manipulations that are presented in Sec. B-5 of App. B.

The mathematical formulation whereby $\Delta\lambda$ is expressed in terms of $\Delta\mathbf{M}$, $\Delta\mathbf{F}$, Φ_0, and λ_0 is first derived for a simple eigenvalue problem that contains only a single operator, say $\mathbf{A}$. The actual reactor problem is treated in the subsequent sections.

The following two equations describe the perturbed and the unperturbed problems defined by the operators $\mathbf{A}$ and $\mathbf{A}_0$, respectively:

$$\mathbf{A}\Phi = \lambda\Phi \tag{4.17a}$$

and

$$\mathbf{A}_0\Phi_0 = \lambda_0\Phi_0 \quad . \tag{4.17b}$$

One introduces Φ_0 into Eq. (4.17a) by writing:

$$\Phi = \Phi_0 + \Delta\Phi \quad . \tag{4.18}$$

Equation (4.17a) then becomes:

$$\mathbf{A}\Phi_0 = \lambda\Phi_0 - (\mathbf{A}\Delta\Phi - \lambda\Delta\Phi) \quad . \tag{4.19}$$

Equations (4.17) through (4.19) depend in general on space and energy. Since the desired quantity, $\Delta\lambda$, is just a number, the space and energy dependencies in corresponding equations are removed by integration. To preserve generality and to add flexibility to the procedure, the equations are multiplied with a weighting function, $\Phi^w = \phi^w(\mathbf{r},E)$, prior to the integration.

The integration over the domain of a product of two functions is called a "scalar product" (see Sec. B-1 of App. B). Multiplication of Eqs. (4.19) and (4.17b) with a weighting function and integrating with respect to space and energy (written in scalar product notation) gives:

$$(\Phi^w,\mathbf{A}\Phi_0) = \lambda(\Phi^w,\Phi_0) - (\Phi^w,[\mathbf{A} - \lambda]\Delta\Phi) \tag{4.20a}$$

and

$$(\Phi^w,\mathbf{A}_0\Phi_0) = \lambda_0(\Phi^w,\Phi_0) \quad . \tag{4.20b}$$

Subtracting Eqs. (4.20) yields:

$$(\Phi^w,\Delta\mathbf{A}\Phi_0) = \Delta\lambda(\Phi^w,\Phi_0) - (\Phi^w,[\mathbf{A} - \lambda]\Delta\Phi) \tag{4.21}$$

with

$$\Delta\mathbf{A} = \mathbf{A} - \mathbf{A}_0 \quad . \tag{4.22}$$

The second term on the right side of Eq. (4.21) contains $\Delta\Phi$. Its first-order contribution can be eliminated with a proper choice of the weighting function Φ^w. Inserting Eqs. (4.13) and (4.22) into the second term on the right side of Eq. (4.21) gives:

$$(\Phi^w,[\mathbf{A} - \lambda]\Delta\Phi) = (\Phi^w,[\mathbf{A}_0 - \lambda_0]\Delta\Phi) + (\Phi^w,[\Delta\mathbf{A} - \Delta\lambda]\Delta\Phi) \quad . \tag{4.23}$$

The first term on the right side of Eq. (4.23) is of "first order" in the difference quantities, whereas the second term is of "second order" since it contains a product of two difference quantities. If the difference of the perturbed and unperturbed systems is small, the second-order term in Eq. (4.23) is small compared to the first-order term. Consequently, if the first-order term can be eliminated, the right side of Eq. (4.23) reduces to only a small second-order term. The first-order term on the right side of Eq. (4.23) can be rewritten—provided Φ^w satisfies certain conditions discussed in Sec. B-3 of App. B—as:

$$(\Phi^w,[\mathbf{A}_0 - \lambda_0]\Delta\Phi) = (\Delta\Phi,[\mathbf{A}_0^* - \lambda_0]\Phi^w) \quad , \tag{4.24}$$

where $\mathbf{A}_0^*$ is the adjoint operator of $\mathbf{A}$ (see App. B). The right side of Eq. (4.24) is equal to zero if Φ^w is chosen to be the adjoint function Φ_0^*, i.e., the solution of the adjoint eigenvalue problem:

$$\mathbf{A}_0^*\Phi_0^* = \lambda_0\Phi_0^* \quad . \tag{4.25}$$

Equation (4.25) is the adjoint problem of the corresponding unperturbed problem, Eq. (4.17b); it has, as shown in Sec. B-4 of App. B, the same eigenvalue, λ_0.

After elimination of the first-order term on the right side of Eq. (4.23) by choosing $\Phi^w = \Phi_0^*$, the unperturbed adjoint function, and after neglect of the second-order term in Eq. (4.23), the right side of Eq. (4.21) reduces to only the first term. Solving for $\Delta\lambda$ gives the first-order perturbation formula for the simplified eigenvalue problem, Eq. (4.17a):

$$\Delta\lambda \simeq \frac{(\Phi_0^*,\Delta\mathbf{A}\Phi_0)}{(\Phi_0^*,\Phi_0)} = \Delta\lambda^{(1)} \quad , \tag{4.26}$$

where $\Delta\lambda^{(1)}$ is the result of "first-order perturbation theory."

Equation (4.26) apparently provides the answer to the question posed in Eq. (4.13). The eigenvalue perturbation and thus the reactivity increment is estimated in a first-order approximation, using only known terms—$\Delta\mathbf{A}$, Φ_0, and Φ_0^*. To obtain this result, it was necessary to eliminate the dependence of Eq. (4.21) on the flux perturbation, $\Delta\Phi = \Phi - \Phi_0$, since Φ and thus $\Delta\Phi$ are not known. The elimination of $\Delta\Phi$ required the introduction of the adjoint function Φ_0^*.

4-2 First-Order Perturbation Theory

The general approach of the first-order perturbation theory for the estimation of eigenvalue perturbations in terms of the operator differences and the unperturbed flux and adjoint flux was discussed in the previous section. The application of the same approach to the eigenvalue problem of reactors, Eq. (4.1), is straightforward. It involves, however, some additional terms relative to the simpler eigenvalue problem, Eq. (4.17a), due to the appearance of an operator, $\mathbf{F}$, on the right side of Eq. (4.1).

As in the simple case, the actual eigenvalue problem is considered as the perturbed problem, and the eigenvalue difference (the negative reactivity increment) is determined using a properly defined unperturbed problem (the subscript zeros):

$$\mathbf{M}\Phi = \lambda\mathbf{F}\Phi \tag{4.27}$$

and

$$\mathbf{M}_0\Phi_0 = \lambda_0\mathbf{F}_0\Phi_0 \quad . \tag{4.28}$$

As before, all perturbed quantities are expressed as the sum of unperturbed quantities and difference terms as shown in Eqs. (4.13), (4.15), and (4.18). Decomposing the flux into an unperturbed part and a perturbation in Eq. (4.27) yields a form of the perturbed eigenvalue problem, which is equivalent to Eq. (4.19):

$$\mathbf{M}\Phi_0 = \lambda\mathbf{F}\Phi_0 - (\mathbf{M} - \lambda\mathbf{F})\Delta\Phi \quad . \tag{4.29}$$

The first term on the right side of Eq. (4.29) is now recast into zero-, first-, and second-order terms:

$$\lambda\mathbf{F}\Phi_0 = \lambda\mathbf{F}_0\Phi_0 + \lambda_0\Delta\mathbf{F}\Phi_0 + \Delta\lambda\Delta\mathbf{F}\Phi_0 \quad . \tag{4.30}$$

Substituting Eq. (4.30), after neglecting the second-order term, into Eq. (4.29) and multiplying the result and the unperturbed eigenvalue problem, Eq. (4.28), with a weighting function Φ^w, yields:

$$(\Phi^w,\mathbf{M}\Phi_0) = \lambda(\Phi^w,\mathbf{F}_0\Phi_0) + \lambda_0(\Phi^w,\Delta\mathbf{F}\Phi_0)$$
$$- (\Phi^w,[\mathbf{M} - \lambda\mathbf{F}]\Delta\Phi) \tag{4.31a}$$

and

$$(\Phi^w,\mathbf{M}_0\Phi_0) = \lambda_0(\Phi^w,\mathbf{F}_0\Phi_0) \quad . \tag{4.31b}$$

Subtracting Eqs. (4.31) gives the interim result:

$$(\Phi^w,[\Delta\mathbf{M} - \lambda_0\Delta\mathbf{F}]\Phi_0) = \Delta\lambda(\Phi^w,\mathbf{F}_0\Phi_0)$$
$$- (\Phi^w,[\mathbf{M} - \lambda\mathbf{F}]\Delta\Phi) \quad . \tag{4.32}$$

The term containing the flux deformation $\Delta\Phi$ can be eliminated in a first-order approximation in the same way as shown above for the simpler problem. First, this term is cast into first-, second-, and third-order contributions:

$$(\Phi^w,[\mathbf{M} - \lambda\mathbf{F}]\Delta\Phi) = (\Phi^w,[\mathbf{M}_0 - \lambda_0\mathbf{F}_0]\Delta\Phi)$$
$$+ (\Phi^w,[\Delta\mathbf{M} - \lambda_0\Delta\mathbf{F} - \Delta\lambda\mathbf{F}_0]\Delta\Phi)$$
$$- \Delta\lambda(\Phi^w,\Delta\mathbf{F}\Delta\Phi) \quad . \tag{4.33}$$

The second- and third-order terms in Eq. (4.33) are neglected in first-order perturbation theory, and the first-order term, which is due to the flux deformation $\Delta\Phi$, is eliminated by choosing the unperturbed adjoint flux Φ_0^* as the weighting function, Φ^w. Revolving the functions in the first scalar product on the right side of Eq. (4.33) and setting Φ^w equal to Φ_0^* gives:

$$(\Delta\Phi,[\mathbf{M}_0^* - \lambda_0\mathbf{F}_0^*]\Phi^w) = (\Delta\Phi, [\mathbf{M}_0^* - \lambda_0\mathbf{F}_0^*]\Phi_0^*) = 0 \quad . \tag{4.34}$$

It is equal to zero since the adjoint flux Φ_0^* is the solution of the adjoint eigenvalue problem:

$$\mathbf{M}_0^*\Phi_0^* = \lambda_0\mathbf{F}_0^*\Phi_0^* \quad . \tag{4.35}$$

After the flux perturbation term in Eq. (4.32) is eliminated as described above, the remainder of Eq. (4.32) can be readily solved for $\Delta\lambda$ (or $\Delta\rho$), which gives the *first-order perturbation formula for reactivity increments* as scalar products, as well as in more explicit notation:

$$\Delta\rho = \frac{(\Phi_0^*,[\lambda_0\Delta\mathbf{F} - \Delta\mathbf{M}]\Phi_0)}{(\Phi_0^*,\mathbf{F}_0\Phi_0)} = -\Delta\lambda \tag{4.36a}$$

$$= \frac{\int_V\int_E \phi_0^*(\mathbf{r},E)(\lambda_0\Delta\mathbf{F} - \Delta\mathbf{M})\phi_0(\mathbf{r},E)\,dE\,dV}{\int_V\int_E \phi_0^*(\mathbf{r},E)\mathbf{F}_0\phi_0(\mathbf{r},E)\,dE\,dV} \quad . \tag{4.36b}$$

See Sec. B-5 of App. B for the explicit form of the leakage contribution.

Applications of the first-order perturbation formula are discussed in Sec. 4-4. Equations (4.36) also provide the basis of the widely used point reactor model (see Sec. 5-2).

In neutronics, the adjoint function $\phi^*(\mathbf{r},E)$ is for the most part called the "adjoint flux," but it is also often called the "importance function." (See Sec. B-4 of App. B and Sec. 7-6 for the interpretation of ϕ^* as the neutron importance function.)

In perturbation theory, the role of the importance function can be formulated as:

> If the flux perturbation $\Delta\phi$ is neglected, the reaction rate differences $\Delta\mathbf{M}\phi_0$ and $\Delta\mathbf{F}\phi_0$ should be multiplied, prior to the space-energy integration, with the respective neutron importance, $\phi_0^*(\mathbf{r},E)$, in order to improve the accuracy of the resultant eigenvalue perturbation.

The first-order perturbation theory formulas are called "stationary" with respect to small changes in the system because the error of the first-order reactivity evolves only "quadratically" with a linearly increasing change. The reactivity as such varies linearly. For a numerical example and applications of first-order perturbation theory to reactor problems, see Sec. 4-4.

4-3 Exact Perturbation Theory

The name "exact perturbation theory" seems to suggest that the perturbation theory approach can be expanded by including higher order perturbation terms, and that the result would converge to the exact solution. Although higher order perturbation terms can, in principle, be calculated, the resulting method is cumbersome and generally not practical.

"Exact perturbation theory," as it appears in the reactor literature, expresses the exact eigenvalue differences in a form similar to that of the first-order perturbation theory formula. This is done only with knowledge of the solution to the perturbed problem. It may seem, therefore, at first glance that nothing is to be gained by this approach.

Since the solution of the perturbed problem is used in exact perturbation theory, the derivation of the corresponding expression for the eigenvalue perturbation is simpler than in first-order perturbation theory. Again, the equations for the perturbed and the unperturbed problems represent the basic set of equations. The derivation is simplified by first converting the unperturbed equation into the corresponding adjoint equations:

$$\mathbf{M}\Phi = \lambda\mathbf{F}\Phi \tag{4.37a}$$

and

$$\mathbf{M}_0^*\Phi_0^* = \lambda_0\mathbf{F}_0^*\Phi_0^* \quad . \tag{4.37b}$$

To obtain scalar products with difference operators, Eqs. (4.37a) and (4.37b) are multiplied by Φ_0^* and Φ, respectively, prior to integration. Revolving the functions of the scalar products in Eq. (4.38b) gives Eq. (4.38c):

$$(\Phi_0^*,\mathbf{M}\Phi) = \lambda(\Phi_0^*,\mathbf{F}\Phi) \quad , \tag{4.38a}$$

$$(\Phi,\mathbf{M}_0^*\Phi_0^*) = \lambda_0(\Phi,\mathbf{F}_0^*\Phi_0^*) \quad , \tag{4.38b}$$

and

$$(\Phi_0^*,\mathbf{M}_0\Phi) = \lambda_0(\Phi_0^*,\mathbf{F}_0\Phi) \quad . \tag{4.38c}$$

The right side of Eq. (4.38c) is rewritten as:

$$(\Phi_0^*,\mathbf{M}_0\Phi) = \lambda_0(\Phi_0^*,\mathbf{F}\Phi) - \lambda_0(\Phi_0^*,\Delta\mathbf{F}\Phi) \quad . \tag{4.38d}$$

Subtracting Eqs. (4.38a) and (4.38d) and solving for $-\Delta\lambda = \Delta\rho$ yields the *exact perturbation formula*[b] *for reactivity increments:*

$$\Delta\rho = \frac{(\Phi_0^*,[\lambda_0\Delta\mathbf{F} - \Delta\mathbf{M}]\Phi)}{(\Phi_0^*,\mathbf{F}\Phi)} = -\Delta\lambda \quad . \tag{4.39}$$

The explicit expression can be found in analogy to Eqs. (4.36b) and (B.50).

The exact formula, Eq. (4.39), differs from the first-order perturbation formula, Eq. (4.36a), by the use of the perturbed flux Φ in both scalar products and by the appearance of $\mathbf{F}$ rather than $\mathbf{F}_0$ in the denominator. Equation (4.36a) is obtained as a first approximation of the exact equation, Eq. (4.39), by approximating Φ with Φ_0 and in the denominator, $\mathbf{F}$ with $\mathbf{F}_0$. The errors of this approximation are second order as previously shown.

An exact formula for the reactivity can be found in the same way as for reactivity increments. If Φ_0 describes a specific critical state, $\lambda_0 = 1$, Eq. (4.39) provides directly the following reactivity:

$$\rho = \rho^{st} = \frac{(\Phi_0^*,[\Delta\mathbf{F} - \Delta\mathbf{M}]\Phi)}{(\Phi_0^*,\mathbf{F}\Phi)} \quad . \tag{4.40}$$

The reactivity, $\rho = 1 - \lambda$, of a state with flux Φ is independent of the specifics of the critical reference state. Thus, the $\mathbf{F}_0$ and $\mathbf{M}_0$ terms in Eq. (4.40) cancel, since

$$(\Phi_0^*,[\mathbf{F}_0 - \mathbf{M}_0]\Phi) = 0 \quad . \tag{4.41}$$

Furthermore, the weighting function need not be the adjoint of any critical problem. Therefore, Eq. (4.40) can also be written as

$$\rho^{st} = \frac{(\Phi^w,[\mathbf{F} - \mathbf{M}]\Phi)}{(\Phi^w,\mathbf{F}\Phi)} \quad , \tag{4.42}$$

with Φ^w as an arbitrary weighting function. If, however, the state with the flux Φ is near a *known* critical state with the adjoint flux Φ_0^*, it can be numerically advantageous to choose this adjoint flux as the weighting function:

$$\rho^{st} = \frac{(\Phi_0^*,[\mathbf{F} - \mathbf{M}]\Phi)}{(\Phi_0^*,\mathbf{F}\Phi)} \quad . \tag{4.43}$$

The *dynamic* reactivity is defined in analogy to Eq. (4.43) if the reactor

---

[b]In an alternative formulation, λ_0 in the numerator is replaced by λ and the $\mathbf{F}$ in the denominator by $\mathbf{F}_0$.

is initially critical, and in analogy to Eq. (4.39) if the initial state is subcritical (compare Secs. 5-1B and 5-1C).

Strictly speaking, Eqs. (4.39) and (4.40) do not contain any information that was not already known after the solution of the two eigenvalue problems, Eqs. (4.37). Indeed, $\Delta\lambda$ can be formed directly as $\lambda - \lambda_0$ rather than through the complicated scalar products of Eq. (4.39). However, perturbed and sometimes also unperturbed eigenvalue problems are often solved with insufficient accuracy. Then, application of the exact perturbation formula provides a tool to improve the accuracy of the reactivity; Eq. (4.39) suggests a way of improving first-order perturbation theory results, as follows.

If Φ and/or Φ_0, and thus λ and λ_0, are calculated with certain numerical or modeling errors, the application of Eq. (4.39) normally yields a more accurate value of the reactivity increment, $\Delta\rho$, than provided by $\lambda - \lambda_0$. Furthermore, the exact perturbation formula can be used to improve first-order perturbation theory results. It suggests that a flux Φ' be found that more closely approximates the perturbed λ mode flux than does Φ_0. Thus, the use of Φ' as an approximate solution of the perturbed problem, together with the substitution of $\mathbf{F}_0$ by $\mathbf{F}$ in the denominator, gives a better estimate of the exact reactivity perturbation than the first-order perturbation formula. This approach is extensively used in reactor dynamics, where a solution of the time-dependent problem is desired that is more accurate than that obtained on the basis of the first-order perturbation reactivity. Thus, the exact perturbation theory formula indicates how to improve practically on the first-order theory results without going to the formal but impractical route of higher order perturbation theories.

4-4 Applications of First-Order Perturbation Theory

Applications of first-order and exact perturbation theories are described and developed in this text, primarily in Chapters 9 and 11. The examples in this section are only presented to illustrate the application of the first-order perturbation formula and to demonstrate the effect of the adjoint flux weighting.

In the simple one-group model, the operators $\mathbf{M}$ and $\mathbf{F}$ degenerate into numbers representing macroscopic cross sections, Eqs. (4.5). Their perturbations are given by Eqs. (4.16). The corresponding neutronics eigenvalue problem, Eq. (4.6) is self-adjoint (see Sec. B-4 of App. B). The adjoint function then equals the flux:

$$\Phi^* = \Phi \quad . \tag{4.44}$$

Furthermore, the flux is just a single number in this simple case; i.e., Φ_0 and Φ_0^* cancel in Eqs. (4.36). The formula for the reactivity increment is then given by

$$\Delta\rho^{(1)} = \frac{\lambda_0 \Delta\nu\Sigma_f - \Delta\Sigma_a'}{\nu\Sigma_{f0}} \tag{4.45a}$$

or

$$\Delta\rho^{(1)} = \frac{\Sigma_{a0}' \, \Delta\nu\Sigma_f}{(\nu\Sigma_{f0})^2} - \frac{\Delta\Sigma_a'}{\nu\Sigma_{f0}} \quad , \tag{4.45b}$$

with Σ_{a0}' defined as:

$$\Sigma_{a0}' = D_0 B_0^2 + \Sigma_{a0} \quad . \tag{4.46}$$

Equation (4.45b) equals the first-order variation of the corresponding reactivity, describing small variations about a reference value. The reference value of ρ is designated as:

$$\rho_0 = 1 - \frac{1}{k_0} = \frac{\nu\Sigma_{f0} - \Sigma_{a0}'}{\nu\Sigma_{f0}} \quad . \tag{4.47}$$

The first-order variation of Eq. (4.47) is then given by

$$\delta\rho^{(1)} = \delta\left(1 - \frac{\Sigma_{a0}'}{\nu\Sigma_{f0}}\right) = \frac{\Sigma_{a0}'\delta\nu\Sigma_f}{(\nu\Sigma_{f0})^2} - \frac{\delta\Sigma_{a0}'}{\nu\Sigma_{f0}} \quad , \tag{4.48}$$

which agrees with Eq. (4.45b).

Since the "exact" reactivity increment can be directly calculated in this simple neutronics model, there is no need to apply a first-order approximation. The exact solution can then be used to show that the exact $\Delta\rho$ differs from $\Delta\rho^{(1)}$ by a second-order term.

The second example deals with a space-dependent perturbation due to a local increase in absorption (e.g., control rod insertion). Let the reactor be modeled as a one-dimensional slab in the x direction. The unperturbed criticality eigenvalue problem for a spatially uniform composition is

$$\left(-D_0 \frac{d^2}{dx^2} + \Sigma_{a0}\right)\phi_0(x) = \lambda_0 \nu\Sigma_{f0}\phi_0(x) \quad , \tag{4.49}$$

with

$$\phi_0(x_b) = 0 \text{ at } x_b = \pm a \quad . \tag{4.50}$$

The space dependence of the unperturbed flux as determined by the boundary conditions, Eq. (4.50), is given by

$$\phi_0(x) = \phi_0 \cos Bx \quad , \tag{4.51}$$

with B^2 being the geometrical buckling, $B^2 = B_x^2$:

$$B^2 = \left(\frac{\pi}{2a}\right)^2 \quad . \tag{4.52}$$

Inserting Eq. (4.51) into Eq. (4.49) yields the eigenvalue,

$$\frac{1}{\lambda_0} = k_0 = \frac{\nu\Sigma_{f0}}{D_0 B^2 + \Sigma_{a0}} = \frac{\nu\Sigma_{f0}}{\Sigma'_{a0}} \quad . \tag{4.53}$$

Equation (4.49) describes a slab-reactor model in terms of a one-group approximation. The one-group operators **M** and **F** are self-adjoint, as discussed in Sec. B-4 of App. B. Since the adjoint operators are the same as the corresponding real operators and since the vacuum boundary conditions are also the same, the adjoint flux equals the flux:

$$\phi_0^*(x) = \phi_0(x) \quad . \tag{4.54}$$

If the change in the reactor model consists only of a change in the macroscopic absorption cross section in a perturbed range around the center (say, from $-x_p$ to $+x_p$), the perturbation reactivity in the one-group approximation is given by ($\Delta\nu\Sigma_f = 0$):

$$\Delta\rho^{(1)} = \frac{-\displaystyle\int_{-x_p}^{x_p} \phi_0(x)\Delta\Sigma_a\phi_0(x)\,dx}{\displaystyle\int_{-a}^{a} \phi_0(x)\nu\Sigma_{f0}\phi_0(x)\,dx} \quad , \tag{4.55}$$

where the integral in the denominator is extended over the entire domain of the reactor.

The integral in the denominator is given by

$$\phi_0^2\,\nu\Sigma_{f0}\int_{-a}^{a} \cos^2 Bx\,dx = \phi_0^2\frac{\nu\Sigma_{f0}}{B}\int_{\frac{-\pi}{2}}^{\frac{\pi}{2}} \cos^2\alpha\,d\alpha$$

$$= \phi_0^2\,\frac{\nu\Sigma_{f0}}{B}\cdot\frac{\pi}{2} = \nu\Sigma_{f0}\cdot a\phi_0^2 \quad . \tag{4.56}$$

If x_p is small compared to the thickness of the reactor, the cosine in the numerator integral can be approximated by unity. The numerator of Eq. (4.55) then becomes:

$$-\phi_0^2\int_{-x_p}^{x_p} \Delta\Sigma_a \cos^2 Bx\,dx \simeq -\,\phi_0^2\,\Delta\Sigma_a\cdot 2x_p \quad . \tag{4.57}$$

Dividing Eqs. (4.57) and (4.56) yields the first-order perturbation reactivity:

$$\Delta\rho^{(1)} \simeq -2 \cdot \frac{\Delta\Sigma_a}{\nu\Sigma_{f0}} \cdot \frac{x_p}{a} \quad . \tag{4.58}$$

Application of Eq. (4.55) to a pure change in Σ_a over the entire separated homogeneous reactor composition gives the following result since the integrals cancel each other:

$$\Delta\rho^{(1)} = -\frac{\Delta\Sigma_a}{\nu\Sigma_{f0}} \quad . \tag{4.59}$$

The three factors in Eq. (4.58) can be interpreted in the following way. The middle factor, $-\Delta\Sigma_a/\nu\Sigma_{f0}$, represents a contribution due to a homogeneous change in Σ_a [note Eq. (4.59)]. The third factor, x_p/a, describes the relative lengths of the $\Delta\Sigma_a$ and the $\nu\Sigma_{f0}$ domains; the resulting product is multiplied by two, which accounts for the fact that the $\Delta\Sigma_a$ change is concentrated in the center of the reactor and not homogeneously distributed over the entire system.

The effect of the adjoint (or importance) weighting in the first-order reactivity formula can be readily demonstrated for the simple case described by Eq. (4.55). If adjoint weighting is omitted, the reactivity increment is expressed purely in terms of reaction rates. In this approximation, Eq. (4.55) reduces to the following equation:

$$\Delta\rho^{(0)} = \frac{-\int_{-x_p}^{x_p} \Delta\Sigma_a \phi_0(x)\, dx}{\int_{-a}^{a} \nu\Sigma_{f0}\phi_0(x)\, dx} \quad , \tag{4.60}$$

where $\Delta\rho^{(0)}$ represents the perturbation increment without application of the adjoint weighting function.

If the reactor is perturbed uniformly ($x_p = a$), the same result as Eq. (4.59) is obtained; i.e., the adjoint weighting has no influence on the result of a uniform perturbation in the simple one-group model.

For a perturbation in a small range, in which the cosine can be approximated by unity, Eq. (4.60) yields:

$$\Delta\rho^{(0)} = -\frac{\pi}{2} \cdot \frac{\Delta\Sigma_a}{\nu\Sigma_{f0}} \cdot \frac{x_p}{a} \quad . \tag{4.61}$$

Thus, the factor of 2 in Eq. (4.58) is replaced by a factor $\pi/2$. In both cases, this factor is larger than unity due to the center location of the perturbation. However, the reaction rate-based formula, Eq. (4.60), accounts for only a part of the total effect. The inclusion of the adjoint weighting shows that the importance of the center location of the perturbation is even more pronounced (a factor of 2 versus 1.57) than is indicated by $\Delta\rho^{(0)}$.

Homework Problems

1. Derive the first-order perturbation formula for

$$\mathbf{M}\Phi = \lambda\mathbf{F}\Phi$$

directly from this equation, i.e., without simultaneously carrying along the unperturbed equation

$$\mathbf{M}_0\Phi_0 = \lambda_0\mathbf{F}_0\Phi_0 \quad .$$

2. Carry out explicitly the manipulations needed to convert

$$\lambda(\Phi_0^*,\mathbf{F}\Phi) - \lambda_0(\Phi,\mathbf{F}_0^*\Phi_0^*)$$

into the terms of the exact perturbation formula. Use the continuous energy formulation.

3. Derive the following four formulas for the exact static reactivity increment:

a.
$$\Delta\rho = \frac{(\Phi_0^*,[\lambda_0\Delta\mathbf{F} - \Delta\mathbf{M}]\Phi)}{(\Phi_0^*,\mathbf{F}\Phi)}$$

b.
$$\Delta\rho = \frac{(\Phi_0^*,[\lambda\Delta\mathbf{F} - \Delta\mathbf{M}]\Phi)}{(\Phi_0^*,\mathbf{F}_0\Phi)}$$

c.
$$\Delta\rho = \frac{(\Phi^*,[\lambda_0\Delta\mathbf{F} - \Delta\mathbf{M}]\Phi_0)}{(\Phi^*,\mathbf{F}\Phi_0)}$$

d.
$$\Delta\rho = \frac{(\Phi^*,[\lambda\Delta\mathbf{F} - \Delta\mathbf{M}]\Phi_0)}{(\Phi^*,\mathbf{F}_0\Phi_0)} \quad .$$

4. Consider a perturbation of $+\Delta\Sigma_a$ for $r < r_a$ in a critical sphere. Assume $r_a \ll R$, with R being the critical radius. Find the corresponding change in the reactivity, using the unperturbed flux (a) from reaction rates and (b) from the first-order perturbation formula for the one-group approximation.

5. Apply the formulas of problem 4 to the following data: $R = 50$ cm; $\Sigma_t = 1.625$/cm; $\Sigma_s = 1.5$/cm; $\Sigma_a = 0.125$/cm; $\Sigma_c = 0.073$/cm; $\Sigma_f = 0.052$/cm; $\nu\Sigma_f = 0.126$/cm; $D = 0.25$ cm. Neglect the extrapolation length.
   a. Find k_{eff}.
   b. Introduce $\Delta\Sigma_c$ as a perturbation for $r < r_a = 5$ cm as a fraction $c\%$ of Σ_c. Find $\Delta\rho(c)$ for c between 0 and 100% from reaction rates and first-order perturbation theory.

c. Apply a one-group diffusion program and numerically calculate $k(c)$; convert it to $\Delta\rho(c)$.
d. Plot the three sets of results from problems 5b and 5c and discuss the comparison.

Review Questions

1. Describe the conceptual idea that leads to the application of perturbation theory.
2. State the neutron multiplication as an eigenvalue problem (in operator notation) and explain the idea behind the formulation of the neutron balance as an eigenvalue problem.

3a. Define the static reactivity in terms of k.
b. Define the reactivity increment in terms of k values.
4. What is first-order perturbation theory?

5a. Give the formula for the reactivity in first-order perturbation theory.
b. Why is the adjoint flux used in this formula?
6. Sketch the error of the formula for the first-order reactivity perturbation as function of the "change" in cross sections.

7a. Give the basic formula for the calculation of the exact reactivity (in operator notation).
b. In which slightly modified form is this formula of important practical value?

REFERENCE

1. K. O. Ott and W. A. Bezella, *Introductory Nuclear Reactor Statics,* American Nuclear Society, La Grange Park, Illinois (1983).

Five

THE POINT KINETICS EQUATIONS

5-1 The Exact Point Kinetics Equations

"Exact" in this context means that a given time-, space-, and energy-dependent neutronics model is condensed or lumped into the form of the point kinetics equation without applying simplifying approximations such as Eqs. (3.35) and (3.36). The exact point kinetics equations are formally about the same as the equations obtained from the one-group model, Eqs. (3.51) and (3.53). However, the derivation of the exact kinetics equations provides improved definitions for the integral kinetics parameters ρ, β, and Λ as well as all quantities involving p, ζ_k, and s.

Exact point kinetics equations were first derived by Henry.[1] Henry's derivation started from the Boltzmann equation, whereas the derivation presented here is based on the diffusion equation (see the discussion in Sec. 3-2A).

In addition to the exact point kinetics equations, there are approximate point kinetics equations that are based on certain simplifying assumptions (see Sec. 5-2). Probably the most common terminology is the use of "point kinetics equations" for the approximate ones, the ones based on the point reactor model presented in Sec. 5-2. They are to be distinguished from the "exact point kinetics equations" presented in this section.

5-1A Flux Factorization and Weighting Functions

The conceptual starting point of the derivation of the exact point kinetics equations from time-, space-, and energy-dependent neutronics equations is the *factorization* of the neutron flux into a purely time-dependent amplitude function, $p(t)$, and a space-, energy-, and time-dependent shape function,[a] $\psi(\mathbf{r},E,t)$:

$$\phi(\mathbf{r},E,t) = p(t) \cdot \psi(\mathbf{r},E,t) \quad . \tag{5.1}$$

---

[a]If the Boltzmann equation is used as in Ref. 1, the shape function also depends on angle.

This factorization of the flux is not an approximation such as the separation, Eq. (3.35). It merely splits one function into two. But then a second equation is required to make the factorization unique. Although flux factorization is general, employment is particularly advantageous if the flux variation with time consists primarily of flux amplitude changes with comparatively small time variations of the shape function. Such conditions are realized for most nuclear reactor transients. Some exceptions are special experiments that use an external neutron source in the form of a pulsed beam (see Chapter 9).

The second equation required to make the factorization unique can be used to shift the major time dependence into the amplitude function by constraining the time variation of the "magnitude" of the shape function. A convenient way to do this is to hold some integral (over space and energy) of the shape function constant in time. The derivation leads to a specific integral of ψ, which is held constant in time in order to obtain the desired form of the point kinetics equations.

Allowing the shape function to depend on time is a first generalization compared to the use of the time-independent shape in the derivation presented in Sec. 3-2. A second generalization used in the derivation does not remove an approximation, but rather exploits a certain freedom of choice; the neutronics equation is multiplied by a weight function, $w(\mathbf{r},E)$, prior to integration with respect to space and energy. Also introducing the flux factorization into the left side of Eq. (3.28) yields:

$$\int_V\int_0^\infty \frac{w(\mathbf{r},E)}{v(E)}\frac{\partial\phi(\mathbf{r},E,t)}{\partial t}\,dE\,dV = \frac{dp(t)}{dt}\int_V\int_0^\infty \frac{w(\mathbf{r},E)\psi(\mathbf{r},E,t)}{v(E)}\,dE\,dV$$
$$+\ p(t)\frac{d}{dt}\int_V\int_0^\infty \frac{w(\mathbf{r},E)\psi(\mathbf{r},E,t)}{v(E)}\,dE\,dV\quad . \tag{5.2}$$

The second term on the right side of Eq. (5.2) that appears with flux factorization can be eliminated by using just this integral to constrain the time variation of the shape function and thus make the factorization unique:

$$\int_V\int_0^\infty \frac{w(\mathbf{r},E)\psi(\mathbf{r},E,t)}{v(E)}\,dE\,dV = \text{constant}\quad . \tag{5.3}$$

Factorizing the neutron flux, applying an arbitrary weighting function, and constraining the variation of the flux shape by Eq. (5.3) do not introduce an approximation. In a sense, the factorization introduces a new degree of freedom (two instead of one function), which needs to be eliminated again by a constraint condition. The weight function only influences the precise split of the flux into an amplitude and shape function as well as the definitions of all other quantities appearing in the point kinetics equations. The end result for the flux,

$$\phi(\mathbf{r},E,t) = p^w(t) \cdot \psi^w(\mathbf{r},E,t) \quad , \tag{5.4}$$

is unique despite the dependence of the magnitude of both factors on the weighting function; the dependence of p and ψ on the choice of the weight function is indicated by the superscript w in Eq. (5.4).

In practice, approximate shape functions are used [e.g., $\psi(\mathbf{r},E,t) \simeq \phi_0(\mathbf{r},E)$ as in Sec. 3-2B and Sec. 5-2; or $\psi(\mathbf{r},E,t)$ is obtained by extrapolation of shape functions at earlier times, as in Chapter 11]. Since the shape function $\psi(\mathbf{r},E,t)$ is not known precisely, the resultant flux and particularly the flux amplitude $p(t)$ will only be approximate solutions. In this practical case then, both will depend on the choice of the weight function. It is thus advantageous to choose the weight function in such a way that it reduces the error resulting from inaccuracies in the shape function. Since the solution of the point kinetics equation is particularly sensitive to an error in the reactivity, a weight function should be chosen that reduces the effect of shape function inaccuracies on the reactivity. The initial adjoint flux, $\phi_0^*(\mathbf{r},E)$, fulfills this objective (compare Chapter 4 and Sec. 5-1B). With the initial adjoint flux as a weighting function,

$$w(\mathbf{r},E) = \phi_0^*(\mathbf{r},E) \quad , \tag{5.5}$$

the constraint condition for the shape function is given by

$$\int_V \int_0^\infty \frac{\phi_0^*(\mathbf{r},E)\psi(\mathbf{r},E,t)}{v(E)} \, dE \, dV = K_0 \quad , \tag{5.6}$$

with K_0 being a constant.

The effect of the choice of the weight function on the normalization of the shape function is briefly illustrated. Suppose an unnormalized shape function $\psi^{un}(\mathbf{r},E,t)$ is known. The space and energy integrals of ψ^{un} will in general depend on time. Two integrals are considered; one without and one with a weighting function:

$$\int_V \int_0^\infty \psi^{un}(\mathbf{r},E,t) \, dE \, dV = \hat{\psi}_1(t) \tag{5.7}$$

and

$$\int_V \int_0^\infty w(\mathbf{r},E)\psi^{un}(\mathbf{r},E,t) \, dE \, dV = \hat{\psi}_w(t) \quad . \tag{5.8}$$

The integrals depend on the weight function, as simple examples readily demonstrate. Dividing ψ^{un} by the right side of Eqs. (5.7) and (5.8) yields shape functions ψ^1 and ψ^w, which have different time dependencies and satisfy different constraint conditions for all time:

$$\int_V \int_0^\infty \psi^1(\mathbf{r},E,t)\, dE\, dV = 1 \text{ for all } t \tag{5.9}$$

and

$$\int_V \int_0^\infty w(\mathbf{r},E)\psi^w(\mathbf{r},E,t)\, dE\, dV = 1 \text{ for all } t \quad . \tag{5.10}$$

The shape functions ψ^1 and ψ^w differ by only a purely time-dependent function. The amplitude function must cancel this difference in the magnitude of the two shape functions in order to yield the unique flux of the left side of Eq. (5.4). In this sense, the amplitude function also depends on the choice of the weight function.

5-1B Derivation of the Exact Point Kinetics Equations for an Initially Critical Reactor

The time-dependent diffusion equation without an independent source in operator notation is given by (see Sec. 3-2A):

$$\frac{1}{v}\frac{\partial \Phi}{\partial t} = (\mathbf{F}_p - \mathbf{M})\Phi + S_d \quad . \tag{5.11}$$

The reactor is assumed to be critical for $t \le 0$:

$$0 = (\mathbf{F}_{p0} - \mathbf{M}_0)\Phi_0 + S_{d0} \quad . \tag{5.12}$$

The stationary delayed neutron source can be readily derived in a completely analogous form to the prompt neutron source from Eqs. (3.29c) and (3.32):

$$S_{d0} = \mathbf{F}_{d0}\Phi_0 \quad . \tag{5.13}$$

The stationary prompt and delayed neutron sources can be combined to obtain the familiar form:

$$0 = (\mathbf{F}_0 - \mathbf{M}_0)\Phi_0 \quad , \tag{5.14}$$

with $\mathbf{F}_0\Phi_0$ explicitly containing the delayed neutrons:

$$\mathbf{F}_0\Phi_0 = \chi_p(E)\int_0^\infty \nu_p\Sigma_f(\mathbf{r},E')\phi_0(\mathbf{r},E')\, dE' + \sum_k \chi_{dk}(E)\int_0^\infty \nu_{dk}\Sigma_f(\mathbf{r},E')\phi_0(\mathbf{r},E')\, dE' \quad . \tag{5.15}$$

The static fission neutron source used in the initial condition of dynamics problems must explicitly contain the static delayed neutron source.

For a description of the degree of off-criticality from the concept of the reactivity, the operator $\mathbf{F}_d$ is added and subtracted within the parentheses in Eq. (5.11):

$$\frac{1}{v}\frac{\partial \Phi}{\partial t} = (\mathbf{F} - \mathbf{M} - \mathbf{F}_d)\Phi + S_d \quad ; \tag{5.16}$$

$\mathbf{F}\Phi$ is given by Eq. (5.15) with the subscript 0 deleted. The use of the simpler form of the total fission operator,

$$\mathbf{F}\Phi = \chi(E) \int_0^\infty \nu\Sigma_f(\mathbf{r},E',t)\phi(\mathbf{r},E',t)\, dE' \quad , \tag{5.17}$$

would lead to an error since the definition of the total $\chi(E)\nu\Sigma_f(\mathbf{r},E')$ in Eq. (5.17) is not consistently based on the sum of prompt and delayed fission neutron sources as in Eq. (5.15). In the framework of purely static problems, the simpler definition, Eq. (5.17), with t deleted, can be applied without encountering inconsistencies resulting from adding and subtracting different delayed neutron sources.

The quantity $\mathbf{F}_d\Phi$ that is added and subtracted in Eq. (5.16) is not the actual delayed neutron source at time t. The latter is given by S_d. The quantity $\mathbf{F}_d\Phi$ may be called the "quasi-stationary delayed neutron source," i.e., a source of delayed neutrons that would be produced in a stationary reactor with fission cross sections and neutron flux as they exist at time t.

Multiplying Eq. (5.16) with the initial adjoint flux and integrating with respect to energy and space yields:

$$\frac{\partial}{\partial t}\left(\Phi_0^*,\frac{1}{v}\Phi\right) = (\Phi_0^*,[\mathbf{F} - \mathbf{M}]\Phi) - (\Phi_0^*, \mathbf{F}_d\Phi) + (\Phi_0^*,S_d) \quad , \tag{5.18}$$

where the initial adjoint flux is the solution of the adjoint problem to Eq. (5.14), i.e.,

$$(\mathbf{F}_0^* - \mathbf{M}_0^*)\Phi_0^* = 0 \quad . \tag{5.19}$$

A frequently used alternative formulation to Eq. (5.18) can be derived using the difference of the operators $\mathbf{F}$ and $\mathbf{M}$ at times $t = 0$ and $t = t$:

$$\Delta\mathbf{F} = \mathbf{F} - \mathbf{F}_0 \tag{5.20a}$$

and

$$\Delta\mathbf{M} = \mathbf{M} - \mathbf{M}_0 \quad . \tag{5.20b}$$

This alternative formulation is obtained by subtracting the vanishing scalar product,

$$(\Phi_0^*,[\mathbf{F}_0 - \mathbf{M}_0]\Phi) = (\Phi,[\mathbf{F}_0^* - \mathbf{M}_0^*]\Phi_0^*) = 0 \quad , \tag{5.21}$$

from Eq. (5.18). This yields:

$$\frac{\partial}{\partial t}(\Phi_0^*,\frac{1}{v}\Phi) = (\Phi_0^*,[\Delta\mathbf{F} - \Delta\mathbf{M}]\Phi) - (\Phi_0^*,\mathbf{F}_d\Phi) + (\Phi_0^*,S_d) \quad . \tag{5.22}$$

The flux factorization, Eq. (5.1), is introduced in Eq. (5.22), and Eqs. (5.18) and (5.22) are divided by the importance-weighted quasi-stationary source of fission neutrons, as produced by the flux shape function, $\psi((\mathbf{r},E,t)$:

$$F(t) = (\Phi_0^*,\mathbf{F}\Psi) \quad . \tag{5.23}$$

Application of the constraint condition, Eq. (5.6), yields the kinetics equation for the flux amplitude function in the form:

$$\Lambda(t)\dot{p}(t) = [\rho(t) - \beta(t)]p(t) + s_d(t) \tag{5.24}$$

with

$$s_d(t) = \frac{(\Phi_0^*,\sum_k \chi_{dk}\lambda_k C_k)}{F(t)} = \frac{F_0}{F(t)}\sum_k \lambda_k \zeta_k(t) \quad . \tag{5.25}$$

The delayed neutron source, $s_d(t)$, appears in the kinetics equation in an adjoint weighted, integrated, and relative form that is referred to as "reduced" in the following text. The number density of the precursors is also "reduced" in the same fashion.

The definitions of the quantities in Eqs. (5.24) and (5.25) as they evolve from the derivation of the exact kinetics equations are the following:

$$\Lambda(t) = \frac{(\Phi_0^*,\frac{1}{v}\Psi)}{(\Phi_0^*,\mathbf{F}\Psi)} = \frac{K_0}{F(t)} \tag{5.26a}$$

and

$$\rho(t) = \frac{1}{F(t)}(\Phi_0^*,[\mathbf{F} - \mathbf{M}]\Psi) \tag{5.26b}$$

or

$$\rho(t) = \frac{1}{F(t)}(\Phi_0^*,[\Delta\mathbf{F} - \Delta\mathbf{M}]\Psi) \tag{5.26c}$$

and

$$\beta(t) = \frac{1}{F(t)}(\Phi_0^*,\mathbf{F}_d\Psi) = \sum_k \beta_k(t) \quad , \tag{5.26d}$$

with

$$\beta_k(t) = \frac{1}{F(t)}(\Phi_0^*,\mathbf{F}_{dk}\Psi) \tag{5.26e}$$

and

$$\zeta_k(t) = \frac{1}{F_0}(\Phi_0^*,\chi_{dk}C_k) \quad . \tag{5.26f}$$

The definition of $\zeta_k(t)$ is obtained from the decomposition of the importance-weighted reduced delayed neutron source; the use of F_0 in the denominator of $\zeta_k(t)$ is discussed below:

$$s_d(t) = \frac{1}{F}(\Phi_0^*,S_d) = \frac{F_0}{F}\sum_k \frac{\lambda_k}{F_0}(\Phi_0^*,\chi_{dk},C_k)$$

$$= \frac{F_0}{F}\sum_k \lambda_k\zeta_k(t) \quad . \tag{5.27}$$

The reduction of the space- and time-dependent precursor balance equation must yield balance equations for the reduced precursors $\zeta_k(t)$ as they are defined in Eq. (5.26f). Thus, Eq. (3.32) must be multiplied not only with $\phi_0^*(\mathbf{r},E)$ but also with $\chi_{dk}(E)$. Introducing the flux factorization and integrating the resultant equation with respect to $\mathbf{r}$ and E yields:

$$\frac{d}{dt}(\Phi_0^*,\chi_{dk}C_k) = -\lambda_k(\Phi_0^*,\chi_{dk}C_k) + (\Phi_0^*,\mathbf{F}_{dk}\Psi)p(t) \quad , \tag{5.28}$$

which is divided by the time-independent quantity,

$$F_0 = (\Phi_0^*,\mathbf{F}_0\Psi_0) \quad , \tag{5.29}$$

rather than by $F(t)$, since only a constant quantity can be taken into the derivative on the left side. This yields the exact balance equation for the reduced precursors,

$$\dot{\zeta}_k(t) = -\lambda_k\zeta_k(t) + \frac{F(t)}{F_0}\beta_k(t)p(t) \quad , \tag{5.30}$$

with $\zeta_k(t)$ and $\beta_k(t)$ given by Eqs. (5.26f) and (5.26e).

To obtain the more familiar form of the precursor balance equations, Eq. (5.28) is divided by the time-independent quantity K_0 instead of F_0, with K_0 expressed as

$$K_0 = F(t)\cdot\Lambda(t) \quad . \tag{5.31}$$

This yields, instead of Eq. (5.30),

$$\dot{c}_k(t) = -\lambda_k c_k(t) + \frac{1}{\Lambda(t)} \beta_k(t) p(t) \quad , \tag{5.32}$$

with

$$c_k(t) = \frac{(\Phi_0^*, \chi_{dk} C_k)}{K_0} = \frac{(\Phi_0^*, \chi_{dk} C_k)}{F(t) \cdot \Lambda(t)} \quad . \tag{5.33}$$

Equations (5.30) and (5.32) differ particularly in the appearance of $1/\Lambda$ in Eq. (5.32) versus $F(t)/F_0$ in Eq. (5.30). The unnecessary introduction of $1/\Lambda$ in Eq. (5.32) has undesirable consequences in the formulation of the important "zero Λ" approximation (see Sec. 8-1). To avoid these difficulties, the exact kinetics equations are used in the form:

$$\dot{p}(t) = \frac{\rho(t) - \beta(t)}{\Lambda(t)} p(t) + \frac{1}{\Lambda_0} \sum_k \lambda_k \zeta_k(t) \tag{5.34a}$$

and

$$\dot{\zeta}_k(t) = -\lambda_k \zeta_k(t) + \frac{F(t)}{F_0} \beta_k(t) p(t) \quad . \tag{5.34b}$$

The more familiar forms,

$$\dot{p}(t) = \frac{\rho(t) - \beta(t)}{\Lambda(t)} p(t) + \sum_k \lambda_k c_k(t) \tag{5.35a}$$

and

$$\dot{c}_k(t) = -\lambda_k c_k(t) + \frac{1}{\Lambda(t)} \beta_k(t) p(t) \quad , \tag{5.35b}$$

are not used in this text for reasons given in Chapter 3 (p. 35).

5-1C Derivation of the Exact Point Kinetics Equations for an Initially Subcritical Reactor

The derivation of the exact point kinetics equations for an initially subcritical reactor is completely analogous to the derivation for the initially critical reactor. The basic difference is in the weighting function. The adjoint flux in a critical reactor can be readily defined as the solution of the problem characterized by the adjoint operator, Eq. (5.19). Similarly, adjoint fluxes may be defined for other homogeneous problems, such as the λ eigenvalue problem [see Eq. (5.41)].

A subcritical reactor, however, with a stationary and finite neutron flux for $t \leq 0$ is not a homogeneous problem since a flux-independent source is required to sustain the finite flux level. In contrast to the adjoint of the critical reactor problem, the adjoint of a subcritical reactor problem with an independent source is not uniquely defined, as discussed below.

The time-dependent neutron balance equation is given by

$$\frac{1}{v}\frac{\partial \Phi}{\partial t} = (\mathbf{F}_p - \mathbf{M})\Phi + S_d + S \quad , \tag{5.36}$$

where

$$S = S(\mathbf{r},E,t) \tag{5.37}$$

denotes the flux-independent source. The other quantities are the same as in Eq. (5.11)

A stationary state is again assumed for $t \leq 0$:

$$0 = (\mathbf{F}_{p0} - \mathbf{M}_0)\Phi_0 + S_{d0} + S_0 \tag{5.38}$$

or

$$0 = (\mathbf{F}_0 - \mathbf{M}_0)\Phi_0 + S_0 \quad , \tag{5.39}$$

where all quantities are independent of time.

Frequently, an adjoint function for a subcritical system is introduced as the solution of the following inhomogeneous problem (see, for example, Ref. 2):

$$(\mathbf{M}_0^* - \mathbf{F}_0^*)\Phi_{det}^* = \Sigma_{det}(\mathbf{r},E) \quad , \tag{5.40}$$

where $\Sigma_{det}(\mathbf{r},E)$ denotes the macroscopic neutron capture or fission cross section of a neutron detector (e.g., a fission chamber). The adjoint function defined by Eq. (5.40) represents the flux response to the injection of source neutrons measured by the special detector, which is characterized by $\Sigma_{det}(\mathbf{r},E)$ (compare Ref. 2). Since many kinds of detectors may be used to measure the flux response to a variation of the independent source, the adjoint as defined by Eq. (5.40) is different for different detectors or detector positions. This detector dependence of the adjoint is indicated by the index *det* in Φ_{det}^*. Obviously Φ_{det}^* is not a uniquely defined function for a given subcritical reactor.

Two other adjoint problems are often defined to generate adjoint weighting functions for the reduction of the space- and energy-dependent kinetics equations to the point kinetics equations:

$$(\mathbf{M}_0^* - \lambda_0\mathbf{F}_0^*)\Phi_{\lambda 0}^* = 0 \tag{5.41}$$

and

$$\left(\mathbf{M}_0^* - \mathbf{F}_0^* + \frac{\alpha_0}{v}\right)\Phi_{\alpha 0}^* = 0 \quad . \tag{5.42}$$

Equations (5.41) and (5.42) yield the initial adjoint λ mode and α mode, respectively.

The comparative advantages of the application of $\Phi_{\lambda 0}^*$ or $\Phi_{\alpha 0}^*$ as a weighting function in reducing the kinetics equations are discussed in detail in Sec. 11-4A. It is shown that normally $\Phi_{\lambda 0}^*$ can be expected to be a better weighting function than $\Phi_{\alpha 0}^*$.

An adjoint flux for the initially subcritical reactor with an independent source that is more closely related to the initial problem than $\Phi_{\lambda 0}^*$ may be defined in the following way: After the inhomogeneous initial problem, Eq. (5.39), has been solved, it may be converted *a posteriori* into a homogeneous problem, with a solution identical to that of Eq. (5.39):

$$0 = (\mathbf{F}_0 - \mathbf{M}_0 + \mathbf{S}_0)\Phi_0 \quad , \tag{5.43}$$

where $\mathbf{S}_0$ is defined by

$$\mathbf{S}_0\Phi_0 = S_0(\mathbf{r},E) \quad . \tag{5.44}$$

The adjoint problem of the "homogenized" initial problem can be written as:

$$0 = (\mathbf{F}_0^* - \mathbf{M}_0^* + \mathbf{S}_0^*)\Phi_{h0}^* \quad . \tag{5.45}$$

The subscript h indicates that Φ_{h0}^* is found from the adjoint of the "homogenized" initial problem.

The definition of the source operator by Eq. (5.44) does not yield a unique result for $\mathbf{S}_0$. The arbitrariness of the source operator $\mathbf{S}_0$ has no impact on the solution of Eq. (5.43) since $\mathbf{S}_0\Phi_0$ is uniquely defined by Eq. (5.44). Different source operators may, however, lead to different adjoint functions, Φ_{h0}^*, as a solution of Eq. (5.45).

The adjoint function as derived from homogenization of the inhomogeneous problem, Eqs. (5.43) to (5.45), has not been practically applied to subcritical kinetics. Also lacking is a general investigation of differences in error cancellation that result from using the various adjoints as weighting functions (for a special investigation based on α mode expansion see Gozani[3]). The λ mode adjoint function of Eq. (5.40) is used as convention, since it leads to a formulation that is closest to the one for the initially critical reactor[4] and thus eliminates the first-order flux errors as shown in Eq. (5.58).

The kinetics equations for the initially subcritical system are derived in the same way as for an initially critical reactor. The initial λ mode adjoint flux, $\Phi^*_{\lambda 0}$, is employed as the weighting function and an independent source is included. By incorporating both generalizations, Eq. (5.18) is replaced by:

$$\frac{\partial}{\partial t}(\Phi^*_{\lambda 0},\frac{1}{v}\Phi) = (\Phi^*_{\lambda 0},[\mathbf{F} - \mathbf{M}]\Phi) - (\Phi^*_{\lambda 0},\mathbf{F}_d\Phi) + (\Phi^*_{\lambda 0},S_d) + (\Phi^*_{\lambda 0},S) \quad ; \tag{5.46}$$

Eq. (5.46) approaches Eq. (5.18) when the initial reactor becomes critical.

Inserting into Eq. (5.46) the flux factorization of Eq. (5.1), using the constraint condition of Eq. (5.6), with ϕ^*_0 replaced by $\phi^*_{\lambda 0}$,

$$\int_V\int_E \frac{\phi^*_{\lambda 0}(\mathbf{r},E)\psi(\mathbf{r},E,t)}{v(E)}\, dE\, dV = K_0 \quad , \tag{5.47}$$

and dividing by

$$F_\lambda(t) = (\Phi^*_{\lambda 0},\mathbf{F}\Psi) \tag{5.48}$$

gives the exact point kinetics equations for an initially subcritical reactor (in ζ notation) as:

$$\dot{p}(t) = \frac{\rho(t) - \beta(t)}{\Lambda(t)}\,p(t) + \frac{1}{\Lambda_0}\sum_k \lambda_k\zeta_k(t) + \frac{s(t)}{\Lambda(t)} \tag{5.49a}$$

and

$$\dot{\zeta}_k(t) = -\lambda_k\zeta_k(t) + \frac{F_\lambda(t)}{F_{\lambda 0}}\,\beta_k(t)p(t) \quad . \tag{5.49b}$$

All quantities in Eqs. (5.49) are defined as for the initially critical reactor, with Φ^*_0 replaced by $\Phi^*_{\lambda 0}$ and $F(t)$ by $F_\lambda(t)$. The additional quantity, the reduced independent source, is defined as:

$$s(t) = \frac{1}{F_\lambda(t)}(\Phi^*_{\lambda 0},S) \quad . \tag{5.50}$$

The time-dependent reactivity in the kinetics equations for the initially subcritical system is given by an expression of the same form as the exact static reactivity:

$$\rho(t) = \frac{(\Phi^*_{\lambda 0},[\mathbf{F} - \mathbf{M}]\Psi)}{(\Phi^*_{\lambda 0},\mathbf{F}\Psi)} \quad . \tag{5.51}$$

The initial value of $\rho(t)$ is equal to the static reactivity of the initially subcritical system, which can be readily shown by revolving the scalar products in Eq. (5.51):

$$\rho(0) = \frac{(\Phi^*_{\lambda 0},[\mathbf{F}_0 - \mathbf{M}_0]\Psi_0)}{(\Phi^*_{\lambda 0},\mathbf{F}_0\Psi_0)}$$

$$= \frac{(\Psi_0,[\mathbf{F}^*_0 - \mathbf{M}^*_0]\Phi^*_{\lambda 0})}{(\Psi_0,\mathbf{F}^*_0\Phi^*_{\lambda 0})} = \rho_0 = \rho_0^{st} \quad . \tag{5.52}$$

The difference between $\rho(t)$ and the initial static reactivity, ρ_0, may be expressed by the differences of the operators in the following form:

$$\delta\rho(t) = \frac{1}{F_\lambda(t)} (\Phi^*_{\lambda 0},[\lambda_0 \Delta\mathbf{F} - \Delta\mathbf{M}]\Psi) \quad , \tag{5.53a}$$

with

$$\rho(t) = \rho_0 + \delta\rho(t) \tag{5.53b}$$

and

$$\rho_0 = 1 - \lambda_0 = \rho_0^{st} \quad . \tag{5.53c}$$

Equations (5.53) can be readily verified:

$$\delta\rho(t) \cdot F_\lambda(t) = (\Phi^*_{\lambda 0},[\lambda_0 \Delta\mathbf{F} - \Delta\mathbf{M}]\Psi)$$

$$= (\Phi^*_{\lambda 0},[\lambda_0 \mathbf{F} - \mathbf{M} - (\lambda_0 \mathbf{F}_0 - \mathbf{M}_0)]\Psi)$$

$$= (\Phi^*_{\lambda 0},[\lambda_0 \mathbf{F} - \mathbf{M}]\Psi)$$

$$= (\Phi^*_{\lambda 0},[(1 - \rho_0)\mathbf{F} - \mathbf{M}]\Psi)$$

$$= (\Phi^*_{\lambda 0},[\mathbf{F} - \mathbf{M}]\Psi) - \rho_0(\Phi^*_{\lambda 0},\mathbf{F}\Psi)$$

$$= [\rho(t) - \rho_0]F_\lambda(t) \quad . \tag{5.54}$$

5-1D Reactivity in the Exact Point Kinetics Equations

A reactivity or a reactivity increment in the amount of β is historically called one dollar = 1\$. One hundredth of a dollar of reactivity is called one cent = 1¢. These reactivity units are normally written behind the amount.

The value of the reactivity dollar differs depending on the approximation applied to calculate β. Equation (5.26d) yields the "exact" value of β. An often applied approximation is $\beta(t) \simeq \beta_0$, the initial value. Then 1\$ is given by β_0.

The reactivity $\rho(t)$, Eq. (5.26c), which appears in the exact kinetics equation, is called "dynamic reactivity." The dynamic reactivity is formed in full analogy to the exact static reactivity[b] (see Sec. 4-3). The essential difference is that ρ^{dyn} is formed with the time-dependent flux $\phi(\mathbf{r},E,t)$ as it physically appears during a transient whereas ρ^{st} is formed with the λ mode, $\Phi_\lambda(\mathbf{r},E)$, of the "perturbed" system; compare Eq. (4.40). The conceptual need for two types of reactivities is discussed in Chapter 9.

If the flux and adjoint flux used in the exact point kinetics equations are inaccurate, the resultant reactivity will also be inaccurate. A substantial part of this inaccuracy results from the zero-order terms, $(\Phi_0^*,\mathbf{F}_0\Psi)$ and $(\Phi_0^*,\mathbf{M}_0\Psi)$. The difference between these two terms must be zero in a critical reactor. With inaccurate adjoint fluxes, an erroneous finite residue results, which leads to an error in the reactivity. The classical conceptual idea for eliminating the error caused by the zero-order terms is to cancel analytically the zero-order terms in the perturbation formula. Then, the first-order perturbation (i.e., ρ), is found directly in terms of the first-order operator changes $\Delta\mathbf{F}$ and $\Delta\mathbf{M}$, and not as the difference between large zero-order terms [compare Eqs. (5.26b) and (5.26c)]. With modern high-speed digital computers, the initial adjoint equation can be solved with such numerical accuracy that the subtracted scalar products, Eq. (5.21), are zero in all the considered digits. In this case, Eqs. (5.26b) and (5.26c) will yield the same numerical result for the reactivity. There are, however, cases in which the initial adjoint equation is not solved accurately enough (see Chapter 11). The reactivity formula based on the operator differences should then be used.

The initial adjoint flux was used in Sec. 5-1B to reduce the error that results from inaccuracies of the time-dependent flux shape. As discussed in Chapter 4, the adjoint flux achieves this by eliminating the effect of the first-order flux change $\Delta\psi(\mathbf{r},E,t)$ on the reactivity. The error of the resulting reactivity formula is then of second order or "stationary" about the reference system.

The stationarity property of the reactivity formula is even more important in dynamics problems than in statics problems since the space- and energy-dependent flux shape varies with time, and since the full time dependence of the flux shape including the delayed neutron distributions is generally more difficult to calculate than a few static flux shapes. It is thus desirable to carry the stationarity property of the exact static reactivity formula over into dynamics by forming the dynamic reactivity in the same way. This obviously can be achieved for the case

---

[b] If necessary, dynamic and static values are distinguished by superscripts: ρ^{dyn} and ρ^{st}.

of an initially critical reactor: Eqs. (5.26b) and (5.26c) are fully analogous to Eqs. (4.43) and (4.40) in their structure as well as in the use of the initial adjoint as the weighting function.

The initial state of a source-driven subcritical reactor, however, is mathematically an inhomogeneous problem. Consequently, there is no "initial" adjoint flux that can be used in true analogy to the exact static reactivity formula. In Sec. 5-1C, the initial adjoint λ mode was applied as the weighting function. Since $\Phi^*_{\lambda 0}$ is not the adjoint flux of the physical initial state, the stationarity property of the resulting reactivity formula is not guaranteed and must be investigated explicitly.

According to Eq. (5.53b), the reactivity can be expressed as the sum of two parts:

$$\rho(t) = \rho_0 + \delta\rho(t) \quad . \tag{5.55a}$$

Due to the use of $\Phi^*_{\lambda 0}$ as the weighting function, the initial reactivity ρ_0 is identical with the static reactivity,

$$\rho_0 = \rho_0^{st} \quad , \tag{5.55b}$$

where $\delta\rho(t)$ describes the reactivity increment about ρ_0 (see Fig. 5-1). The static reactivity ρ_0 is introduced per convention; it is obtained from Eq. (5.53c) without an approximation. Thus, only the stationarity property of the formula for $\delta\rho(t)$ is of interest.

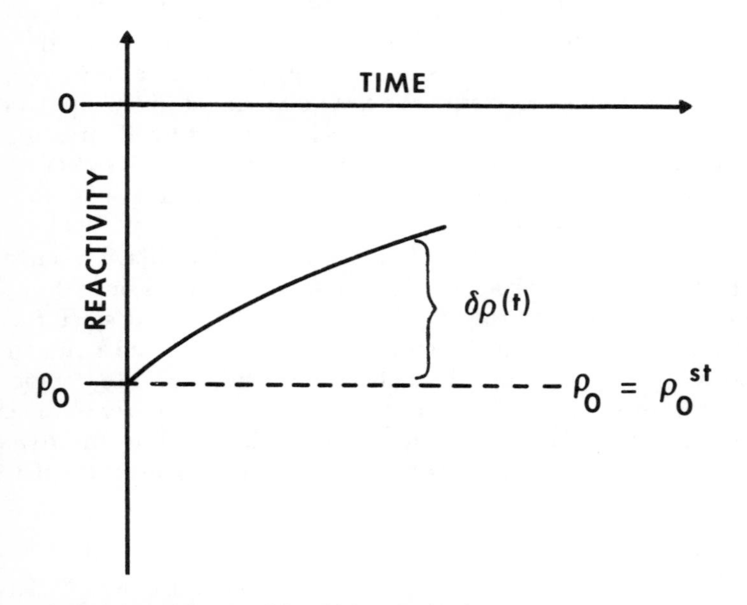

Fig. 5-1. Description of the reactivity $\rho(t)$ in subcritical reactors.

To investigate the stationarity property of Eq. (5.53a), a formula for $\delta\rho$ that still contains the zero-order terms is needed. The proof of Eq. (5.53a) yielded such an interim result [see Eq. (5.54)]:

$$\delta\rho(t) = \frac{1}{F_\lambda(t)}(\Phi^*_{\lambda 0},[\lambda_0\mathbf{F} - \mathbf{M}]\Psi) \quad . \tag{5.56}$$

To partition the numerator of Eq. (5.56) in zero-, first-, and second-order terms, the operators are written as sums of unperturbed operators plus differences, Eq. (5.20); in addition, Ψ is decomposed as

$$\psi(\mathbf{r},E,t) = \psi_0(\mathbf{r},E) + \Delta\psi(\mathbf{r},E,t) \quad , \tag{5.57}$$

where ψ_0 is the initial flux shape. Substituting these sums into Eq. (5.56) yields:

$$\begin{aligned}\delta\rho(t) = \frac{1}{F_\lambda(t)}\{&(\Phi^*_{\lambda 0},[\lambda_0\mathbf{F}_0 - \mathbf{M}_0]\Psi_0) \\ &+ (\Phi^*_{\lambda 0},[\lambda_0\Delta\mathbf{F} - \Delta\mathbf{M}]\Psi_0) \\ &+ (\Phi^*_{\lambda 0},[\lambda_0\mathbf{F}_0 - \mathbf{M}_0]\Delta\Psi) \\ &+ (\Phi^*_{\lambda 0},[\lambda_0\Delta\mathbf{F} - \Delta\mathbf{M}]\Delta\Psi)\} \quad .\end{aligned} \tag{5.58}$$

The first and third terms on the right side of Eq. (5.58) are zero because $\Phi^*_{\lambda 0}$ satisfies the relationship expressed by Eq. (5.47). (Note that in the case of an initially critical reactor, the analogous first term disappears because the operator in parentheses applied to the initial flux is already zero, independent of the weight function.) The second term in Eq. (5.58) is the major part of the reactivity formula. The fourth term contains differences in operators *and* in the flux shape. *Both* of these differences are initially zero and develop after the onset of the transient. Therefore, the fourth term is quadratic[c] in nature. Thus, the reactivity formula of Eq. (5.56) has the same stationarity property as the corresponding formula for the initially critical reactor *if* Ψ_0 is equal to the *initial flux shape*, i.e., calculated from the initial inhomogeneous problem, Eq. (5.39); then $\delta\rho(t)$ is calculated as an increment of the dynamic reactivity.

The stationarity property of the reactivity about an initial subcritical state is different if ρ and $\delta\rho$ are approximated by the static reactivity or

[c]It may easily be shown that the lowest order contribution from the denominator is quadratic in the same sense.

its increment, respectively. In first-order perturbation theory, Ψ is decomposed around the initial λ mode, $\Psi_{\lambda 0}$:

$$\psi(\mathbf{r},E,t) = \psi_{\lambda 0}(\mathbf{r},E) + \Delta\psi_{\lambda}(\mathbf{r},E,t) \quad (5.59)$$

(see Secs. 4-2 and 5-2). If this equation is inserted into Eq. (5.56), the following result is obtained:

$$\begin{aligned} \delta\rho(t) = \frac{1}{F_{\lambda}(t)} \{ & (\Phi^*_{\lambda 0},[\lambda_0 \mathbf{F}_0 - \mathbf{M}_0]\Psi_{\lambda 0}) \\ & + (\Phi^*_{\lambda 0},[\lambda_0 \Delta\mathbf{F} - \Delta\mathbf{M}]\Psi_{\lambda 0}) \\ & + (\Phi^*_{\lambda 0},[\lambda_0 \mathbf{F}_0 - \mathbf{M}_0]\Delta\Psi_{\lambda}) \\ & + (\Phi^*_{\lambda 0},[\lambda_0 \Delta\mathbf{F} - \Delta\mathbf{M}]\Delta\Psi_{\lambda})\} \quad . \end{aligned} \quad (5.60)$$

Again, the first and the third terms disappear for the reasons previously given. The decomposition Eq. (5.59), however, introduces a $\Delta\Psi_{\lambda}$ that does *not* vanish initially. Thus, the fourth term on the right side of Eq. (5.60) is linear in nature. It is, in principle, of the same order of expansion as $\delta\rho(t)$ itself. The neglect of the fourth term in Eq. (5.60) can lead to substantial errors. For further discussion of the reactivity in subcritical reactors, see Sec. 9-2.

A direct effect of application of the energy-dependent weighting function, $\phi^*_0(\mathbf{r},E)$, is that the prompt and delayed neutron emission spectra are not eliminated by the energy integration as in Sec. 3-2B. The importance-weighted[d] integrals of the emission spectra influence the kinetics parameters, particularly β_k (see the discussion below). An additional consequence is that the difference between the importance of neutrons before and after scattering has an effect on the reactivity.

The independent source $S(\mathbf{r},E,t)$ is also weighted by the importance of a neutron injected into the reactor at various locations and with various energies. Obviously, the same source in the center and at the periphery of a reactor have a different effect on the flux response. The neutron importance describes these differences as a function of the source location.

It is emphasized again, at the conclusion of the derivation of the kinetics equations, that the adjoint function weighting was applied in the derivation merely to reduce the reactivity error resulting from an approximately known time-dependent neutron flux; adjoint weighting

---

[d]See Sec. B-4 of App. B for a discussion of the interpretation of the adjoint flux as an importance function.

was *not* applied because of an *a priori* need to weigh any emitted or disappearing neutrons with their respective importance. There is no *a priori* necessity for importance weighting. For a more detailed discussion of the time variations of the flux and its treatment, see Chapter 11.

5-1E Effective Delayed Neutron Fractions and Further Discussion of the Exact Point Kinetics Equations

The "effective" delayed neutron fraction, $\beta(t)$, differs numerically from the corresponding quantity without adjoint flux weighting, i.e., from the one-group β, Eq. (3.50). These differences result from the energy as well as the space dependence of the adjoint flux. The energy dependence of the adjoint flux provides the proper importance weighting for the emission spectra of the delayed neutrons; the space dependence weighs the spatial distribution of the delayed neutron emission. The nature of these differences can be understood, semi-quantitatively, by applying a separation approximation for the initial adjoint flux:

$$\phi_0^*(\mathbf{r},E) \simeq \phi_0^*(\mathbf{r})\varphi^*(E) \quad . \tag{5.61}$$

The approximation of Eq. (5.61) is inserted into the equation for β_k [Eq. (5.26e)]. First consider the simple case of a single fissionable isotope (assuming time-independent ψ and Σ's):

$$\beta_k = \frac{(\Phi_0^*,\mathbf{F}_{dk}\Psi)}{(\Phi_0^*,\mathbf{F}\Psi)} \quad . \tag{5.62}$$

Substituting Eq. (5.61) into Eq. (5.62) yields:

$$\beta_k \approx \frac{\int_E \chi_{dk}(E)\varphi_0^*(E)\, dE \int_V \phi_0^*(\mathbf{r})\nu_{dk} \int_{E'} \Sigma_f(\mathbf{r},E')\psi(\mathbf{r},E')\, dE'\, dV}{\int_E \chi(E)\varphi_0^*(E)\, dE \int_V \phi_0^*(\mathbf{r})\bar{\nu} \int_{E'} \Sigma_f(\mathbf{r},E')\psi(\mathbf{r},E')\, dE'\, dV} \quad . \tag{5.63}$$

Since ν_{dk} is independent of energy (see Sec. 2-2), it is written outside of the E' integral; $\bar{\nu}$ in the denominator is an average value of $\nu(E')$, which originally appeared under the E' integral. The average value $\bar{\nu}$ and ν_{dk} are only weakly dependent on space. If this weak space dependence is neglected, the space and E' integrals in the numerator and denominator cancel and one obtains

$$\beta_k \approx \overline{\beta}_k \gamma_{dk} \tag{5.64a}$$

with

TABLE 5-I
Relative Importance of Delayed and Prompt Fission Neutrons in FBRs

Delayed Neutron Group, k	Relative Importance of Delayed and Prompt Fission Neutrons, γ_{dk}	Delayed Neutron Group, k	Relative Importance of Delayed and Prompt Fission Neutrons, γ_{dk}
1	0.802	4	0.825
2	0.831	5	0.825
3	0.818	6	0.825

$$\overline{\beta}_k = \frac{\nu_{dk}}{\overline{\nu}} \tag{5.64b}$$

and

$$\gamma_{dk} = \frac{\int_E \chi_{dk}(E)\varphi_0^*(E)\, dE}{\int_E \chi(E)\varphi_0^*(E)\, dE} \quad . \tag{5.64c}$$

Thus, for a single fissionable isotope, the "effective" and one-group β's differ by the ratio of the importance weight of the delayed and total fission neutron emission spectra, γ_{dk}.

Table 5-I gives the values of the ratios of the importance weights of delayed and total neutron emission spectra for a typical large fast breeder reactor (FBR). The average importance of the delayed neutrons is ~18% smaller than that of the total number of fission neutrons. The latter importance is close to the importance of prompt neutrons. The reason for the small variation in the importance of the different delay groups and a large difference between delayed and prompt neutron importance is that delayed neutrons in *all* delay groups have only a small chance of causing fission in ^{238}U since most delayed neutrons have energies below the threshold in the ^{238}U fission cross section (see Table 2-IV for the average delayed neutron energies). This is reflected in a lower importance of $\chi_d(E)$ compared to $\chi(E)$. In addition, $\eta(E)$ of ^{239}Pu,

$$\eta(E) = \frac{\nu(E)\sigma_f(E)}{\sigma_a(E)} \quad ,$$

increases with energy, which also gives χ a higher importance than χ_d. These effects cause delayed neutrons to have an ~18% smaller importance than prompt neutrons in fast reactors.

In a thermal reactor, the adjoint spectrum generally increases with decreasing energy because of increasing proximity to the range of the thermal fission. Fast fission is of minimal importance due to the rapid slowing down of the neutrons below the fast fission threshold in ^{238}U. Thus, the increase of $E\varphi_0^*(E)$ with decreasing E prevails in the determination of the average importance of delayed and prompt fission neutrons. Therefore, delayed neutrons have a larger importance than prompt neutrons, in a thermal reactor.

For a reactor with two fissionable isotopes,[e] such as ^{239}Pu and ^{238}U, Eq. (5.65) is obtained instead of Eq. (5.63):

$$\beta_k \simeq \frac{\int_E \chi_{dk}(E)\varphi_0^*(E)\,dE}{\int_E \chi(E)\varphi_0^*(E)\,dE} \times \frac{\int_V \phi_0^*(\mathbf{r})[\nu_{dk9}R_{f9}(\mathbf{r}) + \nu_{dk8}R_{f8}(\mathbf{r})]\,dV}{\int_V \phi_0^*(\mathbf{r})\bar{\nu}R_f(\mathbf{r})\,dV} , \tag{5.65}$$

where R represents an energy-integrated reaction rate. Due to the assumed isotopic independence of the delayed neutron emission spectra, the same factor as in Eq. (5.64c) describes the reduction of the importance of the delayed neutrons compared to the prompt neutrons.

The denominator of Eq. (5.65) is essentially the same as in Eq. (5.63) except that both isotopes contribute to $\bar{\nu}R_f$. In the numerator, however, the ^{238}U effect is strongly amplified by its high delayed neutron yield compared to ^{239}Pu. Removing $R_{f9}(\mathbf{r})$ from the bracket in Eq. (5.65) yields an expression that shows in Table 5-II the strong space, delay group, and system dependence of the contribution of ^{238}U to the delayed neutron production:

$$\tilde{\nu}_{dk}(\mathbf{r}) = \nu_{dk9} + \nu_{dk8}\frac{R_{f8}(\mathbf{r})}{R_{f9}(\mathbf{r})} . \tag{5.66}$$

The relative ^{238}U contribution strongly increases with the decreasing lifetime of the precursor groups from 29 to 315% in the core center. This is explained by the dependence of ν_{dk} for ^{238}U and ^{239}Pu shown in Table 2-III. The bottom line of Table 5-II gives the ratios of the

---

[e]Equations (5.65) through (5.68) and the accompanying discussion may be omitted by less experienced readers.

TABLE 5-II
Comparison of ^{239}Pu and ^{238}U Contributions to the Production Rate of Delayed Neutrons

	Core Center	Core/Blanket Interface
	Percentage Addition of ^{238}U-Produced Delayed Neutrons to ^{239}Pu Contribution	
Delayed Neutron Group, k	$100 \frac{\nu_{dk8}}{\nu_{dk9}} \frac{R_{f8}}{R_{f9}}$	
1	29	17
2	44	26
3	71	42
4	105	62
5	182	108
6	315	185
	R_{f8}/R_{f9}	
	0.142	0.0836

fission rates in ^{238}U and ^{239}Pu at an FBR core center and at the core/blanket interface, where the neutron spectrum is much softer than in the center. The reduction of the relative fission rate across the fast reactor core amounts to more than a factor of 1.5. This reduction is reflected also in the relative delayed neutron contributions as shown in columns 2 and 3.

The adjoint flux weighting in Eq. (5.65) amplifies the increase of the $\tilde{\nu}_d$ values toward the core center since the importance of neutrons increases toward the center of the core. Thus, spatial weighting with the adjoint function increases the effective delayed neutron fraction in fast reactors since it provides a higher weight for the higher ^{238}U contributions near the center of the core.

The total reduction of β resulting from the *energy* dependence of $\phi^*(\mathbf{r},E)$ is generally larger than the increase caused by the spatial weighting. For long-lived delayed neutron groups, however, the increase due to spatial adjoint flux weighting could be larger than the decrease due to spectral adjoint flux weighting.

Equation (5.65) can be written in terms of one-group β's as:

$$\beta_k = \gamma_{dk}\left(\beta_{k9}\frac{S_{f9}^*}{S_f^*} + \beta_{k8}\frac{S_{f8}^*}{S_f^*}\right) \tag{5.67}$$

with

$$S_{fi}^* = \int_V \phi_0^*(\mathbf{r}) R_{fi}(\mathbf{r})\, dV \quad . \tag{5.68}$$

The expression in parentheses in Eq. (5.67) describes the effect of spatial adjoint flux weighting whereas γ_{dk} contains the effect of spectral weighting.

The integral kinetics parameters [see Eqs. (5.26)] have $F(t)$ in the denominator. This denominator is conventionally applied to obtain the definition of the dynamic reactivity in the same form as the static reactivity. After dividing by $\Lambda(t)$, the denominator obviously cancels in all terms. Thus, the solution, $p(t)$, of the point kinetics equations does not depend on $F(t)$ or any other denominator that may have been used. The integral kinetics parameters appear only as ratios, e.g., ρ/Λ and β/Λ. The individual quantities ρ, β, and Λ are, in principle, arbitrary quantities due to the arbitrariness of their denominator. The direct consequence of the "conventional" nature of the denominators in the individual definitions of ρ, β, and Λ is that the integral kinetics quantities cannot be measured individually. Kinetics experiments can only yield ratios of kinetics parameters (see Chapter 9).

The time dependence of the total reactor power, $P(t)$, may be calculated approximately[f] by means of the stationary fission-to-heat conversion factor C_{fsh}:

$$P(t) = p(t) \cdot \frac{1}{C_{fsh}} \int_V \int_0^\infty \Sigma_f(\mathbf{r},E,t)\psi(\mathbf{r},E,t)\, dE\, dV \quad . \tag{5.69}$$

The integral in this equation depends on time. The time variation of the integral over the power *distribution* is not very strong in reactor transients, so that $p(t)$ may be considered an approximation to the time dependence of the relative power $P(t)/P_0$.

5-2 The Point Reactor Model

The integral kinetics parameters ρ, β_k, and Λ are defined as integrals over space and energy of integrands that depend on the time-dependent shape function, $\psi(\mathbf{r},E,t)$. If $\psi(\mathbf{r},E,t)$ could be calculated and used to find $\rho(t)$, $\beta_k(t)$, and $\Lambda(t)$ from their "exact" definitions, Eqs. (5.26), one could use them in the exact point kinetics equations to determine the amplitude function $p(t)$. However, generation of such shape functions would re-

---

[f]Equation (5.69) disregards the delay of part of the fission power (see Sec. 10-1A).

quire solutions of full $(\mathbf{r},E,t)$-dependent problems. Approximate methods for such calculations comprise what is called "space-energy-dependent dynamics" (see Chapter 11). There is, however, a definite need for simpler approximate methods. The most commonly used approximate approach is called the "point reactor model."

The kinetics equations for the point reactor model are normally referred to as "point kinetics equations," without an adjective. They can be obtained from the exact point kinetics equations by introducing one major and two minor simplifications. The major simplification consists of neglecting the time dependence of the flux shape; that is, the initial flux shape,

$$\psi(\mathbf{r},E,t) \simeq \phi_0(\mathbf{r},E) \quad , \tag{5.70a}$$

is used in forming the integral kinetics parameters. If $p_0 = 1$, the initial flux shape is equal to the initial flux $\phi_0(\mathbf{r},E)$. The time dependence of the neutron flux is thus assumed separable from the $(\mathbf{r},E)$ dependence.

There are two minor simplifications. The arbitrary denominator $F(t)$, Eq. (5.23), is replaced by its initial value

$$F(t) \rightarrow F_0 = (\Phi_0^*,\mathbf{F}_0\Psi_0) \quad , \tag{5.70b}$$

and the delayed fission neutron operator $\mathbf{F}_d$ is approximated by the initial operator:

$$\mathbf{F}_d \simeq \mathbf{F}_{d0} \quad . \tag{5.70c}$$

These three simplifications have the following consequences:

1. Substitution of Eqs. (5.70a) and (5.70b) into Eq. (5.26c) converts the exact reactivity formula into the "first-order perturbation theory formula" (see Sec. 4-2) if the reactor is initially critical[g]:

$$\rho(t) \rightarrow \rho^{(1)}(t) = \frac{1}{F_0}(\Phi_0^*,[\Delta\mathbf{F} - \Delta\mathbf{M}]\Phi_0) \quad . \tag{5.71}$$

2. Substitution of $F(t)$ by F_0 in Eq. (5.26a) yields:

$$\Lambda(t) \rightarrow \Lambda_0 = \frac{K_0}{F_0} \tag{5.72a}$$

and

$$\frac{F_0}{F(t)} \rightarrow 1 \quad . \tag{5.72b}$$

[g]The equations for an initially subcritical reactor are discussed near the end of this section.

3. The approximation of Eq. (5.70c) leads to the neglect of the (weak) time dependence of the delayed neutron fractions,

$$\beta_k(t) \rightarrow \beta_{k0} = \frac{1}{F_0}(\Phi_0^*, \mathbf{F}_{dk0}\Phi_0) \quad . \tag{5.73}$$

Implementation of these approximations and replacements into the exact point kinetics equations, Eqs. (5.34) or (5.35), yields what is usually referred to as the *point kinetics equations:*

$$\dot{p}(t) = \frac{\rho(t) - \beta}{\Lambda} p(t) + \frac{1}{\Lambda}\sum_k \lambda_k \zeta_k(t) \tag{5.74a}$$

and

$$\dot{\zeta}_k(t) = -\lambda_k \zeta_k(t) + \beta_k p(t) \quad , \tag{5.74b}$$

or in c notation,

$$\dot{p}(t) = \frac{\rho(t) - \beta}{\Lambda} p(t) + \sum_k \lambda_k c_k(t) \tag{5.75a}$$

and

$$\dot{c}_k(t) = -\lambda_k c_k(t) + \frac{\beta_k}{\Lambda} p(t) \quad . \tag{5.75b}$$

Note that Eqs. (5.74) contain the initial values of β, β_k, and Λ. This is not indicated by an index 0 in order to avoid indices in these very frequently used initial integral kinetics parameters. The index 0 will only be used if a distinction from corresponding time-dependent values is required, for example, in Eq. (5.26a).

The neglect of the time dependence of the flux shape in the point reactor model is usually understood to be an assumption for the "gross neutron flux." The "fine structure" of the flux, which is often accounted for by "effective cross sections," is allowed to depend on time. The most important "fine structure" of the flux, which must be considered time dependent, is the spectrum deformation caused by Doppler broadening of resonances. This resonance broadening is a function of temperature and the temperature depends on time during a transient. The time-dependent spectral deformation due to Doppler broadening is combined with the resonance cross sections and averaged over group intervals. The results are "effective" group cross sections, which depend on temperature and thus on time. Reaction rates are formed by multiplication of the effective, time-dependent, group cross section with the time-independent gross flux, i.e.,

$$\sigma_g[T(t)]\phi_g(\mathbf{r}, t = 0) \quad .$$

Time-dependent flux fine structure caused by the motion of control rods can be accounted for in a similar way. This is commonly done in reactor dynamics.

The adjoint weighting flux in the kinetics equations for an initially subcritical reactor is the initial λ mode, $\Phi^*_{\lambda 0}$. To preserve the important stationarity property of the reactivity formula around the initial state (see Sec. 5-1D), the reactivity change after $t = 0$ must be calculated by

$$\delta\rho(t) = \frac{1}{F_{\lambda 0}} (\Phi^*_{\lambda 0},[\lambda_0 \Delta\mathbf{F} - \Delta\mathbf{M}]\Psi_0) \quad . \tag{5.76}$$

This equation does not represent the usual first-order perturbation theory reactivity since Ψ_0 is the source-driven actual flux and not the initial λ mode flux.

Generally, point kinetics is a good approximation when the flux shape is largely invariable during a transient. This is the case for tightly coupled smaller reactors, or for transients in large reactors if they involve fairly uniform perturbations. Transients in large reactors—for example, 1000-MW(e) light water reactors in which distant areas of the core are only loosely coupled to each other—that are caused by nonuniform perturbations generally require treatment by a space-energy-dependent kinetics method.

Homework Problems

1. Calculate β for a fast reactor with 85% fissions in ^{239}Pu and 15% in ^{235}U. Use the ν_{dk} values from Table 2-III and the γ_k values from Table 5-I.

2. Calculate a burnup-dependent β for an LWR with 2% fissions in ^{238}U and initially 98% in ^{235}U. As ^{239}Pu is produced, it takes over part of the fission rate. Let f_{239} be the fraction of the total fission rate that comes from fissioning ^{239}Pu. Use $\gamma_k = 1.08$.
   a. Find $\beta(f_{239})$ for $f_{239} \leq 50\%$.
   b. Plot your results as a function of f_{239}.

3. In Chapter 8, the zero Λ approximation is introduced and applied. Carry out the limit $\Lambda \to 0$ in the point kinetics equations
   a. in ζ notation,
   b. in c notation.

4. *Calculation of the Reduced Source.* An (α,n) source emitting 10^5 neutron/s is moved toward a swimming pool reactor core in a vertical

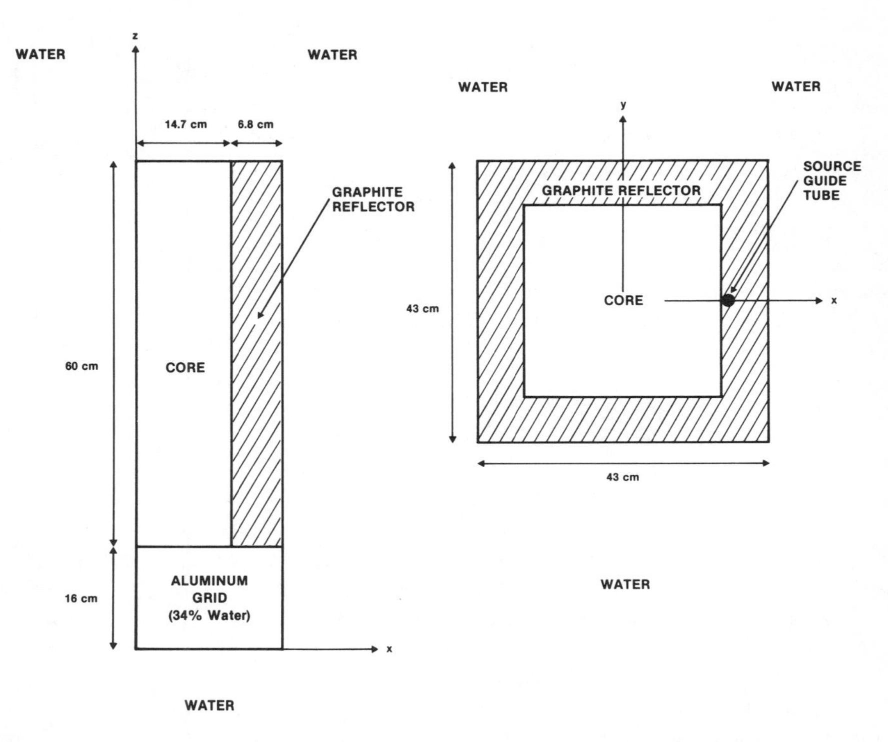

Fig. 5-2. Top and side views (with dimensions) of a swimming pool reactor, PUR-1.

guide tube. While the source is lowered, its multiplication effect depends on its position. In the definition of the reduced source,

$$s(z) = \frac{(\Phi_0^*, S[z])}{(\Phi_0^*, \mathbf{F}_0 \Phi_0)} ,$$

this variation is automatically accounted for by the adjoint flux weighting.

a. Calculate the flux $\phi_0(x,y,z)$ and thus $\phi^*(x,y,z)$ in a one-group approximation for a typical swimming pool reactor (such as Purdue's PUR-1). The dimensions and the one-group data are given in Fig. 5-2 and Table 5-III.

b. Apply flux separation:

$$\phi(x,y,z) = \phi_0 f_x(x) f_y(y) f_z(z) ,$$

assuming the same shape in the x and y directions. Calculate $f_x(x)$ and $f_z(z)$ with a suitable diffusion code (e.g., FOG), with a

TABLE 5-III
PUR-1 One-Group Macroscopic Cross Sections

Region	D	Σ_a	$\nu\Sigma_f$
Core	0.4559	0.0218	0.0327
Water	0.1850	0.0137	0.0
Aluminum Grid	0.5096	0.0082	0.0
Graphite	0.6612	0.0023	0.0

Note: Use $B_x^2 + B_y^2 = 0.01/\text{cm}^2$ for the axial calculation and $B_y^2 + B_z^2 = 0.01/\text{cm}^2$ for the x dimension calculation.

transverse buckling describing the lateral leakage, and normalize it to unity at $x = 0$. Search for control absorption (by variation of Σ_a in the core) such that $k = 0.99$.

c. Determine the reduced source $s(z)$, with z varying along the vertical source guide tube for z from $+c$ to $z = 0$ (core midplane). Use as an independent source an (α,n) point source emitting 10^5 neutron/s. Location of the source guide tube is at $x = a/2$ and $y = 0$.

d. Plot $s(t)$ and discuss the results.

5. *Source Multiplication.*

a. Insert the reduced source, $s(z)$, into the source multiplication formula

$$p = \frac{s}{-\rho} \quad ,$$

and convert the formula for p into one for the adjoint weighted fission source (note that $p\Phi_0 = \Phi$):

$$S_f^* = (\Phi_0^*, \mathbf{F}_0\Phi) = (\Phi_0^*, S_f) \quad .$$

b. Apply this formula and the results of problem 4 to find $S_f^*(z)$; z being the elevation of the source above the midplane, within the guide tube.

c. Calculate the power for the various source locations (assume $\Sigma_f = \dfrac{1}{2.43}\nu\Sigma_f$ with $\nu\Sigma_f$ given in Table 5-III).

d. Plot and discuss the results.

Review Questions

1. Factorize the flux and give the rationale for the factorization.
2. Discuss explicitly the time derivative and find the constraint condition for the flux shape function, using a general weighting function.

3a. Which weighting function is commonly chosen?
 b. Give the rationale for this choice.

4. What does "exact" mean in the context of exact point kinetics equations?
5. Describe in words (and possibly some formulas) the derivation of (a) the differential equation for the flux amplitude and (b) the differential equation for the precursors. Emphasize the features different from the derivation of the one-group point kinetics equation.
6. Give the exact point kinetics equations (a) in ζ notation and (b) in c notation.
7. Are ρ, β, and Λ uniquely defined? If not, why?
8. What is the direct impact of adjoint weighting on scattering and on $\chi(E)$ and $\chi_d(E)$?
9. What is dynamic reactivity and why is it different from the static reactivity?

10a. What is the major difference between β effective and the corresponding quantity of one-group point kinetics?
 b. Explain the sign of these differences qualitatively for thermal and fast reactors.

11a. In the derivation of the exact point kinetics equations, what complication arises for an initially subcritical reactor?
 b. What is the commonly used weighting function?
 c. Give the resulting equation for the flux amplitude.

12. What is the basic assumption in the point reactor model?
13. Give the other two alterations introduced into the exact point kinetics equations to obtain the common form of kinetics equations for the point reactor model.
14. What are the two kinds of flux shape variations commonly accounted for in the point reactor model?

REFERENCES

1. A. F. Henry, "The Application of Reactor Kinetics to the Analysis of Experiments," *Nucl. Sci. Eng.*, **3,** 52 (1958).

2. G. I. Bell and Samuel Glasstone, *Nuclear Reactor Theory,* p. 257, Van Nostrand Reinhold Co., New York (1970).
3. T. Gozani, "Consistent Subcritical Fast Reactor Kinetics," p. 109 in *Dynamics of Nuclear Systems,* D. L. Hetrick, Ed., University of Arizona Press, Tucson (1972).
4. D. A. Meneley, K. O. Ott, and E. S. Wiener, "Fast Reactor Kinetics—The QX1 Code," ANL-7769, Argonne National Laboratory (Mar. 1971).

Six

SOLUTION OF BASIC KINETICS PROBLEMS

The practical solution of almost all kinetics problems is complicated by the individual behavior of the various groups or families of delayed neutrons. Therefore, computer programs are required to obtain the numerical solution of a given problem. Analytical solutions of approximate model problems, however, are extremely valuable. These solutions often yield analytical relations between basic kinetics parameters and thus help in the *understanding* of the results. Such analytical relations can also provide an approximate and completely independent check of numerical solutions.

The analytical treatment presented is kept as simple as possible so that the analytical results can be readily understood. Such a degree of simplicity generally requires the application of one or both of the following two approximations:

1. reduction of the six groups of delayed neutrons to a single "effective" group (see Sec. 6-1)
2. simplification of the kinetics equations by proper use of the small neutron generation time in fast and light water reactors (LWRs) (see Secs. 8-1 and 8-2).

The results obtained on the basis of these two approximations are applicable to all LWRs and fast reactors. The second approximation is more accurate in fast reactors due to the smaller neutron generation time.

How the reactivity is found is unimportant for these model investigations. The results are therefore quite general and independent of the sophistication of the model applied for computing the reactivity from changes in the composition of the reactor system. The slight time dependencies of β and Λ (compare Sec. 5-2 and Chapter 11) are neglected; that is, they are assumed to be constants. For simplicity, the investigation is restricted to the most important class of problems; namely, those for which the reactor is in a stationary state for $t \le 0$. Thus, the kinetics equations that are solved in this chapter are (in ζ notation):

$$\dot{p}(t) = \frac{\rho(t) - \beta}{\Lambda} p(t) + \frac{1}{\Lambda} \sum_k \lambda_k \zeta_k(t) + \frac{1}{\Lambda} s(t) \qquad (6.1a)$$

and

$$\dot{\zeta}_k(t) = -\lambda_k \zeta_k(t) + \beta_k p(t) \quad . \qquad (6.1b)$$

To be consistent with the initial (stationary) state, the constant values of β and Λ are set equal to their initial values:

$$\Lambda = \Lambda_0 \text{ and } \beta = \beta_0 \quad . \qquad (6.1c)$$

This approximation also requires that $F(t)$, Eq. (5.23), be replaced by F_0.

6-1 Kinetics for Small and Large Time Values

Reducing the complexity of kinetics with six groups of delayed neutrons to simple but meaningful kinetics approximations is vital for obtaining sufficiently transparent and useful analytic results. This reduction in complexity can only be achieved by finding appropriate approximations of the delayed neutron source. The investigation of transients at small and at large times suggests such approximations.

6-1A The Stationary Solution: Source Multiplication Formulas

Prior to discussing approximate treatments for short time transients, some relevant information is obtained by considering the point kinetics equations for a stationary state. The stationary kinetics equations are obtained by setting the time derivatives in Eqs. (6.1) equal to zero and removing the generation time:

$$\dot{p} = 0 = (\rho_0 - \beta)p_0 + \sum_k \lambda_k \zeta_{k0} + s_0 \qquad (6.2a)$$

and

$$\dot{\zeta}_k = 0 = -\lambda_k \zeta_{k0} + \beta_k p_0 \quad . \qquad (6.2b)$$

In Eq. (6.2a), ρ_0 and s_0 are equal to zero if the reactor is critical.

The above equations show that *the stationary solution is independent of the generation time;* it appears neither in the balance equation for the flux nor in the precursor equation.

The stationary precursor balance, Eq. (6.2b), is discussed first, it expresses the fact that the stationary decay rate of reduced precursors $\lambda_k \zeta_{k0}$ is equal to their stationary production rate, $\beta_k p_0$. Summing Eq. (6.2b) over all delayed neutron groups gives the following form for the stationary reduced delayed neutron source in a critical or subcritical reactor:

$$s_{d0} = \sum_k \lambda_k \zeta_{k0} = \beta p_0 \quad . \tag{6.3}$$

According to Eq. (3.47), s_{d0}, in the simple one-group description, is defined as the ratio of integrated delayed and total fission neutron sources:

$$s_{d0} = \sum_k \frac{\lambda_k \hat{C}_{k0}}{\hat{S}_{f0}} = \frac{\hat{S}_{d0}}{\hat{S}_{f0}} \quad . \tag{6.4}$$

Setting $p_0 = 1$ and combining Eqs. (6.3) and (6.4) gives the following expression for the stationary delayed neutron source in critical as well as in subcritical reactors:

$$\hat{S}_{d0} = \beta \cdot \hat{S}_{f0} \quad , \tag{6.5}$$

i.e., *the stationary integrated delayed neutron source is equal to* β *times the integrated total fission neutron source.*

The same equation, Eq. (6.5), is obtained if adjoint flux weighting is employed. The only difference is that all quantities, including β, are defined with adjoint flux weighting (see Sec. 5-1).

In a *subcritical system,* insertion of Eq. (6.3) into the amplitude balance equation, Eq. (6.2a), eliminates the delayed neutrons from the steady-state equation:

$$0 = \rho_0 p_0 + s_0 \quad . \tag{6.6}$$

The integrated relative source, s_0, is given by Eq. (3.48) with $t = 0$, i.e.,

$$s_0 = \frac{\hat{S}_0}{\hat{S}_{f0}} \quad , \tag{6.7}$$

which is the basis for the approximate, while unweighted, source multiplication formulas. Setting $p_0 = 1$ (i.e., $\psi_0 = \phi_0$) and combining Eqs. (6.6) and (6.7) yields the *static (unweighted) source multiplication formula between* $\hat{S}$ *and* $\hat{S}_f$*:*

$$\hat{S}_{f0} = \frac{1}{-\rho_0} \hat{S}_0 \quad . \tag{6.8a}$$

The factor in front of the independent source is called the "source multiplication factor." Other source multiplication formulas, for stationary and nonstationary states, are derived and applied throughout this text. Further static source multiplication formulas are introduced below. By adding $\hat{S}_0$ on both sides of this equation and introducing k_0, the following form is obtained for the *static (unweighted) source multiplication formula between* $\hat{S}$ *and* $\hat{S}_{tot}$*:*

$$\hat{S}_{tot,0} = \frac{\hat{S}_0}{1 - k_0} = \frac{\hat{S}_0}{-\Delta k_0} , \tag{6.8b}$$

with $\hat{S}_{tot,0}$ denoting the total neutron source, i.e., $\hat{S}_{f0} + \hat{S}_0$.

Solving Eq. (6.6) directly for p_0, the stationary flux level, the *source multiplication formula* is obtained in the form

$$p_0 = \frac{s_0}{-\rho_0} , \tag{6.8c}$$

i.e., the source multiplication formula of Eq. (6.8a) also holds between the stationary flux amplitude p_0 and the stationary "reduced" source s_0.

If adjoint weighting is employed, the sources as well as ρ_0 and ζ_0 are defined differently (see Sec. 5-1) and have different values. The adjoint weighted source multiplication formulas are much more accurate than the unweighted ones of Eqs. (6.8). The *static adjoint weighted source multiplication formulas* are:

$$(\Phi_0^*,\mathbf{F}_0\Phi_0) = -\frac{1}{\rho_0}(\Phi_0^*,S_0) , \tag{6.9a}$$

$$(\Phi_0^*,[\mathbf{F}_0\Phi_0 + S_0]) = -\frac{1}{\Delta k_0}(\Phi_0^*,S_0) , \tag{6.9b}$$

and

$$p_0 = -\frac{1}{\rho_0}\frac{(\Phi_0^*,S_0)}{(\Phi_0^*,\mathbf{F}_0\Psi_0)} . \tag{6.9c}$$

In a *critical system,* ρ_0 and s_0 are equal to zero. Equation (6.2a) then becomes:

$$0 = -\beta p_0 + \sum_k \lambda_k \zeta_{k0} = -\beta p_0 + s_{d0} ; \tag{6.10}$$

thus,

$$p_0 = \frac{s_{d0}}{\beta} , \tag{6.11}$$

i.e., the source multiplication formula between p_0 and the independent source s_0 also holds between p_0 and s_{d0} with the source multiplication factor given by $1/\beta$; i.e., β plays the role of a negative reactivity in this source multiplication formula. Equation (6.11) indicates the very important fact that *a critical reactor may be viewed as being subcritical with the stationary delayed neutron source sustaining a finite flux level.* The corresponding source multiplication factor is determined by the reactivity increment below prompt critical. The full physical significance of Eq. (6.11) is investigated and illustrated in Sec. 7-1.

A similar source multiplication formula can be obtained from Eq. (6.2a) for a subcritical reactor, $s_0 > 0$:

$$p_0 = \frac{s_0 + s_{d0}}{\beta - \rho_0} \quad . \tag{6.12}$$

Equation (6.12) shows that *the stationary flux amplitude in a subcritical reactor can also be obtained from independent plus delayed neutron sources by source multiplication with a source multiplication factor determined by the reactivity increment below prompt critical.*

Equation (6.11) is a special case of the more general formula of Eq. (6.12). The source multiplication formulas described above are used subsequently to understand certain important features of transients as well as certain experiments involving reactivity measurements.

6-1B Kinetics for Small Time Values

The investigation of kinetics at small times provides an understanding of the onset of transients and also provides information that suggests approximations of the delayed neutron source for small time intervals.

Practically, any perturbation of an initially stationary state of a reactor starts gradually even though it may occur very fast. A falling fuel subassembly, the runaway of a control rod, or a sudden onset of coolant boiling initially causes only gradual insertions of reactivity. In actual safety investigations, the realistic time dependence of the reactivity needs to be used. In survey investigations, however, reactivity insertions are commonly idealized by "steps" or "ramps," which for an initially critical reactor, are given by:

$$\text{reactivity step:}\quad \rho(t) = \begin{cases} 0 & \text{for } t < 0 \\ \rho_1 & \text{for } t \geq 0 \end{cases} \tag{6.13a}$$

and

$$\text{reactivity ramp:}\quad \rho(t) = \begin{cases} 0 & \text{for } t \leq 0 \\ at & \text{for } t \geq 0 \end{cases} \quad , \tag{6.13b}$$

with a denoting the reactivity insertion rate. Step and ramp reactivity insertions may be combined in the form of a "limited" ramp or "terminated" ramp:

$$\text{limited reactivity ramp: } \rho(t) = \begin{cases} 0 & \text{for } t \leq 0 \\ at & \text{for } 0 \leq t \leq t_1 \\ \rho_1 & \text{for } t \geq t_1 \end{cases} \quad . \tag{6.13c}$$

Idealized reactivity insertions are particularly appropriate for a basic understanding of transient phenomena and are applied for the analytical investigations presented in this chapter.

Neither the neutron flux nor the delayed neutron source are able to follow immediately a reactivity step insertion into an initially critical reactor. Their change requires time, and they still have their initial values right after the reactivity insertion. The only "quantity" that changes directly with the reactivity is $\dot{p}$:

$$\dot{p}(0) = \frac{\rho_1 - \beta}{\Lambda} p_0 + \frac{1}{\Lambda} \sum_k \lambda_k \zeta_{k0} = \frac{\rho_1}{\Lambda} p_0 \tag{6.14}$$

and

$$\dot{\zeta}_k(0) = -\lambda_k \zeta_{k0} + \beta_k p_0 = 0 \quad . \tag{6.15}$$

The step change of the right side of Eq. (6.14) leads to a step change in $\dot{p}$. Thus, $p(t)$ starts to change linearly at $t = 0$ whereas $\zeta_k(t)$ has a zero slope at $t = 0$.

The slopes of p and ζ at $t = 0$ suggest the following two approximations of the delayed neutron source for small times:

1. *Constant delayed neutron source (CDS) approximation:*

$$s_d(t) = s_{d0} \quad . \tag{6.16}$$

2. *Precursor accumulation approximation:* Since the ζ's start to vary "later" than p, an improvement of the constant delayed neutron source approximation can be obtained by setting ζ_k equal to ζ_{k0} in the precursor balance and allowing $p(t)$ to vary with time:

$$\dot{\zeta}_k(t) = -\lambda_k \zeta_{k0} + \beta_k p(t) \quad . \tag{6.17}$$

If $p(t)$ increases with time, this approximation means that newly formed precursors are accumulated and contribute to the delayed neutron source. The fraction that has decayed at time t, however, is still negligibly small. This approximation is called the "precursor accumulation (PA) approximation."

With $\lambda_k \zeta_{k0} = \beta_k p_0$, Eq. (6.17) can be written as:

$$\dot{\zeta}_k(t) = \beta_k[p(t) - p_0] \quad . \tag{6.18}$$

The delayed neutron source thus becomes

$$s_d(t) = s_{d0} + \sum_k \beta_k \lambda_k I(t) \quad , \tag{6.19}$$

where the integral over the additional flux amplitude is abbreviated by $I(t)$, i.e.,

$$I(t) = \int_0^t [p(t') - p_0]\, dt' \quad . \tag{6.20}$$

The combination in which the delayed neutron parameters appear in Eq. (6.19) suggests the introduction of an average λ:

$$\bar{\lambda} = \frac{1}{\beta} \sum_k \beta_k \lambda_k \quad . \tag{6.21}$$

By using Eqs. (6.3) and (6.21), the two short time approximations of the delayed neutron source can be written as

$$\text{CDS approximation: } s_d(t) = \beta p_0 \tag{6.22}$$

and

$$\text{PA approximation: } s_d(t) = \beta p_0 + \beta \bar{\lambda} I(t) \quad . \tag{6.23}$$

The delayed neutron data appear only in form of the total β in the CDS approximation. In the PA approximation, the β_k weighted average value of the decay constants λ_k appears in addition. Thus, in the PA approximation, the decay of newly formed precursors enhances the delayed neutron source.

Typical values for $\bar{\lambda}$ are:

$$\bar{\lambda} \simeq 0.4 \text{ s}^{-1} \text{ for } {}^{235}\text{U-fueled thermal reactors} \tag{6.24a}$$

and

$$\bar{\lambda} \simeq 0.6 \text{ s}^{-1} \text{ for large mixed-oxide-fueled fast reactors} \quad . \tag{6.24b}$$

Insertion of the two simple expressions for the delayed neutron source into the amplitude equation, Eq. (6.1a), gives the following two kinetics equations for small times:

kinetics equation in CDS approximation:

$$\dot{p} = \frac{\rho - \beta}{\Lambda} p + \frac{\beta p_0}{\Lambda} + \frac{s(t)}{\Lambda} \tag{6.25}$$

kinetics equation in PA approximation:

$$\dot{p} = \frac{\rho - \beta}{\Lambda} p + \frac{1}{\Lambda} \left\{ \beta p_0 + \beta \bar{\lambda} \int_0^t [p(t') - p_0] \, dt' \right\} + \frac{s(t)}{\Lambda} \quad . \tag{6.26}$$

The two short time approximations (CDS and PA) do not allow for the full time dependence (buildup *and* decay) of precursors. In both cases, the precursor decay is restricted to the rate of decay in the stationary state. The decay of newly formed precursors is considered only in the delayed neutron source and then only in the PA approximation.

There is a definite need to augment the PA approximation by accounting for the decay of precursors in the associated balance equations. The simplest model to treat such decay is the "one-delay-group kinetics" model. The equations are:

$$\dot{p} = \frac{\rho - \beta}{\Lambda} p + \frac{1}{\Lambda} \lambda \zeta + \frac{s}{\Lambda} \tag{6.27a}$$

and

$$\dot{\zeta} = -\lambda \zeta + \beta p \quad . \tag{6.27b}$$

A suitable collapsing procedure must be found to reduce the six-delay-group model into a one-delay-group model with appropriate definitions for λ and ζ. The situation is similar to condensing the energy scale into groups, particularly into a model with one-group cross sections.

A straightforward collapsing procedure would consist of merely summing up the precursor balance equations:

$$\frac{d}{dt} \sum_k \zeta_k(t) = -\sum_k \lambda_k \zeta_k(t) + \sum_k \beta_k p(t) \quad . \tag{6.28}$$

The form of the left side of Eq. (6.28) describes the rate of change of the total number of precursors. This form suggests that ζ be defined as the total number of precursors:

$$\zeta(t) = \sum_k \zeta_k(t) \quad . \tag{6.29}$$

Equating the first terms on the right side of Eqs. (6.27b) and (6.28) and inserting ζ from Eq. (6.29) defines a condensed λ as:

$$\lambda(t) = \frac{\sum_k \lambda_k \zeta_k(t)}{\sum_k \zeta_k(t)} \quad . \tag{6.30}$$

The one-delay-group λ is thus obtained as a function of time. It is the $\zeta_k(t)$ weighted average of the λ_k and depends, therefore, on the entire solution of the problem. This is similar to the space-energy collapse of spatially independent cross sections in which the collapsed cross sections depend on space. As in the case of the energy groups, the problem dependence of λ might be weak, at least in a certain domain; for example, the group cross sections are virtually independent of space in the inner area of the core where a fundamental mode spectrum prevails. To find out if there is a situation for λ that is analogous, $\lambda(t)$ is evaluated at first for short times.

The initial value of λ may be calculated readily from the initial ζ_{k0} values given by Eq. (6.2b):

$$\lambda(0) = \frac{\sum_k \lambda_k \zeta_{k0}}{\sum_k \zeta_{k0}} = \frac{\sum_k \beta_k}{\sum_k \frac{\beta_k}{\lambda_k}} = \overline{\left(\frac{1}{\lambda}\right)}^{-1} = \overline{\lambda^{in}} \quad , \tag{6.31}$$

with $\overline{\lambda^{in}}$ denoting the average decay constant obtained by averaging the inverse constants $1/\lambda_k$.

The straightforward collapsing procedure yields $\overline{\lambda^{in}}$ as "initial" one-decay-group λ. It is the ζ_{k0} weighted average of λ_k. Since the initial condition, Eq. (6.2b), results from a stationary state prior to $t = 0$, then $\overline{\lambda^{in}}$ correctly describes the average decay constant of the total number of precursors in a stationary state, i.e.,

$$\overline{\lambda^{in}} = \lambda_{\text{stationary}} \quad . \tag{6.32}$$

One might expect $\overline{\lambda^{in}}$ to be a good approximation for a one-delay-group λ for short times because the ζ_k changes with zero slope after the onset of a transient [compare Eq. (6.15)]. However, another short time consideration, the PA approximation, yielded $\bar{\lambda}$, Eq. (6.21). To investigate the consistency of these two findings—$\overline{\lambda^{in}}$ versus $\bar{\lambda}$—numerical values are required. The numerical values of $\bar{\lambda}$ and $\overline{\lambda^{in}}$ are very different from each other in both thermal and fast reactors. The value of $\bar{\lambda}$ is strongly influenced by the short-lifetime precursor groups with large λ_k values:

$$\bar{\lambda} = \frac{1}{\beta}(\beta_1 \lambda_1 + \ldots + \beta_6 \lambda_6) \quad . \tag{6.33}$$

According to Table 2-II, λ_6 is ~300 times larger than λ_1; thus λ_6 influences $\bar{\lambda}$ much stronger than λ_1. The opposite is true for $\overline{\lambda^{in}}$:

$$\frac{1}{\overline{\lambda^{in}}} = \overline{\lambda^{-1}} = \frac{1}{\beta}\left(\frac{\beta_1}{\lambda_1} + \ldots + \frac{\beta_6}{\lambda_6}\right) \quad , \tag{6.34}$$

where the terms in parentheses with small λ_k values dominate. Consequently, $\overline{\lambda^{in}}$ can be expected to be much smaller than $\bar{\lambda}$.

Typical values of $\overline{\lambda^{in}}$ for both ^{235}U-fueled thermal as well as mixed U-Pu-fueled fast reactors are $\sim 0.08\ s^{-1}$. Comparison with $\bar{\lambda}$, Eqs. (6.24), shows that $\overline{\lambda^{in}}$ is about six to eight times smaller than $\bar{\lambda}$.

Figure 6-1 shows $\lambda(t)$ as defined by Eq. (6.30) for a transient following a reactivity step insertion of $\rho_1 = 0.5\$$; $\lambda(t)$ increases and levels off asymptotically at an asymptotic value λ_{as}, which is discussed in Sec. 6-1C. Thus, the straightforward derivation of the one-delay-group kinetics equation yields a λ that depends strongly on time and is quite different from $\bar{\lambda}$, the value suggested by Eq. (6.21).

The previous discussion poses the following question: Why does the approximation of the delayed neutron source, Eq. (6.23), and the straightforward condensation of the balance equations for the six precursor groups lead to discrepant results for the small time approximation of the one-delay-group λ? The qualitative answer is already indicated in

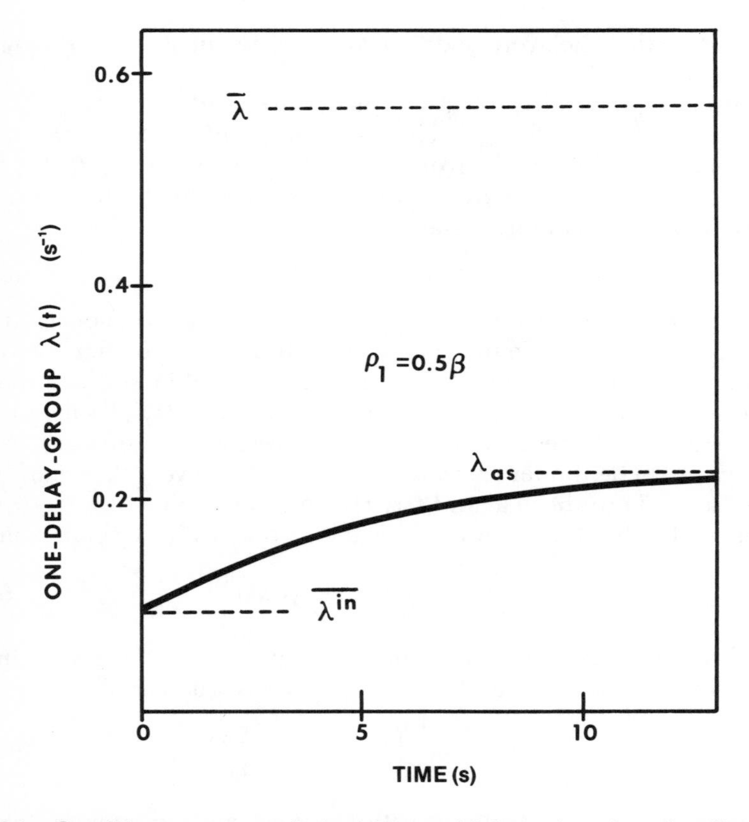

Fig. 6-1. One-delay-group λ, as obtained from straightforward condensation, Eq. (6.30), as a function of time for a step-induced transient of $\rho_1 = 0.5\$$.

the formulation of the question: $\bar{\lambda}$ results from the approximate description of the delayed neutron source for small t, whereas the straightforward condensation leads to a one-delay-group λ, which is appropriate to describe the balance of the total precursor population.

The total number of precursors is by itself virtually of no interest. The neutron flux is of primary interest, and the precursors enter the balance equation for the neutron flux only in the form of the delayed neutron source. It is therefore better to obtain the one-delay-group value of λ from an appropriate approximation of the delayed neutron source, after the onset of a transient, and not from the description of the total number of precursors.

To provide a definition of a one-delay-group λ that has a better basis than the PA approximation leading to Eq. (6.23), an analytical expression is needed for the delayed neutron source that holds over not too small an interval of time. Such an expression cannot be obtained from a Taylor expansion of the full solution of the kinetics equations, e.g., after a reactivity step insertion. Expansions of this type hold only over a very small interval in time (see Fig. 6-3 and the corresponding discussion in this section). An expansion that holds over a much larger time interval can, however, be obtained by solving the differential equation for $\zeta_k(t)$ and expanding only the exponential functions, which depend on the precursor decay constants; the flux is not expanded since it may vary much faster than exponential functions, which describe the beta-decay of the precursors.

The precursor balance equation,

$$\dot{\zeta}_k(t) = -\lambda_k \zeta_k(t) + \beta_k p(t) \quad , \tag{6.35}$$

is formally solved (see Sec. C-1 of App. C) by treating $p(t)$ as a known function[a]:

$$\zeta_k(t) = \zeta_{k0} \exp(-\lambda_k t) + \beta_k \int_0^t \exp[-\lambda_k(t - t')]\, p(t')\, dt' \quad . \tag{6.36}$$

The total "number" of precursors and the delayed neutron source thus become:

$$\sum_k \zeta_k(t) = \sum_k \zeta_{k0} \exp(-\lambda_k t) + \sum_k \beta_k \int_0^t \exp[-\lambda_k(t - t')]\, p(t')\, dt' \tag{6.37}$$

[a]The insertion of Eq. (6.36) into the first of the kinetics equations yields an integro-differential equation for $p(t)$, which is sometimes used in the literature.

and

$$s_d(t) = \sum_k \lambda_k \zeta_{k0} \exp(-\lambda_k t) + \sum_k \lambda_k \beta_k \int_0^t \exp[-\lambda_k(t - t')]p(t')\, dt' \quad . \tag{6.38}$$

Expanding the exponential functions, retaining only the first two terms, and inserting $\zeta_{k0} = \beta_k p_0/\lambda_k$ yields:

$$\sum_k \zeta_k(t) = p_0 \left(\frac{\beta}{\overline{\lambda^{in}}} - \beta t\right) + \beta \int_0^t p(t')\, dt' - \beta\overline{\lambda}\, p^{(1)}(t) + \ldots \tag{6.39}$$

and

$$s_d(t) = p_0(\beta - \beta\overline{\lambda}t) + \beta\overline{\lambda} \int_0^t p(t')\, dt' - \beta\overline{\lambda^2}\, p^{(1)}(t) + \ldots \quad , \tag{6.40}$$

with $\overline{\lambda^2}$ defined as the β_k weighted average of λ_k^2. The corresponding expressions in the one-delay-group model are

$$\zeta(t) = p_0 \left(\frac{\beta}{\lambda} - \beta t\right) + \beta \int_0^t p(t')\, dt' - \beta\lambda p^{(1)}(t) + \ldots \tag{6.41}$$

and

$$s_d(t) = p_0(\beta - \beta\lambda t) + \beta\lambda \int_0^t p(t')\, dt' - \beta\lambda^2 p^{(1)}(t) + \ldots \quad , \tag{6.42}$$

with

$$p^{(1)}(t) = \int_0^t (t - t')p(t')\, dt' \quad . \tag{6.43}$$

Comparing Eqs. (6.39) through (6.42) yields two conclusions:

1. If a one-delay-group λ is desired for a good description of $\sum_k \zeta_k(t)$, the comparison of Eqs. (6.39) and (6.41) yields

$$\lambda = \overline{\lambda^{in}} \quad . \tag{6.44}$$

2. If a one-delay-group λ is desired for a good description of the delayed neutron source, the comparison of Eqs. (6.40) and (6.42) yields

$$\lambda = \overline{\lambda} \quad . \tag{6.45}$$

In each case, agreement is obtained for the first three terms of the expansion, i.e., only the $p^{(1)}$ term and further terms are not described accurately. For sufficiently small times, the $p^{(1)}$ and higher terms are small. Thus the choice $\lambda = \bar{\lambda}$ should give good agreement of one- and six-delay-group kinetics results (see the quantitative comparisons of kinetic results in Secs. 6-2A and 6-2C).

Figure 6-2a shows a comparison of one-delay-group *delayed neutron sources* using $\bar{\lambda}$ and $\overline{\lambda^{in}}$ with the six-delay-group solution for a 0.5\$-step-induced transient. The results calculated with $\lambda = \bar{\lambda}$ agree fully with the six-delay-group $s_d(t)$ for the first 50 ms. The agreement is still acceptable for another 200 ms. In contrast, the $s_d(t)$ curve, which is calculated with $\lambda = \overline{\lambda^{in}}$, is discrepant in the entire time domain. The good agreement

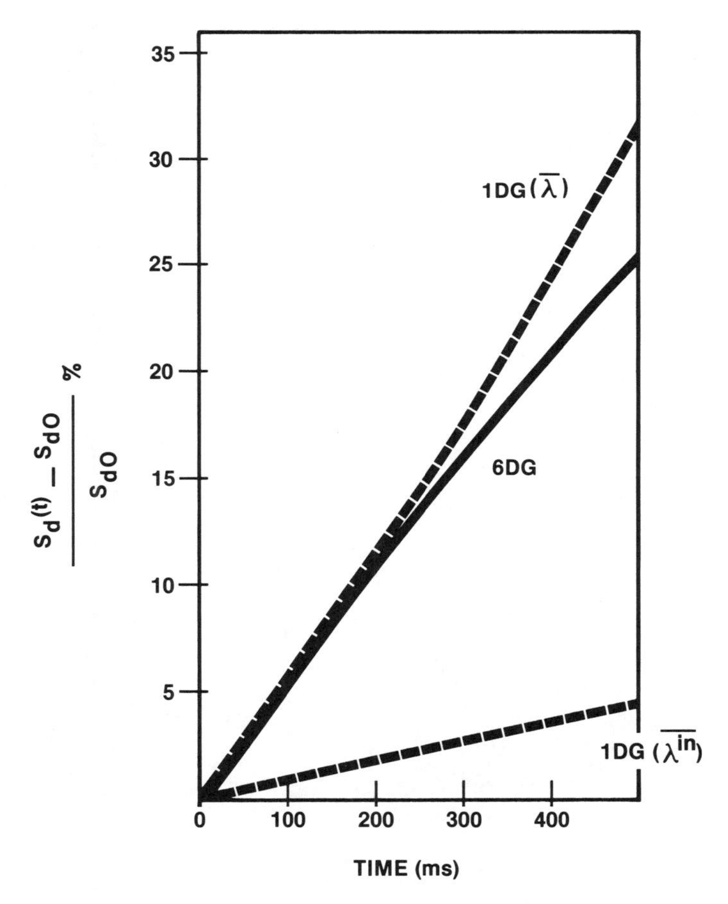

Fig. 6-2a. Comparison of the six-delay-group neutron source with the one-delay-group neutron source obtained with $\lambda = \bar{\lambda}$ and $\lambda = \overline{\lambda^{in}}$ for a transient following a 0.5\$ reactivity step insertion.

obtained with $\lambda = \bar{\lambda}$ supports the theoretical evidence for this choice of the one-delay-group λ. Figure 6-2b shows that the description of the total $\zeta(t)$ requires the use of $\overline{\lambda^{in}}$ as the one-delay-group value.

A brief investigation of the convergence of the straightforward Taylor expansion solution of the point kinetics equations is of general methodological interest. For the purpose of this investigation, it suffices to expand the one-delay-group solutions, and to consider only a step-induced transient.

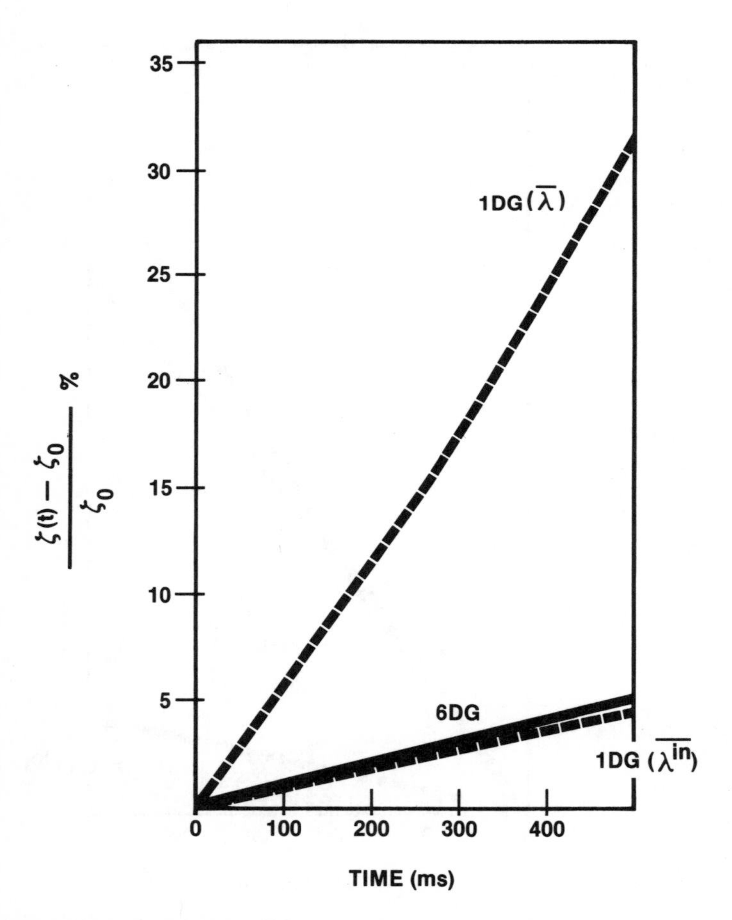

Fig. 6-2b. Comparison of the reduced precursors calculated with six delayed groups and with one delayed group, obtained with $\lambda = \bar{\lambda}$ and $\lambda = \overline{\lambda^{in}}$ for a transient following a 0.5$ reactivity step insertion.

The flux amplitude function and the reduced precursor concentrations are expanded as

$$p(t) = p_0 + p_1 t + p_2 t^2 + \ldots \tag{6.46}$$

and

$$\zeta_k(t) = \zeta_{k0} + \zeta_{k1} t + \zeta_{k2} t^2 + \ldots \quad . \tag{6.47}$$

Insertion of these equations into Eqs. (6.27) readily yields:

$$\zeta_{k1} = 0 \quad , \tag{6.48}$$

$$\zeta_{k2} = \frac{1}{2} \beta_k p_1 \quad , \tag{6.49}$$

$$p_1 = \frac{\rho_1}{\Lambda} p_0 \quad , \tag{6.50}$$

and

$$p_2 = \frac{\rho_1(\rho_1 - \beta)}{2\Lambda^2} p_0 \quad . \tag{6.51}$$

Equation (6.50) shows that the flux rises with a very steep slope after a reactivity step if Λ is very small, e.g., with $\rho_1 = 2 \times 10^{-3}$ and $\Lambda = 4 \times 10^{-7}$ s, the linear flux ($p_0 + p_1 t$) doubles in 0.2 ms. In addition, the curvature is very large due to the Λ^2 term in the denominator of p_2 in Eq. (6.51).

The appearance of increasing powers of Λ in the denominator of the coefficients results in very poor convergence of the Taylor expansion. The same poor convergence is exhibited by Taylor expansions around any point during a transient. Figure 6-3 shows a comparison of two-, three-, and four-term Taylor expansions with the exact one-delay-group solution. The lack of convergence is obvious. The methodological impact of this poor convergence is that numerical methods for the solution of the kinetics equations need very small time steps if extrapolations or interpolations over a time step are based on Taylor expansions. The expansions of the exponential functions in Eqs. (6.39) to (6.42) provide expressions that hold over very large time intervals compared to the validity of the Taylor expansion of the entire solution (compare Figs. 6-2 and 6-3).

6-1C Asymptotic Transients and Inhour Formulas

After insertion of a small reactivity into a *critical* reactor with a low initial power density, an asymptotic transient develops, characterized by a "stable" or "asymptotic" period. A constant reactivity is required for

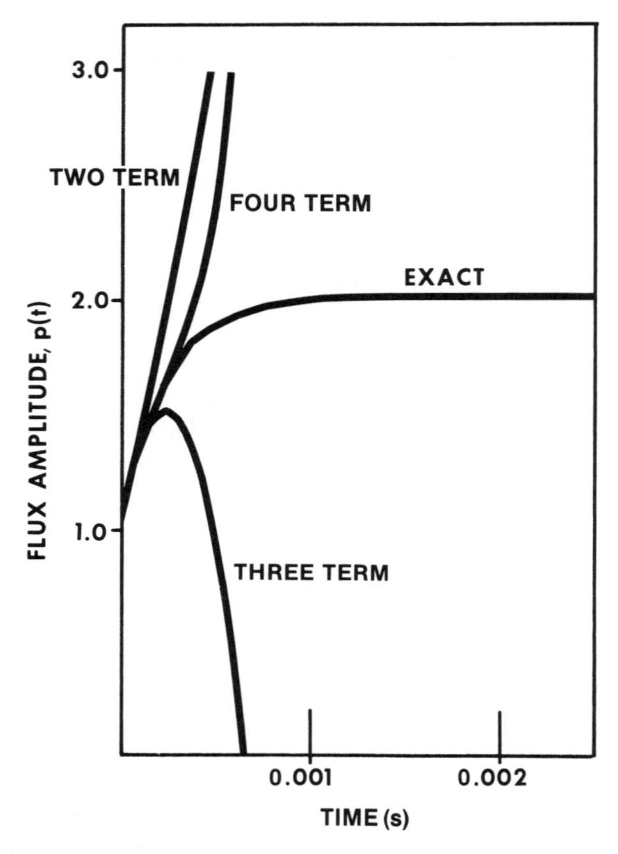

Fig. 6-3. Comparison of two-, three-, and four-term Taylor expansions with the one-delay-group reference solution.

establishing a transient with a stable period. The power density should be small enough so that no noticeable reactivity feedback disturbing the constant reactivity develops. The asymptotic transients are discussed in this section as well as the relation between the reactivity and stable periods, the inhour formula. Important approximate relations are also discussed. The approximations introduced are either approximations of the delayed neutron source or are made possible by the smallness of the neutron generation time.

The asymptotic transient following a constant reactivity insertion[b] in a reactor without feedback is of a purely exponential form, i.e., the

[b]The β and Λ values are also assumed to be constant.

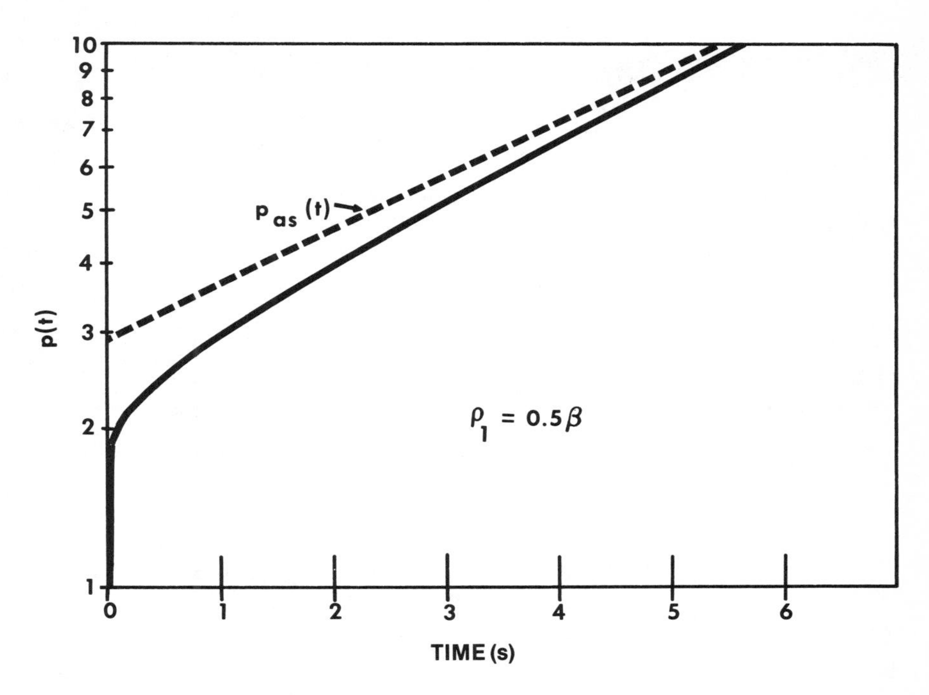

Fig. 6-4. A semi-logarithmic plot of the flux amplitude following a 0.5$ reactivity step insertion.

period is constant. The purely exponential form is realized after a transition time during which the period depends on time. Figure 6-4 shows a typical transient ($\rho_1 = 0.5\$$) calculated with six groups of delayed neutrons exhibiting an obvious transition phase between the stationary state and the asymptotic transient characterized by a constant period. The flux amplitude in the asymptotic part of the transient may be written as

$$p_{as}(t) = p_{as} \exp(\alpha t) \quad , \tag{6.52}$$

with α equal to the inverse stable period.

In general, the inverse period depends on time. The general definition of the inverse period $\alpha(t)$ is:

$$\alpha(t) = \frac{\dot{p}(t)}{p(t)} = \text{inverse period} \quad . \tag{6.53}$$

The inverse period, Eq. (6.53), is often called the "instantaneous inverse period," particularly in older literature, in order to emphasize the fact that it is not a "stable inverse period." The term $\alpha(t)$ is the slope of the flux amplitude in a semi-logarithmic plot such as Fig. 6-4 at time t, i.e.,

$$\frac{d}{dt} \ln \frac{p(t)}{p_0} = \frac{\dot{p}}{p} = \alpha(t) \quad . \tag{6.54}$$

The flux amplitude can be expressed in terms of the inverse period by solving the differential equation, Eq. (6.53):

$$p(t) = p_0 \exp\left[\int_0^t \alpha(t')\, dt'\right] \quad . \tag{6.55}$$

If the period is asymptotically constant, as in the case depicted in Fig. 6-4, $\alpha(t)$ may be split into an asymptotic and a transitory part:

$$\alpha(t) = \alpha + \delta\alpha(t) \quad , \tag{6.56}$$

with $\delta\alpha(t)$ converging to zero. Equation (6.55) then yields

$$p(t) = p_0 \exp\left[\int_0^t \delta\alpha(t')\, dt'\right] \exp(\alpha t) \rightarrow p_{as} \exp(\alpha t) \quad . \tag{6.57}$$

The first integral in the exponent of Eq. (6.57) converges since $\delta\alpha(t)$ rapidly approaches zero after the transition phase. The converged integral changes the value of p_0 into p_{as}.

A purely exponential variation of the flux amplitude is only possible if all other components of the solution vary proportionally to the flux. If the delayed neutron source were to have a different time variation, the flux would be affected.

The asymptotic amplitudes of the precursors in terms of the asymptotic behavior of the flux are obtained by formally solving the respective balance equations. Instead of the solution, Eq. (6.36), it is convenient for this derivation to write $\zeta_k(t)$ in the form of a convolution integral over the entire flux history without singling out an initial state:

$$\zeta_k(t) = \beta_k \int_{-\infty}^t p(t') \exp[-\lambda_k(t - t')]\, dt' \quad . \tag{6.58}$$

To obtain the asymptotic ζ_k, the term $p_{as}(t')$ from Eq. (6.57) is substituted for $p(t')$:

$$\zeta_{k,as}(t) = \beta_k p_{as} \int_{-\infty}^t \exp[\alpha t' - \lambda_k(t - t')]\, dt' \quad . \tag{6.59}$$

Integrating Eq. (6.59) gives

$$\zeta_{k,as}(t) = \frac{\beta_k p_{as}}{\alpha + \lambda_k} \exp(\alpha t) = \zeta_{k,as} \exp(\alpha t) \quad , \tag{6.60}$$

with

$$\zeta_{k,as} = \frac{\beta_k p_{as}}{\alpha + \lambda_k} \quad . \tag{6.61}$$

The asymptotic reduced delayed neutron source is given by

$$s_d(t) = \sum_k \frac{\beta_k \lambda_k}{\alpha + \lambda_k} p_{as} \exp(\alpha t) \quad . \tag{6.62}$$

Inserting Eq. (6.60) into the first of the point kinetics equations and cancelling the exponential function and p_{as} gives

$$\alpha = \frac{\rho - \beta}{\Lambda} + \frac{1}{\Lambda} \sum_k \frac{\beta_k \lambda_k}{\alpha + \lambda_k} \quad . \tag{6.63}$$

Solving for ρ yields

$$\rho = \alpha\Lambda + \beta - \sum_k \frac{\beta_k \lambda_k}{\alpha + \lambda_k} \quad . \tag{6.64}$$

The β on the right side of this equation is usually combined with the sum over k. This then yields the standard form of the so-called "inhour" equation, which gives ρ as a function of α:

$$\rho = \alpha\Lambda + \sum_k \frac{\beta_k \alpha}{\alpha + \lambda_k} \quad . \tag{6.65}$$

The reactivity corresponding to $\alpha = 1$ inverse hour is called "1 inhour." Therefore, Eq. (6.65) is historically called the "inhour equation."

The reactivity "1 inhour" is often used as the reactivity unit in reactor experiments or operations involving very small amounts of reactivity. The delayed neutron data in Sec. 2-3 along with the β_k definitions of Sec. 3-2B give the approximate value,

$$\rho(\alpha = 1/3600\ \text{s})$$

$$= 1\ \text{inhour} \simeq 0.0035\beta \simeq \begin{cases} 2.5 \times 10^{-5} & \text{(LWR)} \\ 1.0 \times 10^{-5} & \text{(FBR)} \end{cases} \quad . \tag{6.66}$$

Note that the asymptotic solutions of Eqs. (6.52) and (6.60) are independent of the initial conditions, except for the general normalization, i.e., the proportionality to p_0. The independence of the asymptotic solution from the initial conditions is common to all systems of linear differential equations. This general rule is easily understood if the asymptotic forms, Eqs. (6.52) and (6.60), are inserted in the equation system and cancelling $\exp(\alpha t)$. What remains is a homogeneous *system of equations* for p_{as} and $\zeta_{k,as}$, i.e., not a system of *differential* equations. This homogeneous system of equations has, for the proper value of α, a unique solution except for a free normalization factor. Initial conditions have no effect on this solution.

To obtain additional insight into the relation between ρ, α, β, and Λ as provided by the inhour formula, three extreme cases are examined:

1. $\alpha \ggg \lambda_k$, very rapid transients
2. $\alpha \gg \lambda_k$, rapid transients
3. $\alpha \ll \lambda_k$, very slow transients.

By exploiting the inequalities, simplifications can be made that are, in effect, approximations of the delayed neutron source. The resulting approximate inhour equations are therefore compared with the corresponding result of the one-delay-group kinetics approximation. The one-delay-group inhour equations corresponding to Eqs. (6.64) and (6.65) are

$$\rho = \alpha\Lambda + \beta - \frac{\beta\lambda}{\alpha + \lambda} \tag{6.67}$$

and

$$\rho = \alpha\Lambda + \frac{\beta\alpha}{\alpha + \lambda} \quad . \tag{6.68}$$

In the first limit, $\alpha \ggg \lambda_k$, one deals with very rapid transients. The resultant approximation is called "prompt kinetics." The inverse period α is assumed to be so large that it suffices to retain only the first term of a Taylor expansion of the sum in Eq. (6.65) in powers of λ_k/α; i.e., the λ_k are completely negligible compared to α. The resulting inhour equation is

$$\rho = \alpha\Lambda + \beta \quad . \tag{6.69}$$

If the reactivity is given, one can solve for α and obtain the "prompt" inverse period:

$$\alpha_p = \frac{\rho - \beta}{\Lambda} = \text{prompt inverse period} \quad . \tag{6.70}$$

Equation (6.69) can also be obtained from Eq. (6.64), where $\alpha \ggg \lambda_k$ leads to the complete neglect of the sum on the right side. Comparison with Eq. (6.62) shows that neglecting this sum means physically *the total neglect of the delayed neutron source.* Therefore, the six- and one-delay-group inhour equations, Eqs. (6.65) and (6.68), lead to the same prompt kinetics approximation.

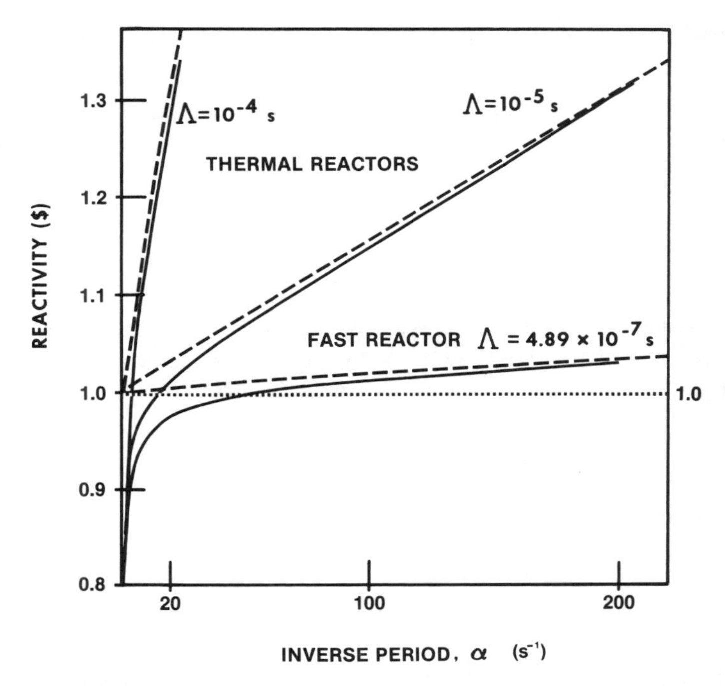

Fig. 6-5. The relation of stable inverse periods (solid lines) and prompt inverse periods (dashed lines) with the reactivity, ρ, for three generation times.

Graphically,[c] the asymptotic relation, Eq. (6.69), for $\rho(\alpha)$ appears as a straight line through $\rho = \beta$ with Λ as the slope. Figure 6-5 shows $\rho(\alpha)$ from Eq. (6.69) and the six-delay-group formula for three different lifetimes with ρ around prompt critical:

1. $\Lambda = 4.9 \times 10^{-7}$ s for a typical large mixed-oxide (Pu-U)-fueled reactor
2. $\Lambda = 1 \times 10^{-5}$ s for a typical light-water-moderated reactor
3. $\Lambda = 10^{-4}$ s for a more dilute reactor, using, for example, D_20 or graphite as the moderator.

The most striking feature in Fig. 6-5 is the strong increase of α with increasing ρ in the superprompt critical domain.

In the limit $\alpha \gg \lambda_k$, one deals with *rapid transients.* By taking into account the second term of the Taylor expansion of the sum in Eq. (6.65),

---

[c]Only part of the *positive branch* of the complicated and multiple valued function $\rho(\alpha)$ is presented. The complete function $\rho(\alpha)$ is discussed in Sec. 6-2D in the context of the roots of the characteristic equation for analytic solutions of certain kinetics problems.

an expression is obtained that describes the approach toward the asymptotic relation, Eq. (6.69); this rapid transient approximation yields:

$$\rho = \alpha\Lambda + \beta - \frac{1}{\alpha}\sum_k \beta_k\lambda_k \tag{6.71a}$$

or

$$\rho = \alpha\Lambda + \beta\left(1 - \frac{\overline{\lambda}}{\alpha}\right) \quad . \tag{6.71b}$$

Considering *two* terms of the Taylor expansion of the sum in Eq. (6.65) is equivalent to considering *one* term of the sum in Eq. (6.64) and thus *one* term in the delayed neutron source, Eq. (6.62):

$$s_d(t) = \sum_k \beta_k\lambda_k \cdot \frac{1}{\alpha} \cdot p_{as} \exp(\alpha t) \quad \text{for } \alpha \gg \lambda_k \quad , \tag{6.72a}$$

which may be written in an integral form as:

$$s_d(t) = \beta \cdot \overline{\lambda} \cdot p_{as} \int_{-\infty}^{t} \exp(\alpha t')\, dt' \quad . \tag{6.72b}$$

A comparison of Eq. (6.72b) with Eq. (6.59) shows that Eq. (6.72b) does not contain the exponential term, $\exp[-\lambda_k(t - t')]$, which describes the decay of newly formed precursors. Thus, the limit $\alpha \gg \lambda_k$ is equivalent to neglecting the decay of newly formed precursors in their balance equations [see also Eqs. (6.17) to (6.23)]. Therefore, the one-delay-group λ is the same as it appeared for short-term transients in Section 6-1B. However, the precise physical interpretations of both approximations differ.

For *short-term transients,* the decay of newly formed precursors is negligible because the time interval of the transient considered does not provide enough time for newly formed precursors to change through decay. The decay of the stationary precursor population is, however, included. In other words, in the rapid transient approximation of an exponentially increasing neutron flux, "nearly all" of the precursors have been produced so recently that their total number practically has not yet changed through decay.

Apart from this small difference in the interpretation, the approximate inhour equation, Eqs. (6.71), is equivalent to the precursor accumulation approximation, Eqs. (6.17) to (6.23). The application of the same approximation, $\alpha \gg \lambda_k$, to the one-delay-group inhour equations, Eqs. (6.67) and (6.68), yields the same approximate inhour equation as derived from the six-delay-group kinetics, Eq. (6.71b), *if* λ *is chosen to equal* $\overline{\lambda}$.

Figure 6-6 illustrates how the $\bar{\lambda}$ term in Eq. (6.71b) describes the approach into the prompt kinetics limit, Eq. (6.69). Only one set of curves is shown in this illustration, since the subtracted term does not depend on Λ. The comparison presented in Fig. 6-6 shows that the prompt kinetics approximation augmented by the $\bar{\lambda}/\alpha$ term in Eq. (6.71b), which subtracts a hyperbola from the straight line of the prompt period approximation of Eq. (6.69), describes the inhour equation very well for all reactors in the entire reactivity range above $\approx$0.95\$. Thus, the approximate inhour equation, Eq. (6.71b), represents a useful approximation around prompt critical and in the entire superprompt critical reactivity range.

In the limit of very slow transients, α is much smaller than the λ's ($\alpha \ll \lambda_k$). For $\rho = 0$, obviously α is also equal to zero. The increase of α with increasing ρ in the range of very small reactivities can be approximately described by an inhour formula obtained from neglecting α in

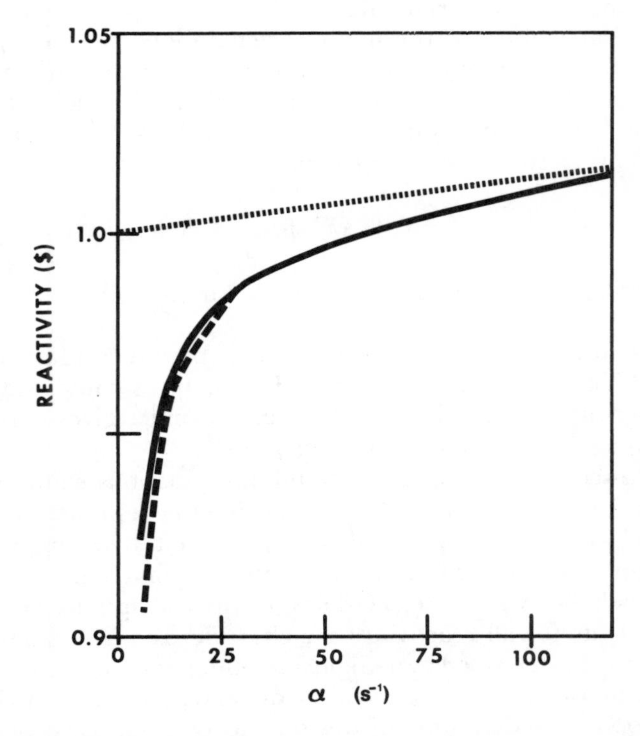

Fig. 6-6. The stable inverse period as a function of the reactivity: exact values (solid lines); large alpha approximation of Eq. (6.71b) (dashed line); and prompt period approximation of Eq. (6.69) (dotted line).

the denominators of Eq. (6.65):

$$\rho = \alpha\Lambda + \alpha \sum_k \frac{\beta_k}{\lambda_k} \tag{6.73}$$

or

$$\rho = \alpha[\Lambda + \beta(\overline{1/\lambda})] = \alpha\,[\Lambda + \beta/(\overline{\lambda^{in}})] \quad . \tag{6.74}$$

With the same approximation, the one-delay-group kinetics [Eq. (6.68)] yields

$$\rho = \alpha(\Lambda + \beta/\lambda) \quad . \tag{6.75}$$

Thus, the appropriate λ in the one-delay-group kinetics for *very slow transients* is $\overline{\lambda^{in}}$ and not $\overline{\lambda}$. This is not surprising since a reactor during very slow transients is *nearly stationary.* The precursor decay in a stationary reactor was previously found to be described exactly by a one-delay-group $\lambda = \overline{\lambda^{in}}$. The same λ, then, approximately describes the precursor decay during very slow transients.

The contribution of the neutron generation time in Eq. (6.74) is negligible for almost all reactors except very large graphite-moderated reactors. In addition, $\overline{\lambda^{in}}$ is about the same for ^{235}U-fueled thermal as well as for ^{238}U-Pu-fueled fast reactors (see Sec. 6-1B). These two facts lead to the very convenient approximation,

$$\alpha = \frac{\rho}{\beta}\overline{\lambda^{in}} \simeq \frac{\rho}{\beta}\,0.08\ \mathrm{s}^{-1} \quad \text{for } \frac{\rho}{\beta} \lesssim 3 \times 10^{-2} \quad . \tag{6.76}$$

For example, if $\rho = 0.0125\beta$ (= 1.25¢), Eq. (6.76) yields a period of 1000 s.

Figure 6-7 shows the asymptotic inverse period as a function of the reactivity. The linear approximation, Eq. (6.76), is compared with the six-delay-group result. This simple linear formula gives a reasonably good estimate of the stable period for $\rho \lesssim 3$¢.

The discussion of approximate inhour equations showed that the decay of the six precursor groups may be described, in the two extremes of the positive α range, by a single decay constant λ: During rapid transients $\overline{\lambda}$ and during very slow transients, the much smaller $\overline{\lambda^{in}}$ were found to yield fairly accurate one-delay-group descriptions. Since asymptotic transients for all reactivities are characterized by a single period, they may also be described by a single time-independent one-delay-group λ. The value of this $\lambda = \lambda_{as}$ certainly depends on the reactivity or the stable period. The definition of this one-delay-group λ_{as} is obtained by equating the corresponding inhour equations, Eqs. (6.65) and (6.68), which yields:

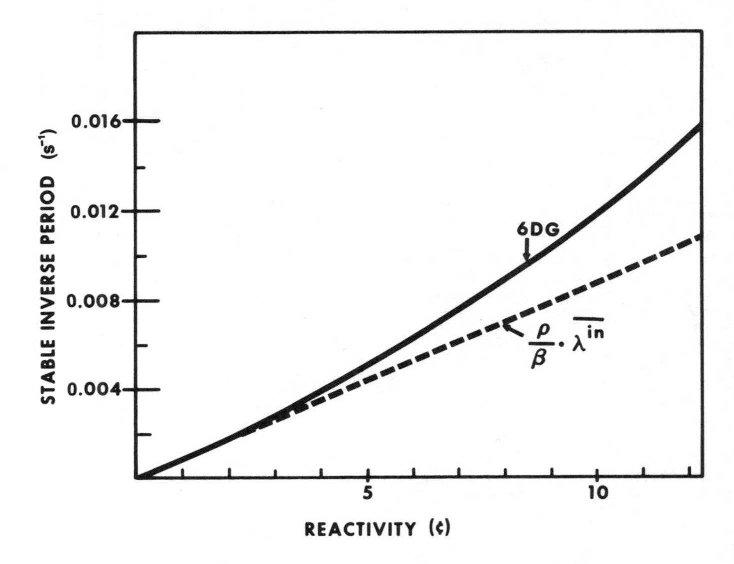

Fig. 6-7. Stable inverse period as a function of ρ for small reactivities (solid line) and a linear approximation, Eq. (6.76) (dashed line).

$$\lambda_{as} = \frac{\beta}{\sum_k \frac{\beta_k}{\alpha + \lambda_k}} - \alpha \quad . \tag{6.77}$$

Figure 6-8 shows the transition of $\lambda_{as}(\rho)$ from its low reactivity limit $\overline{\lambda^{in}}$ into the high reactivity limit $\bar{\lambda}$. The curve $\lambda_{as}(\rho)$ is calculated from the definition of Eq. (6.77) with $\alpha(\rho)$ found from the inhour equation, Eq. (6.65).

The inhour equation is identical with the characteristic equation of the system of (point kinetics) differential equations. The characteristic equation and its roots are further discussed in Sec. 6-2D in the context of the analytic solution of transients with constant reactivity. Specifically investigated are the values of p_{as} as well as the time required for establishing the asymptotic form of the transient. These quantities cannot be obtained by investigating only the asymptotic solution.

6-1D Summary of Delayed Neutron Source Approximations

Several approximations of the point kinetics equations have been introduced and discussed earlier in this chapter. The common feature of these approximations is the—more or less drastic—simplification of the delayed neutron source. The following list presents a sequence of

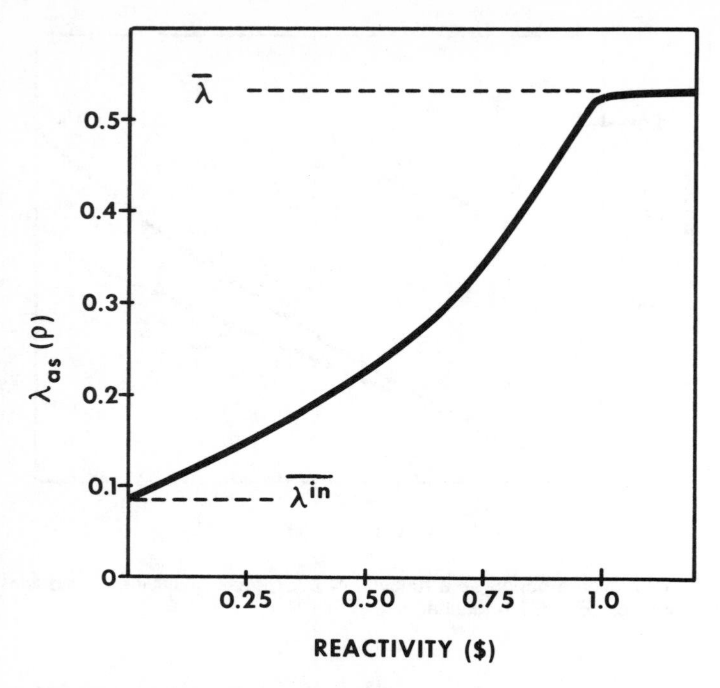

Fig. 6-8. One-delay-group constant, λ_{as}, as a function of ρ.

point kinetics models, which result from approximating the delayed neutron source with decreasing accuracy (note that β and Λ are assumed to be constant here, as well as in most other applications in this text):

six-delay-group kinetics:

$$\dot{p} = \frac{\rho - \beta}{\Lambda} p + \frac{1}{\Lambda} \sum_k \lambda_k \zeta_k(t) \quad , \tag{6.78a}$$

two-delay-group kinetics:

$$\dot{p} = \frac{\rho - \beta}{\Lambda} p + \frac{1}{\Lambda} [\lambda^{(1)}\zeta^{(1)}(t) + \lambda^{(2)}\zeta^{(2)}(t)] \quad , \tag{6.78b}$$

one-delay-group kinetics ($\bar{\lambda}$ kinetics):

$$\dot{p} = \frac{\rho - \beta}{\Lambda} p + \frac{1}{\Lambda} \bar{\lambda}\zeta(t) \quad , \tag{6.78c}$$

precursor accumulation (PA) approximation:

$$\dot{p} = \frac{\rho - \beta}{\Lambda} p + \frac{1}{\Lambda} [\beta p_0 + \bar{\lambda}\beta I(t)] \quad , \tag{6.78d}$$

constant delayed neutron source (CDS) approximation:

$$\dot{p} = \frac{\rho - \beta}{\Lambda} p + \frac{1}{\Lambda} \beta p_0 \quad , \tag{6.78e}$$

prompt kinetics approximation; neglect of explicit delayed neutron source:

$$\dot{p} = \frac{\rho - \beta}{\Lambda} p \quad , \tag{6.78f}$$

and

kinetics without delayed neutrons:

$$\dot{p} = \frac{\rho}{\Lambda} p \quad . \tag{6.78g}$$

The sequence of these approximations is reviewed in the following in an inverted order, starting with the least accurate one. This allows us to introduce and explain the various kinetics features in the simplest possible way, thus identifying the cause/effect relationships on the most rudimentary levels. Most of the results that are referred to in this discussion are presented elsewhere in Chapter 6.

Kinetics Without Delayed Neutrons

As problems of reactor *statics* do not require the consideration of delayed neutrons, the *stationary states in subcritical systems are correctly described* by Eq. (6.78g). This description assumes the form of source multiplication formulas [see, for example, Eqs. (6.86) and (6.87)]. The transient between stationary subcritical states is described per Eq. (6.78g) as a prompt transition. This is only partially correct.

Historically, this approximation was used for semiquantitative considerations of transients with $\rho \gg \beta$, but reactivity insertions of such magnitude are not feasible in nuclear reactors.

Prompt Kinetics

Prompt kinetics is a useful approximation for the treatment of superprompt-critical transients (see Chapters 8 and 10). It allows us to treat problems with nonlinear feedback effects analytically and thus come to simple relations for important safety characteristics such as the relation between reactivity insertion and energy release in idealized transients. The delayed neutron source can be approximately accounted for in the form of a modified initial condition (pseudo-initial condition).

The CDS Approximation

The CDS approximation is the lowest level that allows a correct interpretation and quantitative description of the *prompt jump*. It is a useful tool to understand the kinetics behavior shortly after any sudden change in the system, where "shortly" means prior to a considerable change in the delayed neutron source, ~0.1 s in the example of Table 6-II. These sudden flux changes can be quantified in terms of a source multiplication formula; see, for example, Eq. (6.92b). The CDS approximation is widely applied in the simplified analyses of reactivity measurements that employ rapid changes. Examples are rod-drop or source-jerk reactivity measurement methods; see Sec. 9-3.

The PA Approximation

The PA approximation is a first-order extension of the CDS approximation toward larger times, to ~0.5 s after a sudden change in the system as shown in the example of Table 6-II. It also gives a first-order correction to the prompt kinetics form of the inhour equation.

The One-Delay-Group Kinetics

The inclusion of one group of delayed neutrons leads to the lowest level of a kinetics that can formally be applied for the entire time scale. However, as a "correct" one-delay-group would depend considerably on time, a one-delay-group kinetics with a constant λ (especially $\lambda = \bar{\lambda}$) is restricted in its accuracy to short times (e.g., smaller than ~1 s after a sudden change).

Kinetics with Two Groups of Delayed Neutrons

The two-delay-group kinetics was a popular approximation in the time before large computers became available. It yields an improved value for the stable period and also allows a description of transitory behavior between the prompt jump and the asymptotic transition that eventually follows a reactivity step that was not possible with one delay group. It disappeared years ago from practical applications and is mentioned here merely for historical reasons.

Six-Delay-Group Kinetics

This is the normally applied version of the kinetics equations.

Kinetics with More Than Six Delay Groups

More than six delay groups are occasionally used in a case where significant contributions to the delayed neutron source come from more than one fissioned nuclide. Examples would be the high-burnup fuel in LWRs with comparable contributions from ^{235}U and ^{239}Pu, or from ^{239}Pu and ^{238}U in fast reactors. Then several delay groups are added for the λ values that are considerably different for the two nuclides and thus require separate precursor balance equations. Table 2-II shows that the delay groups 3 to 6 have considerably different λ values for ^{235}U and ^{239}Pu. This then suggests the use of 10 delay groups; groups 1 and 2 are combined for both nuclides, but groups 3 through 6 are treated separately.

These complications can be avoided by using the readjusted ν_{dk} values that match the nuclide-dependent results with a single set of λ values (as given in Table 2-III). Then the six-delay-group kinetics is adequate in all cases.

6-2 Transients with Constant Reactivity

Many reactor transients are caused by a relatively fast insertion of a fixed amount of reactivity—positive or negative. Typical examples are rod-drop or rod-jerk reactivity measurements, rapid insertion of a shutdown rod (scram), falling of a fuel subassembly (loading accident), or hypothetical runaway of a control rod. In all cases, the reactivity is increased or decreased quickly and then remains constant. This type of a reactivity insertion can be described approximately by a reactivity step, Eq. (6.13a). In addition to reactivity steps, step changes in the intensity of an independent source in subcritical systems are of interest. The resulting transients, which follow a step change in the system, are analyzed in the following sections with models of increasing sophistication in order to provide an understanding of the basic features of the transient.

6-2A The Prompt Jump in a Subcritical Reactor

If an originally stationary *subcritical* reactor is subjected to a step-type change, within the subcritical range, the eventual result will be a modified stationary flux. In Sec. 6-1A, it was shown that the delayed neutron terms drop out of the kinetics equations in any *stationary* state of the reactor, critical or subcritical. Thus, the approximation of complete neglect of delayed neutrons, Eq. (6.78g), can be usefully employed since it correctly describes the stationary states. The delayed neutrons can only have an effect on the transition between the two stationary states (see below).

Let ρ_0 and s_0 denote the initial values of reactivity and source and ρ_1 and s_1 the values after the respective step change. The following combinations of step changes in the subcritical reactor are investigated:

1. $s_0 \rightarrow s_1$, step change in source
2. $\rho_0 \rightarrow \rho_1$, reactivity step
3. $\rho_0 \rightarrow \rho_1$ and $s_0 \rightarrow s_1$, reactivity and source steps.

The kinetics equation without delayed neutrons is given by

$$\dot{p} = \frac{\rho}{\Lambda} p + \frac{s}{\Lambda} \quad . \tag{6.79}$$

The general solution for the combined step changes can be readily found[d]:

$$p(t) = p_0 \exp\left(\frac{\rho_1}{\Lambda} t\right) - \frac{s_1}{\rho_1}\left[1 - \exp\left(\frac{\rho_1}{\Lambda} t\right)\right] \quad . \tag{6.80}$$

It is particularly instructive to have the prompt inverse period appearing explicitly in analytic solutions. Consequently, a special notation for the prompt inverse period is introduced:

$$\alpha_p = \frac{\rho - \beta}{\Lambda} \quad . \tag{6.81}$$

The α_p value without delayed neutrons is denoted by

$$\alpha_p^0 = \frac{\rho}{\Lambda} \quad , \tag{6.82}$$

where the superscript 0 indicates the neglect of β.

With this notation, the solution, Eq. (6.80), is written as:

$$p(t) = p_0 \exp(\alpha_{p1}^0 t) - \frac{s_1}{\rho_1}[1 - \exp(\alpha_{p1}^0 t)] \tag{6.83}$$

with

$$\alpha_{p1}^0 = \alpha_p^0(\rho = \rho_1)$$

and

$$\alpha_{p0}^0 = \alpha_p^0(\rho = \rho_0) \quad . \tag{6.84}$$

The solutions for the special cases listed above follow directly from the respective simplification of the general solution, Eq. (6.83):

step change in source:

$$p(t) = p_0 \exp(\alpha_{p0}^0 t) - \frac{s_1}{\rho_0}[1 - \exp(\alpha_{p0}^0 t)] \quad . \tag{6.85a}$$

step change in reactivity:

$$p(t) = p_0 \exp(\alpha_{p1}^0 t) - \frac{s_0}{\rho_1}[1 - \exp(\alpha_{p1}^0 t)] \quad . \tag{6.85b}$$

---

[d]See, for example, Sec. C-1 of App. C, also see Fig. 6-13 for the interpretation of the two terms of $p(t)$ as it results from Eq. (6.79).

step changes in source and reactivity:

$$p(t) = p_0 \exp(\alpha_{p1}^0 t) - \frac{s_1}{\rho_1}[1 - \exp(\alpha_{p1}^0 t)] \quad . \tag{6.85c}$$

According to the source multiplication formula, Eq. (6.8c), p_0 is given by

$$p_0 = \frac{s_0}{|\rho_0|} \quad . \tag{6.86}$$

All transients described by Eqs. (6.85) have one feature in common: They start at $p_0 = s_0/|\rho_0|$ and approach "promptly" (i.e., within several prompt periods) a new level, e.g., $p_1 = s_1/|\rho_1|$. *The rapid or prompt transition that follows a step change in the multiplication properties, ρ, or independent source intensity, s, is called a "prompt jump."*

In Fig. 6-9, a typical transient is depicted as it would occur after step changes in a subcritical reactor without delayed neutrons.

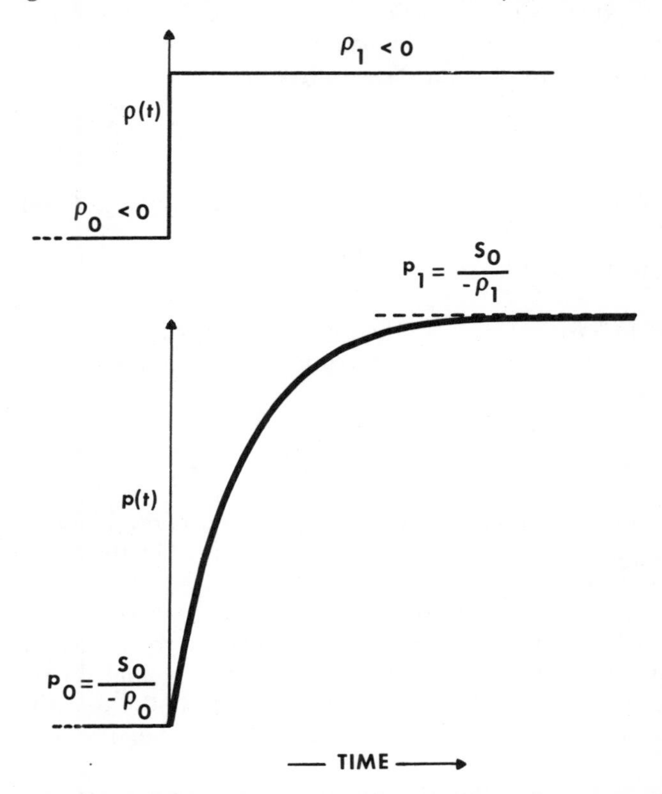

Fig. 6-9. A typical transient following a reactivity step change in a subcritical reactor without delayed neutrons.

The time period for the prompt jump transition such as the one depicted in Fig. 6-9 may be very short, e.g., $\sim 10^{-3}$ s for $|\rho_1| = 5 \times 10^{-3}$ and $\Lambda = 5 \times 10^{-6}$ s. The simple kinetics without delayed neutrons already provides a correct *qualitative* understanding of the important kinetics phenomenon, the prompt jump, as the prompt response of the neutron flux to a step-type change in the multiplication or source characteristics of a reactor. Although the kinetics without delayed neutrons does not give the correct value of the flux immediately after the prompt jump, it does give the correct asymptotic flux in the new stationary state, i.e.,

$$p_1 = \frac{s_1}{|\rho_1|} \quad . \tag{6.87}$$

The following investigation of kinetics with delayed neutrons shows that the actual value of the flux change during the prompt jump is smaller than that obtained from Eq. (6.87).

A higher degree of sophistication in the description of the prompt jump phenomenon is obtained by accounting for the existence of delayed neutrons in the form of a constant source $s_d = s_{d0}$ as in Eq. (6.78e):

$$\dot{p} = \frac{\rho - \beta}{\Lambda} p + \frac{s_{d0} + s}{\Lambda} \quad . \tag{6.88}$$

The solution of Eq. (6.88) has the same form as Eq. (6.83), i.e.,

$$p(t) = p_0 \exp(\alpha_{p1} t) + \frac{s_{d0} + s_1}{\beta - \rho_1} [1 - \exp(\alpha_{p1} t)] \quad . \tag{6.89}$$

The main difference between Eqs. (6.83) and (6.89) is that the flux approaches an intermediate level during the prompt jump, the so-called "prompt jump flux," p_{pj}:

$$p_{pj} = \frac{s_{d0} + s_1}{\beta - \rho_1} \quad ; \tag{6.90}$$

p_{pj} is different from the final flux level $s_1/|\rho_1|$. A second difference is that the prompt jump transition in Eq. (6.89) is described by the actual prompt period, α_{p1}, and not by its truncated value α_{p1}^0.

The source multiplication formula, Eq. (6.90), is very similar to the one given by Eq. (6.12). The only difference is that the reactivity appears in Eq. (6.90) with its value right after the step change in the system rather than with its initial value as in Eq. (6.12). In both cases, the delayed neutron source is the one established during stationary operation at $t < 0$.

To see how accurately the different approximations, particularly the one with a constant delayed neutron source, describe the prompt jump phenomenon, a numerical comparison is presented in Fig. 6-10. A step increase in reactivity to $\rho_1 = -\beta$ is applied to a subcritical system with $\rho_0 = -3\beta$. The results of Eqs. (6.89) and (6.83) are compared for short times with six-delay-group results. Equation (6.83) obviously yields a wrong value for the prompt jump whereas Eq. (6.89) describes the prompt jump phenomenon very accurately. Equation (6.89) cannot, however, yield the increase in the flux due to the increase in the precursor population since the delayed neutron source is assumed constant.

The good agreement between Eq. (6.89) and the six-delay-group results for short times shows that the simple model with a constant delayed neutron source provides a correct interpretation as well as an accurate quantitative description of the prompt jump phenomenon:

The prompt jump phenomenon consists of a prompt adjustment of the flux to a step change in the neutron multiplication capability or independent source.

The flux after the prompt jump is given by a source multiplication formula with the as-yet unchanged delayed neutron source (together with an independent source if applicable) in the numerator and the reactivity difference below prompt critical in the denominator.

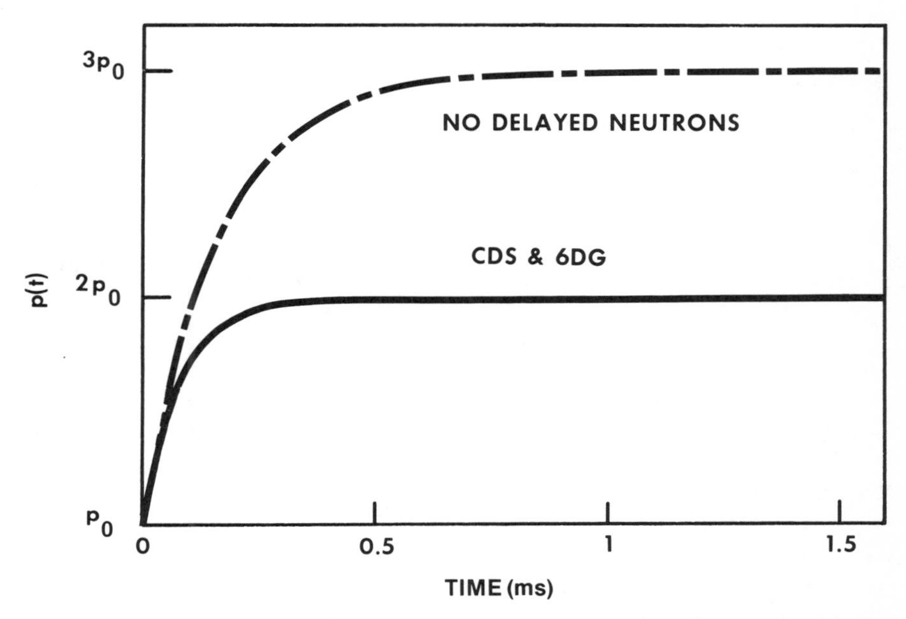

Fig. 6-10. Comparison of the results of various models for short times following a reactivity-step-induced transient in a subcritical reactor.

The *rapid* transition of the prompt jump must be followed by a second transition phase during which the delayed neutron source, which is virtually unchanged during the prompt jump, establishes a balance with the new flux. For example, if the flux increased during the prompt jump as in Fig. 6-10, the production rate of precursors increases proportionally to the new flux, i.e., it is twice the original production rate. The decay rate, however, is still given by the virtually unchanged initial decay rate. Thus the precursor population increases and with it the flux. Asymptotically the flux approaches a level that results from the static source multiplication in the new stationary state:

$$p_1 = \frac{s_1}{|\rho_1|} \quad , \tag{6.91a}$$

or in its alternative formulation [compare Eqs. (6.8c) and (6.12)]:

$$p_1 = \frac{s_{d1} + s_1}{\beta + |\rho_1|} \quad . \tag{6.91b}$$

This gives the following sequence of source multiplication formulas for subcritical step-induced transients:

initial static source multiplication:

$$p_0 = \frac{s_{d0} + s_0}{\beta + |\rho_0|} \quad , \tag{6.92a}$$

prompt jump source multiplication:

$$p_{pj} = \frac{s_{d0} + s_1}{\beta + |\rho_1|} \quad , \tag{6.92b}$$

final static source multiplication:

$$p_1 = \frac{s_{d1} + s_1}{\beta + |\rho_1|} \quad . \tag{6.92c}$$

The prompt jump source multiplication differs from the initial static one by containing the new ρ_1 and s_1, and it differs from the asymptotic source multiplication by still containing the initial delayed neutron source.

Figures 6-11 and 6-12 show the transition of the transient of Fig. 6-10 into the asymptotic state, calculated with six-delay-group kinetics. In Fig. 6-11, the flux amplitude $p(t)$ is plotted on a logarithmic time scale in order to demonstrate clearly the two transitions after a step change in the system: the prompt jump up to a level on which the flux stays for many prompt periods and the subsequent slow adjustment of the delayed neutron source. Figure 6-12 shows the flux on a linear time scale where the prompt jump virtually coincides with the ordinate if the

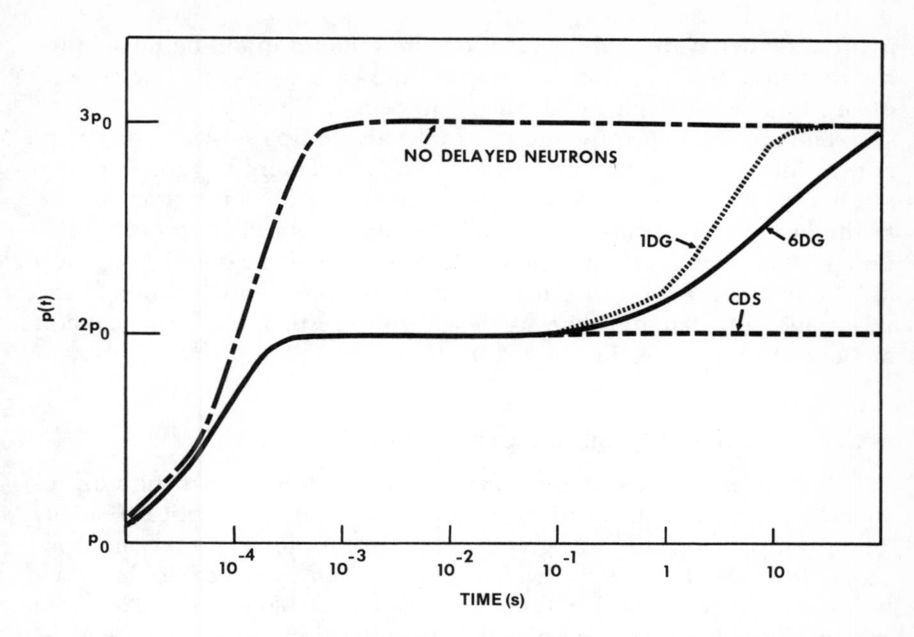

Fig. 6-11. A reactivity-step-induced transient in a subcritical reactor on a logarithmic time scale.

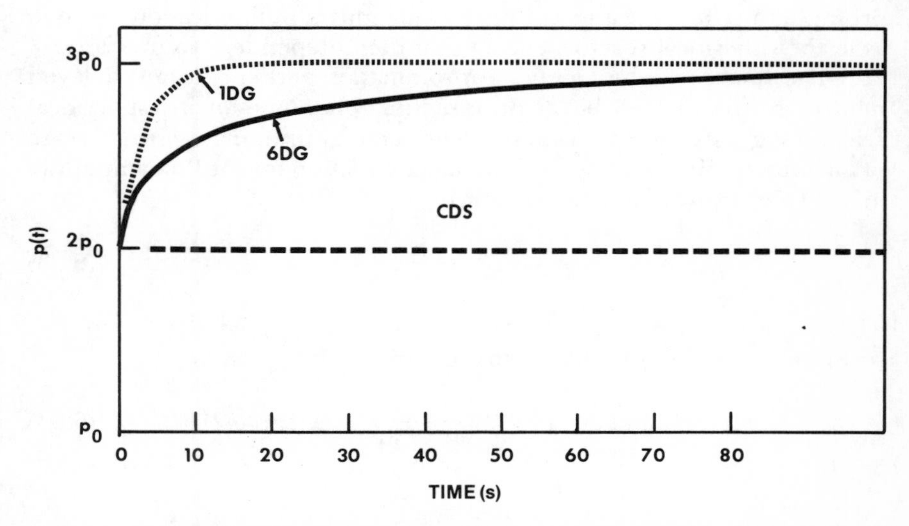

Fig. 6-12. A reactivity-step-induced transient in a subcritical reactor for long times on a linear scale.

asymptotic behavior is also shown on the same graph. The linear plot clearly shows the flux rise "right after" the prompt jump. The "slope" of this flux rise is explicitly discussed in Secs. 8-1 and 10-3.

The one-delay-group kinetics results are compared with the reference solution in Figs. 6-11 and 6-12. The agreement is good for very small and very large times; the good agreement for large times is due to the fact that the transients considered asymptotically approach a stationary state that does not depend on the choice of the one-delay-group λ. The one-delay-group results, however, converge much faster to the new stationary state than the six-delay-group results; see the discussion of the transitory roots in Sec. 6-2D.

6-2B The Prompt Jump in a Critical Reactor

The investigations in the previous section showed that the flux responds with a prompt jump to a step change in a subcritical reactor. The prompt jump flux was given by the prompt jump source multiplication formula, Eq. (6.92b), which described an intermediate level between the two stationary states p_0 and p_1, Eqs. (6.92a) and (6.92c). The applicability of source multiplication formulas in a *supercritical* reactor, resulting from a reactivity step insertion ρ_1 into a *critical* reactor, is not at all obvious. The astonishing fact, however, is that as long as ρ_1 is smaller than β, i.e., as long as the reactor is not superprompt-critical, the flux response to a step change in ρ is also a prompt jump, and the prompt jump flux, p_{pj}, is given by the same source multiplication formula as in the subcritical reactor, except that the independent source is zero.

The success of the kinetics approximation with a constant delayed neutron source to describe the prompt jump phenomenon in a subcritical reactor suggests the application of the same approximation in the case of an initially critical reactor. The balance equation for the flux amplitude in the CDS approximation is given by:

$$\dot{p} = \alpha_p p + \frac{\beta}{\Lambda} p_0 \quad , \tag{6.93}$$

where α_p is the inverse prompt period[e] after the reactivity insertion. For constant α_p, the solution [compare Eq. (6.89)] is:

$$p(t) = p_0 \exp(\alpha_p t) + p_0 \frac{\beta}{\beta - \rho_1} [1 - \exp(\alpha_p t)] \quad . \tag{6.94}$$

---

[e]The term α_p corresponds to the α_{p1} of Eq. (6.89). Since only one reactivity value appears in this treatment, α_p is not indexed by 1.

Equation (6.94) shows that the *flux responds with a prompt jump also in a critical reactor* for $\rho_1 < \beta$; the intermediately established flux level is given by:

$$p_{pj} = \frac{\beta}{\beta - \rho_1} p_0 \tag{6.95a}$$

or

$$p_{pj} = \frac{s_{d0}}{\beta - \rho_1} = \text{prompt jump source multiplication} \quad , \tag{6.95b}$$

where $\rho_1 < \beta$ may be positive or negative. The flux change that occurs during the prompt jump is:

$$\Delta p_{pj} = \frac{\rho_1}{\beta - \rho_1} p_0 \quad . \tag{6.96}$$

Equations (6.89) and (6.94) provide an instructive interpretation. The first term of Eqs. (6.89) and (6.94) describes a rapid decrease of the initial flux, which would occur if all sources including the delayed neutron precursors were suddenly removed. The second term describes an equally rapid buildup of new flux through source multiplication. This interpretation of the analytic solutions, Eqs. (6.89) and (6.94), is illustrated in Fig. 6-13.

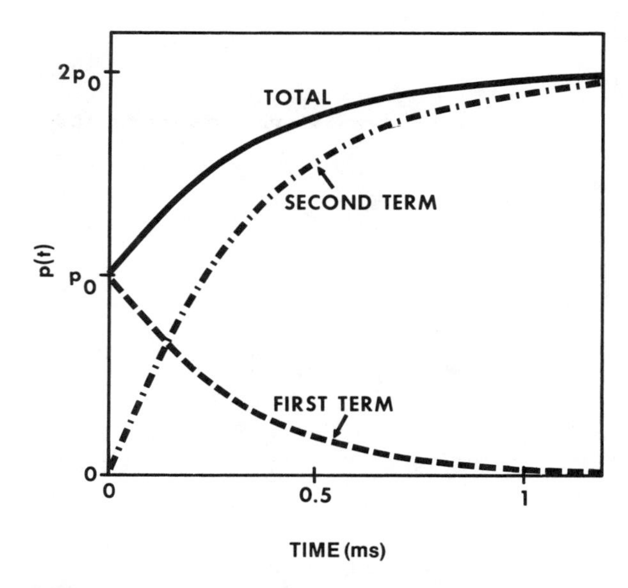

Fig. 6-13. Illustration of the two terms of the analytic solution of the kinetics equation in the CDS approximation.

Figures 6-14 and 6-15 show a step-induced transient, $\rho_1 = 0.5\beta$, initiated in a critical reactor. The results are calculated with six delayed neutron groups and are also compared with the CDS and one-delay-group results, and are presented in both figures for small and large time scales in analogy to Figs. 6-11 and 6-12. The transients show the same basic characteristics exhibited in subcritical transients. The flux promptly adjusts itself to the new source multiplication in an initial transition. In a second transition, the delayed neutron source adjusts itself to the changed flux such that an asymptotic transient is established. In the case of an initially critical reactor, however, the asymptotic transient consists of an exponential increase or decrease and not of a new stationary state. Again,

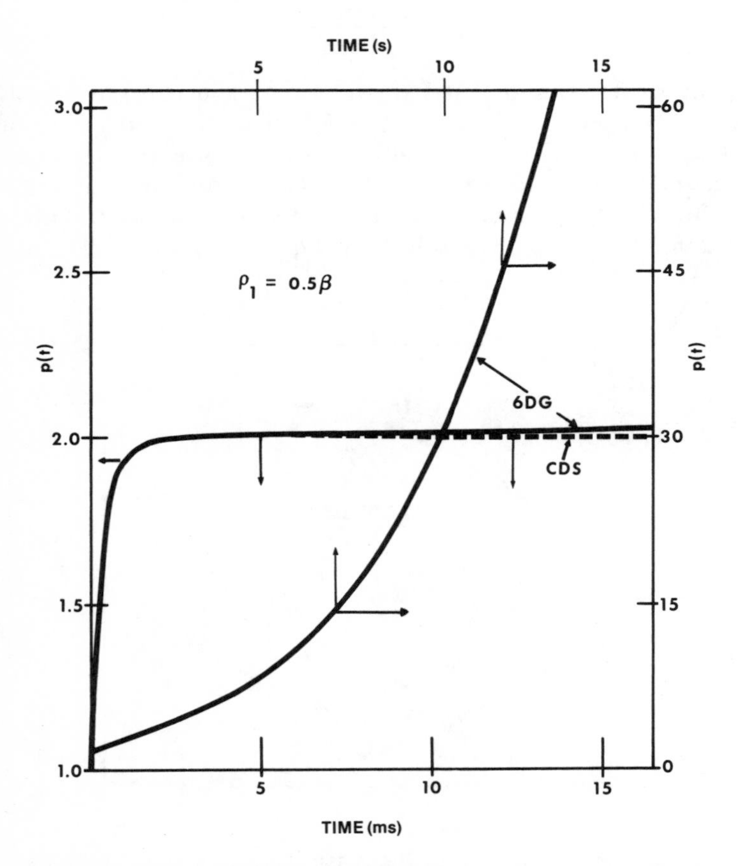

Fig. 6-14. Transient following a positive reactivity step, $\rho_1 = 0.5\beta$, in a critical reactor for short and long times: six delay group (solid line); constant delayed neutron source (dashed line). Note the difference in the abscissas and ordinates for the two curves.

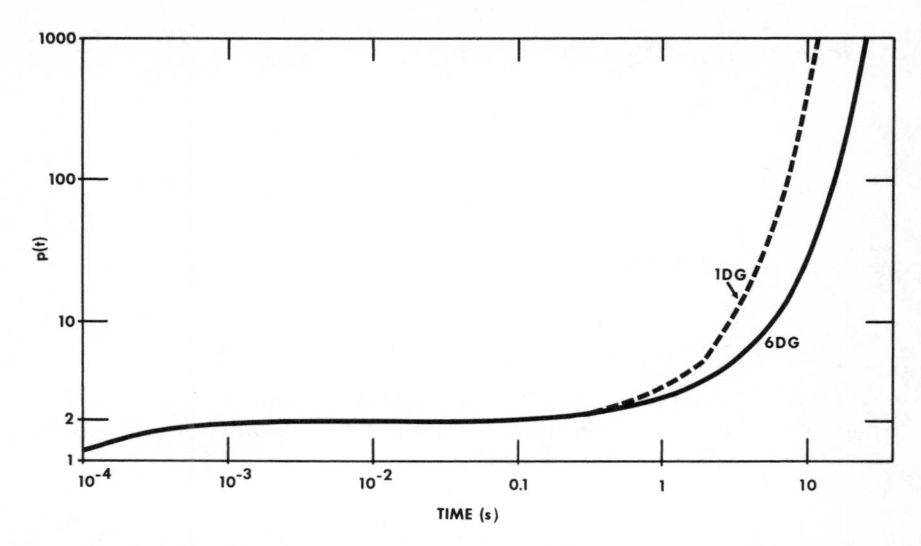

Fig. 6-15. The flux amplitude, $p(t)$, for the step-induced transient of Fig. 6-14 plotted on a log-log scale.

the simple model with a constant delayed neutron source accurately describes the time dependence during the prompt jump, but cannot describe the further flux rise after the prompt jump.

6-2C Kinetics with One Group of Delayed Neutrons

The prompt jump phenomenon was accurately described in the previous two sections by the simple kinetics model in which the delayed neutron source was assumed constant. The description of the transient after the prompt jump clearly requires a model in which the subsequent changes in the delayed neutron source are accounted for. The simplest model of this kind is the PA approximation, in which the newly formed precursors are accumulated, Eq. (6.78d). The analytical solution of Eq. (6.78d) is not, however, easier than the solution of the kinetics equations with one delay group, Eq. (6.78c). Consequently, the one-delay-group model is considered here. Since the PA model is subsequently used for important estimates and since the analytical solution can be found along with the one-delay-group solution with only little additional effort, the PA model is also treated explicitly.

Only transients initiated in a critical reactor are treated in this section; the solution for subcritical problems is fully analogous.

The kinetics equations in the two models considered are given by ($\lambda = \overline{\lambda}$):

one-delay-group kinetics (1DG):

$$\dot{p} = \frac{\rho_1 - \beta}{\Lambda} p + \frac{1}{\Lambda} \lambda \zeta \tag{6.97a}$$

and

$$\dot{\zeta} = -\lambda \zeta + \beta p \tag{6.97b}$$

precursor accumulation approximation (PA):

$$\dot{p} = \frac{\rho_1 - \beta}{\Lambda} p + \frac{\beta}{\Lambda} [p_0 + \lambda I(t)] \quad . \tag{6.98}$$

To cast Eqs. (6.97) and (6.98) into the same mathematical form, the equations are differentiated with respect to time; elimination of ζ yields second-order differential equations:

$$\ddot{p} + (\lambda - \alpha_p)\dot{p} - \frac{\lambda}{\Lambda} \rho_1 p = 0 \quad \text{(1DG)} \tag{6.99}$$

and

$$\ddot{p} - \alpha_p \dot{p} - \frac{\lambda}{\Lambda} \beta p = -\frac{\lambda}{\Lambda} \beta p_0 \quad \text{(PA)} \quad , \tag{6.100}$$

where ρ_1 denotes the step reactivity inserted and α_p the corresponding absolute inverse prompt period.

The "initial" conditions for both cases are:

$$p(0) = p_0 \tag{6.101a}$$

and

$$\dot{p}(0+) = \frac{\rho_1}{\Lambda} p_0 \quad , \tag{6.101b}$$

where $\dot{p}(0+)$ is the slope after insertion of the reactivity step. Equation (6.101b) is derived by assuming a stationary state for $t = 0$, finding $\zeta(0)$ from Eq. (6.97b) and inserting it into Eq. (6.97a). Equation (6.98) yields (6.101b) directly since $I(0) = 0$.

Since Eqs. (6.99) and (6.100) are linear differential equations with constant coefficients, the solution is composed of exponential functions with exponents found as roots of the characteristic equations:

$$\alpha^2 + (\lambda - \alpha_p)\alpha - \frac{\lambda}{\Lambda} \rho_1 = 0 \quad \text{(1DG)} \quad , \tag{6.102}$$

with

$$\alpha_{1,2} = -\frac{\lambda - \alpha_p}{2} \pm \left[\left(\frac{\lambda - \alpha_p}{2} \right)^2 + \frac{\lambda}{\Lambda} \rho_1 \right]^{1/2} \quad \text{(1DG)} \quad . \tag{6.103}$$

Analogous equations are obtained for the PA approximation.

It may easily be verified that the term in parentheses is much larger than the term that follows (except when ρ_1 is very close to β). Consequently, the square root in Eq. (6.103) can be approximated by the first two terms of a Taylor expansion, i.e.:

$$\frac{\lambda - \alpha_p}{2}\left[1 + \frac{2}{(\lambda - \alpha_p)^2}\frac{\lambda}{\Lambda}\rho_1\right] = \frac{\lambda - \alpha_p}{2} + \frac{\lambda\rho_1}{\Lambda(\lambda - \alpha_p)} \quad . \tag{6.104}$$

This gives for the two roots:

$$\alpha_1 \simeq \frac{\lambda\rho_1}{\Lambda(\lambda - \alpha_p)} \simeq -\frac{\lambda\rho_1}{\Lambda\alpha_p} = \frac{\lambda\rho_1}{\beta - \rho_1} \tag{6.105a}$$

and

$$\alpha_2 \simeq \alpha_p - \lambda\frac{\beta}{\beta - \rho_1} \quad , \tag{6.105b}$$

where λ has been neglected in the denominators of Eqs. (6.105) since it is much smaller than α_p. For the same reason, α_2 may be further approximated as

$$\alpha_2 \simeq \alpha_p \quad . \tag{6.105c}$$

The general solution of Eq. (6.99) is then given by

$$p(t) = A_1 \exp(\alpha_1 t) + A_2 \exp(\alpha_2 t) \quad , \tag{6.106}$$

with A_1 and A_2 determined by the initial conditions, Eqs. (6.101):

$$A_1 + A_2 = p_0$$

and

$$\alpha_1 A_1 + \alpha_2 A_2 = \frac{\rho_1}{\Lambda} p_0 \quad . \tag{6.107}$$

Neglecting terms that are much smaller than those that contain $1/\Lambda$ gives for A_1 and A_2:

$$A_1 \simeq \frac{\beta}{\beta - \rho_1} p_0$$

and

$$A_2 \simeq -\frac{\rho_1}{\beta - \rho_1} p_0 \quad . \tag{6.108}$$

Thus, the solution of the one-delay-group kinetics equation for a step-reactivity-induced transient is given by:

$$p(t) = p_0 \left[\frac{\beta}{\beta - \rho_1} \exp(\alpha_1 t) - \frac{\rho_1}{\beta - \rho_1} \exp(\alpha_p t) \right] \quad , \tag{6.109}$$

with α_1 given by Eq. (6.105a). The second term decreases very rapidly after the start of the transient and thus raises the flux in a prompt jump. The first term is essentially constant *during* the prompt jump and describes an exponential increase *after* the prompt jump. The solution, Eq. (6.109), is compared below with other approximations.

In the same way as above, one obtains for the precursor accumulation approximation:

$$\alpha_1 \simeq \frac{\lambda\beta}{\beta - \rho_1} \tag{6.110a}$$

and

$$\alpha_2 \simeq \alpha_p \quad . \tag{6.110b}$$

Since Eq. (6.100) is inhomogeneous, the sum of the exponential functions as on the right side of Eq. (6.106) gives only the solution of the corresponding homogeneous problem. A particular solution of the inhomogeneous equation must be added before subjecting the result to the initial conditions. The simplest particular solution of the inhomogeneous equation is obviously

$$p_{\text{inhom}}^{\text{part.}} = p_0 = \text{constant} \quad . \tag{6.111}$$

The sum of Eqs. (6.106) and (6.111) gives the solution of the inhomogeneous problem:

$$p(t) = p_0 + A_1 \exp(\alpha_1 t) + A_2 \exp(\alpha_2 t) \quad . \tag{6.112}$$

Application of the initial conditions yields:

$$p_0 + A_1 + A_2 = p_0$$

and

$$\alpha_1 A_1 + \alpha_2 A_2 = \frac{\rho_1}{\Lambda} p_0 \quad . \tag{6.113}$$

After neglecting small terms as in Eq. (6.108), the solution of Eq. (6.113) is obtained as:

$$A_1 \simeq \frac{\rho_1}{\beta - \rho_1} p_0$$

and

$$A_2 \simeq - A_1 \quad . \tag{6.114}$$

TABLE 6-I
Comparison of Different Approximations of a Flux Transient Following a $\rho_1 = 0.5\$$ Reactivity Step

$\bar{\lambda}t =$	0.01	0.05	0.10	0.20	0.40
CDS	2.000	2.000	2.000	2.000	2.000
PA	2.020	2.105	2.221	2.491	3.226
1DG	2.020	2.102	2.209	2.442	2.982
6DG	2.020	2.097	2.190	2.366	2.701
error CDS (%)	−1.0	−4.8	−8.7	−15.4	−26.0
error PA (%)	0.0	0.4	1.4	5.3	19.4
error 1DG (%)	0.0	0.2	0.9	3.2	6.7

The flux transient in which the decay of the newly formed precursors is neglected is thus given by

$$p(t) = p_0\left[1 + \frac{\rho_1}{\beta - \rho_1}\exp(\alpha_1 t) - \frac{\rho_1}{\beta - \rho_1}\exp(\alpha_p t)\right] , \quad (6.115)$$

with α_1 given by Eq. (6.110a).

Table 6-I presents a comparison of results of the various approximations discussed above. The transient was induced by a 0.5$ reactivity step. The results are presented as a function of $\bar{\lambda}t$. The six-delay-group (6DG) results, which are obtained as a function of t, are converted to a function of $\bar{\lambda}t$ with $\bar{\lambda} = 0.565\ s^{-1}$, the $\bar{\lambda}$ of the six-delay-group data and the fast reactor model used here.

The basic features of the various approximations may easily be identified in Table 6-I. The CDS solution stays constant after the prompt jump; the precursor accumulation model yields too high a flux since the decay of the newly formed precursors is neglected; the one-delay-group model (with $\bar{\lambda}$) correctly describes, according to Eq. (6.109), the first-order decay of the delayed neutron source—it therefore holds over a much larger time interval than the two simpler models.

Since all three approximate models are applied below to obtain simple estimates of key dynamics quantities (such as total energy release), it is important to have an idea of the approximate range of applicability of these "small time" models. If, arbitrarily, a deviation of 5% is used as a limit of the range of applicability, the values given in Table 6-II are obtained from Table 6-I. The specific values given in Table 6-II vary for different reactor types or transients, but the order of magnitude and the basic trend are the same.

TABLE 6-II
Approximate Limits of Applicability of "Small Time" Kinetics Models for 0.5$ Reactivity Step Transients

Kinetics Model	Limits of Applicability ($\approx$5% Deviation)
CDS model	$\lesssim$90 ms
PA model	$\lesssim$350 ms
1DG model ($\lambda = \bar{\lambda}$)	$\lesssim$600 ms

6-2D Kinetics with Six Groups of Delayed Neutrons

The previous investigations showed that the transient behavior at small times can be fully understood and accurately described with just one group of delayed neutrons. For example, see Fig. 6-11 and Tables 6-I and 6-II. However, from the investigation of asymptotic transients, it followed that six delayed neutron groups are normally needed.

This is even more so for the transitory range between short-time and asymptotic behavior. The one-delay-group solution, Eq. (6.109), yields for the entire range

$$p(t) = p_{as}(t)\left\{1 - \frac{\rho_1}{\beta}\exp[-(\alpha_1 - \alpha_p)t]\right\} \quad , \tag{6.116}$$

where the braces describe the transition into the asymptotic solution, $p_{as}(t)$. Since $-\alpha_p \gg \alpha_1$, the time constant for this transition is practically given by the inverse prompt period, α_p. Thus, in the one-delay-group model, the asymptotic transient is established after several prompt periods. However, the investigation of the solution with six groups of delayed neutrons in this section shows that the prompt period does *not* determine the duration of the transitory phase between short-time and asymptotic behavior. Six groups of delayed neutrons are needed to describe this transition and thus the most important transient phenomena from a practical standpoint.

The analytic solution of the six-delay-group kinetics is derived in the same way as for one delay group. Equation (6.106) is replaced by an equation with seven terms:

$$p(t) = \sum_{n=1}^{7} A_n \exp(\alpha_n t) \quad . \tag{6.117}$$

The characteristic equation that yields the seven exponents α_n is the inhour equation [see the derivation of Eq. (6.65)]:

$$\rho = \alpha\Lambda + \alpha\sum_k \frac{\beta_k}{\alpha + \lambda_k} \quad . \tag{6.118}$$

The common procedure to get a semi-quantitative idea of the values of the seven roots of this equation is to plot the right side of Eq. (6.118) as a function of α, say $\tilde{\rho}(\alpha)$:

$$\tilde{\rho}(\alpha) = \alpha\Lambda + \alpha\sum_k \frac{\beta_k}{\alpha + \lambda_k} \quad . \tag{6.119}$$

The intersections of the different branches of $\tilde{\rho}(\alpha)$ with a *given* ordinate value ρ yields the seven roots of the characteristic equation, $\alpha_n(\rho)$.

It is difficult to display $\tilde{\rho}(\alpha)$ in a single diagram such that the key features of the complicated function are quantitatively exhibited. Three different diagrams are therefore employed to demonstrate the three key features of $\tilde{\rho}(\alpha)$:

1. The stable period (largest root) for positive as well as negative reactivities is best shown on a linear alpha scale (Fig. 6-16). The behavior

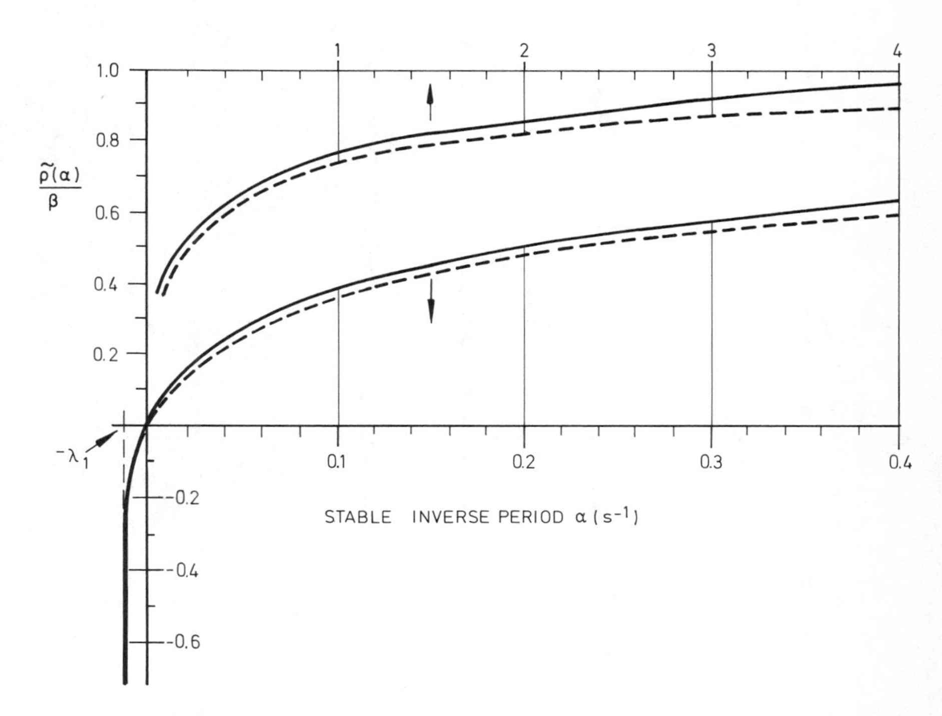

Fig. 6-16. Largest eigenvalue of the characteristic equation $\rho = \tilde{\rho}(\alpha)$, for large light water reactors (LWRs) (solid lines) and fast breeder reactors (FBRs) (dashed lines).

of this branch for small and large positive α has been discussed in the context of asymptotic transients (see Sec. 6-1C). Figure 6-16 shows that the stable period for negative reactivities is limited by the smallest precursor decay constant, λ_1, i.e., the neutron population in a subcritical reactor cannot die away faster than the delayed neutron group with the longest lifetime.

2. The function $\tilde{\rho}(\alpha)$ has six singularities at $\alpha = -\lambda_k$. The λ_k values and thus the singularities are about equally spaced on a logarithmic scale. Therefore, a logarithmic scale is employed around the range of the singularities of $\tilde{\rho}(\alpha)$ in Fig. 6-17.

3. To show a larger picture, which includes the roots near the positive *and* negative prompt period, a linear scale is employed in Fig. 6-18, but over a much larger range of α than in Fig. 6-16. The singularities of the six-delay-group function cannot be presented on the large scale of Fig. 6-18. They all disappear in the vertical dashed line.

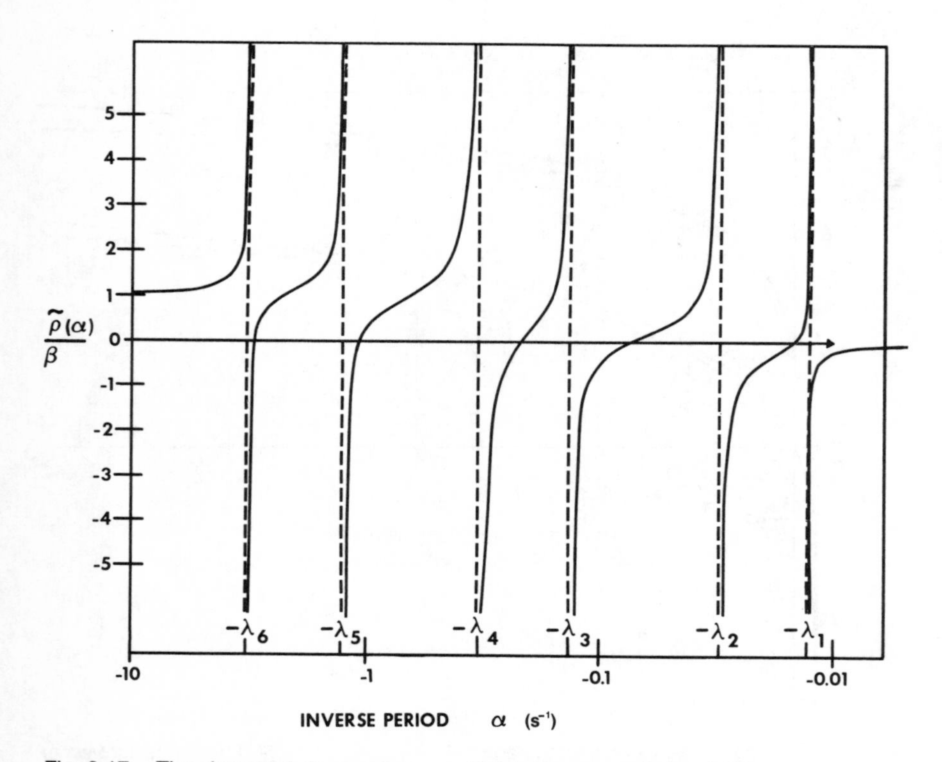

Fig. 6-17. The characteristic equation $\rho = \tilde{\rho}(\alpha)$ in the range of singularities.

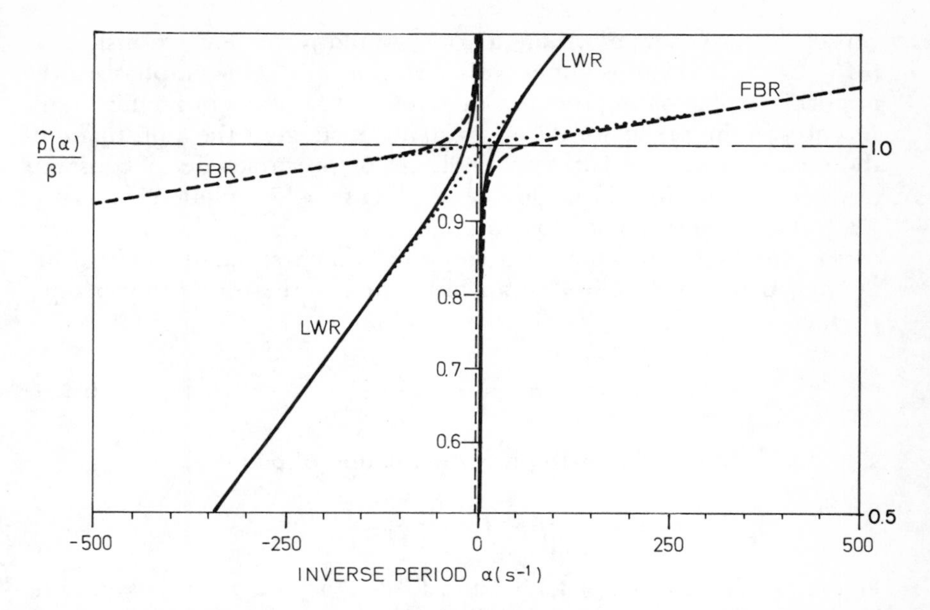

Fig. 6-18. The asymptotic behavior of the characteristic equation $\rho = \tilde{\rho}(\alpha)$ for large LWRs and FBRs.

Figure 6-16 shows the branch of the characteristic equation that yields the stable period. The stable period branch approaches $-\infty$ for $\alpha \rightarrow -\lambda_1 = -0.0129\ \mathrm{s}^{-1}$; i.e., stable inverse periods for negative reactivities are fairly small and are bounded by the decay constant of the longest living precursor group:

$$\left.\begin{array}{r} -0.0129\ \mathrm{s}^{-1} < \alpha < 0 \\ 80\ \mathrm{s} \gtrsim -\dfrac{1}{\alpha} < \infty \end{array}\right\} \text{ for } \rho < 0 \quad . \tag{6.120}$$

The stable periods for small negative reactivities are substantially larger than 80 s. It may therefore take many minutes to reduce the reactor power in a shutdown procedure well below the level achieved by the prompt jump response to the negative reactivity insertion. Only minimal differences exist between fast and thermal reactors in these subcritical transients, since the period is largely determined by the decay time of $^{87}\mathrm{Br}$, the longest living precursor isotope.

Figure 6-17 shows that there are five roots that describe the transition effects resulting from the adjustment of the delayed neutron source (e.g., for $\rho = 0.5\$$). These five roots are bounded by the six delayed neutron decay constants. If an isotope-dependent set of precursor decay constants were used, more singularities would result, e.g., 12 instead of 6 for two fissionable isotopes with different λ_k's. This emphasizes the importance of avoiding the introduction of statistically insignificant differences in the precursor decay constants; it suggests the adjustment of the delayed neutron data to a single set of precursor decay constants (see Sec. 2-3 and Ref. 17 in Chapter 2). Figure 6-17 is plotted for a large FBR; it looks very similar for an LWR.

Figure 6-18 shows the two asymptotic branches, i.e., the positive and the negative branches. Both branches approach the prompt inverse period, α_p:

$$\alpha_p = \frac{\rho - \beta}{\Lambda} \quad . \tag{6.121}$$

The dotted line in Fig. 6-18 gives the relation of ρ and α_p:

$$\frac{\rho}{\beta} = 1 + \frac{\Lambda}{\beta}\alpha_p \quad . \tag{6.122}$$

Figure 6-18 is plotted for LWRs and FBRs. Each of the two asymptotic branches contributes a root for all reactivities.

For all reactivities below β, the negative root is very large and for all reactivities above β, the positive root is very large. Figure 6-18 shows that the large roots may be approximated by the prompt period if ρ is not too close to β:

$$\begin{aligned} \alpha_7 &\lesssim \alpha_p \;\; \text{for } \rho < \beta \\ \alpha_1 &\gtrsim \alpha_p \;\; \text{for } \rho > \beta \quad . \end{aligned} \tag{6.123}$$

Furthermore, the one- and six-delay-group results are very close together for these large positive and negative inverse periods:

$$\alpha(\text{1DG}) \simeq \alpha(\text{6DG}) \;\; \text{for } |\alpha| \text{ very large} \quad . \tag{6.124}$$

Equation (6.123), which holds for six as well as one group of delayed neutrons, was used above in the analytic solution of the one-delay-group transients.

To emphasize the fact that the characteristic equation has three widely different kinds of roots, the general analytic solution, Eq. (6.117), is rewritten as follows:

$$\begin{aligned} p(t) = A_1 \exp(\alpha_1 t) \Bigg\{ 1 &+ \frac{A_7}{A_1} \exp[(\alpha_7 - \alpha_1)t] \\ &+ \sum_{n=2}^{6} \frac{A_n}{A_1} \exp[(\alpha_n - \alpha_1)t] \Bigg\} \quad . \end{aligned} \tag{6.125}$$

The factor in front of the brace is the asymptotic solution, $p_{as}(t)$. The braces describe the *entire* transition into the asymptotic solution.

For $\rho < \beta$, the root α_7 is very large and negative; thus, the first exponential function in the braces decreases very rapidly with increasing t. This represents the analytic description of the *prompt jump* in the six-delayed-group solution. Since $\alpha_7 \simeq \alpha_p$ for six as well as one delayed neutron group [Eq. (6.124)], the same prompt jump transition is obtained in both models. The fact that the flux level after the prompt jump is also the same in both models cannot be deduced readily from consideration of only the roots. The previous investigations, however, showed that there was no change in p_{pj} due to increasing the sophistication of the model from "constant delayed neutron source" to "precursor accumulation" to "$\bar{\lambda}$ model." Thus, it can be expected that six delayed groups yield the same value, which is supported by numerical results previously presented in, for example, Table 6-I.

The sum over the contributions of the five intermediate roots in Eq. (6.125) thus describes the comparatively slow (compared to $1/\alpha_7$) transition that leads to the asymptotic time behavior of the delayed neutron population. Earlier, the question of how long it takes to complete this delayed adjustment was raised. An estimate of this time (or the corresponding time constant) can be obtained from the time constant of the slowest decaying exponential function in the braces of Eq. (6.125); namely, $\alpha_2 - \alpha_1$. This gives a

$$\text{delayed adjustment period} \approx \frac{1}{\alpha_1 - \alpha_2} \quad . \tag{6.126}$$

If $|\alpha_2|$ is much smaller than α_1, which is satisfied for transients around and above $\rho = 0$, the following simpler formula is obtained:

$$\text{delayed adjustment period} \approx \frac{1}{\alpha_1} \quad ; \tag{6.127}$$

thus, even for moderately rapid transients, the asymptotic transient is established after several stable periods.

Equation (6.125) also shows why the one-delay-group model principally fails to describe the delayed transition: the one-delay-group model has no transitory roots and therefore no sum in the braces as in Eq. (6.125).

The basic results of the discussion of the six-delay-group kinetics of step-induced transients is illustrated in several diagrams.

Figure 6-19 shows the prompt jump transition for $\rho_1 = 0.5\$$ with its time constant $|\alpha_p|$ for both one and six delay groups. The results are in total agreement for the entire domain of the lower time scale. The upper curves show the continuation of the transient over a much larger

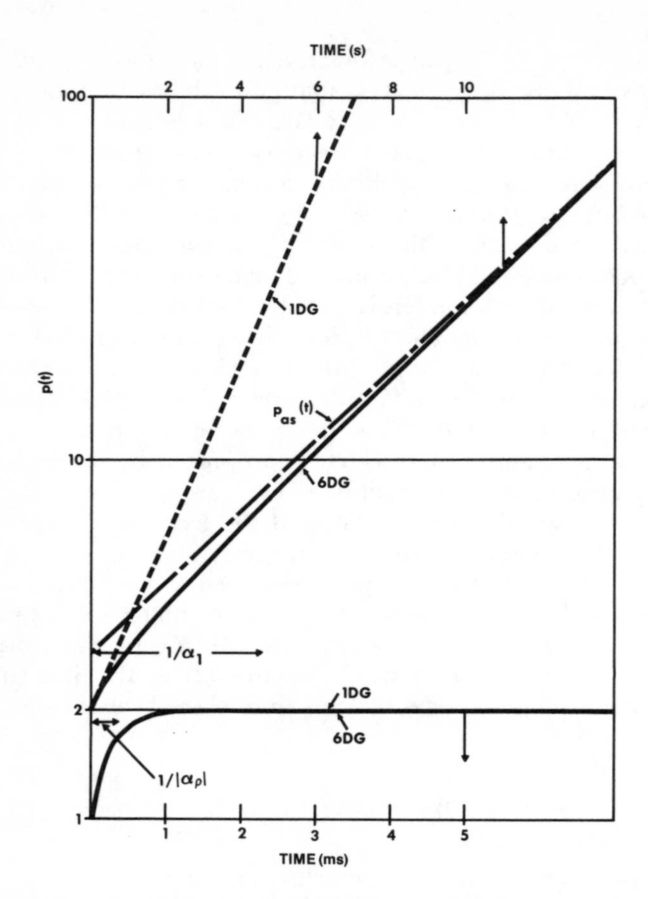

Fig. 6-19. The transition phases in a delayed supercritical transient ($\rho_1 = 0.5\$$).

time interval (upper scale). The prompt jump disappears on the ordinate. The one-delay-group solution is purely exponential after the completion of the prompt jump. The six-delay-group solution, however, branches slowly away from the one-delay-group solution and into a different and much later established purely exponential rise. The period for this delayed neutron adjustment, $1/\alpha_1 \approx 4$ s, can clearly be identified.

Figure 6-20 shows the asymptotic behavior of the one- and six-delay-group treatment of a transient for larger reactivities, i.e., $\rho_1 = 1\$$ and $1.1\$$. The agreement is much better than for $\rho_1 = 0.5\$$, as was suggested by the previous investigations.

Figure 6-21 shows the two transitions for negative reactivity inser-

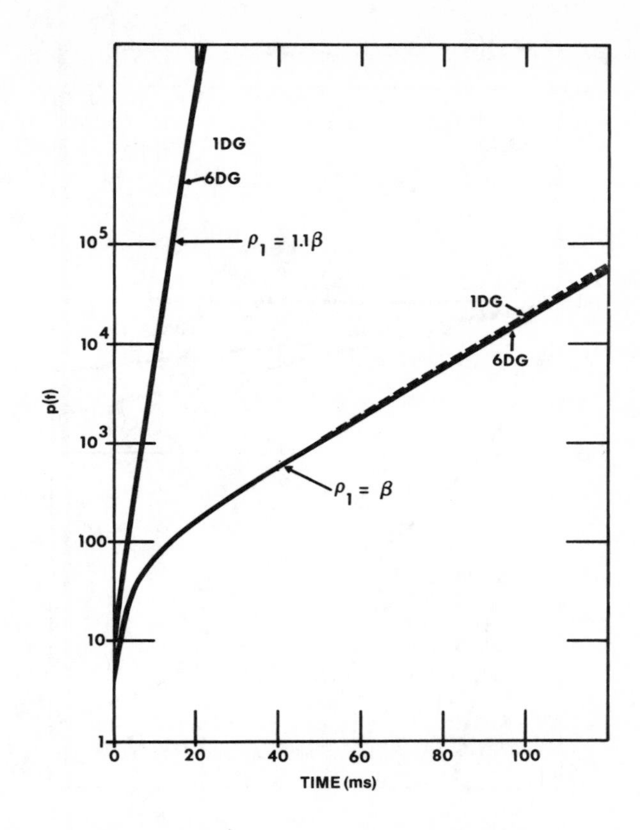

Fig. 6-20. Comparison of prompt and superprompt-critical transients ($\rho_1 = \beta$ and 1.1β) calculated with one and six delayed neutron groups.

tions, $\rho_1 = -1\$$. The prompt jump is again described equally well in both approximations. The transition into the asymptotic transient takes much longer for negative reactivities than for positive ones. The reasons are that the stable period is much larger and that the roots are much closer together. It thus takes many more periods, $(\alpha_1 - \alpha_2)^{-1}$, to clearly establish the dominance of the asymptotic mode. The one-delay-group model is asymptotically very inaccurate since the reactivity is even farther away from $+\beta$ than in Fig. 6-19.

In conclusion, it is emphasized that practical kinetics applications employ the full information about delayed neutrons (normally six groups). Only transient phenomena at very short times, such as prompt jumps and superprompt-critical transients, can be sufficiently explained by considering just one delay group.

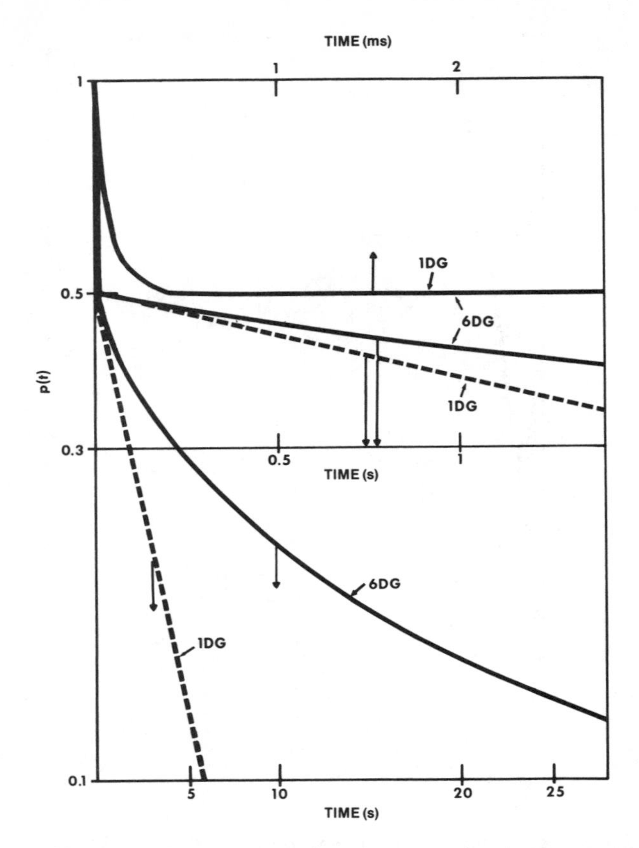

Fig. 6-21. Comparison of a subcritical transient ($\rho_1 = -1\$$) calculated with one and six delayed neutron groups.

Homework Problems

1. *Inhour Equation.*
   a. Find the stable and prompt-period branches for ^{235}U as fuel and $\Lambda = 10^{-4}$, 10^{-5}, and 4×10^{-7} s (data given in Table 2-III).
   b. Find $\rho(\alpha)$ in the one-delay-group approximation with $\lambda = \bar{\lambda}$.
   c. Plot both results outside of the range of the singularities, with a shadowed area indicating the singularities.
   d. Discuss the comparison.

2. Find $\rho(\alpha)$ for $\alpha < 0.1$/s (for the same three Λ values as problem 1a). Plot and discuss the comparison of the results with the ap-

proximate formula given in the text for very small α values. Extend the comparison to negative α.

3. *Source Drop into a Critical Reactor.* Use a point kinetics program to find the corresponding solution to the source drop problem; $\beta_k = 1.07\ \beta_k^{\text{phys}}$ (^{235}U), $\nu = 2.43$, ν_{dk} from Table 2-III, and $\Lambda = 3 \times 10^{-6}$ s.
   a. Find $p(t)$ for the following source values: $s = 10^{-2}$, 10^{-1}, 3×10^{-1}, and 1.0. Let t go to 1 s.
   b. Plot the results with a computer plot program for 0 to 6 ms, and for 0 to 1 s.
   c. Interpret the results quantitatively in terms of a prompt jump and a later linear rise. Find the slope in terms of s, β, and some average $\lambda = \langle\lambda\rangle$. Find $\langle\lambda\rangle$ and compare it with $\overline{\lambda^{in}}$ and $\bar{\lambda}$.
   d. Find the analytical one-delay-group solution in the prompt jump approximation ($\Lambda \rightarrow 0$) and compare it with your numerical solution.
4. Figure 6-22 shows a printout of counts registering a transient that follows two source steps in a critical reactor (PUR-1). Use a point kinetics program with β_k values 1.07 times the ν_{dk}/ν values given in

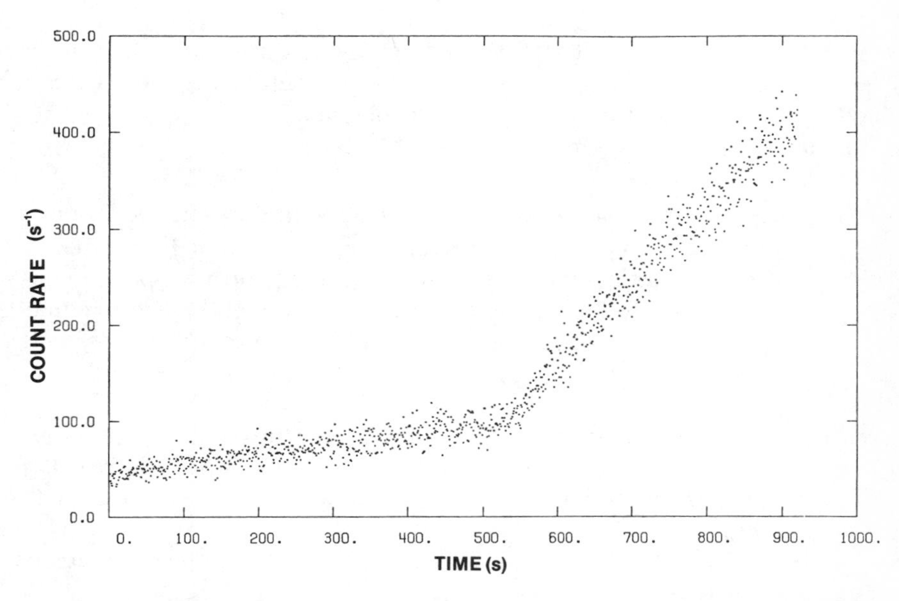

Fig. 6-22. Counting rates in counts per second for a two-step source drop in a critical reactor (PUR-1).

Table 2-III for ^{235}U; use the isotope-independent λ_k values and $\Lambda = 3 \times 10^{-6}$ s.

a. Prepare the input for the source step problem. Find $p(t)$ and match it with the first part of the count rate by using various s values.

b. If your point kinetics program allows a source change during the transient, change s to match the second part. If not, restart at the new flux level with stationary "initial" conditions (the first part of the transient is so slow that most precursor groups are practically saturated).

5. *Transients with Constant Reactivity.* Find numerically (with a suitable code) the solution of the point kinetics equations for the following reactivity insertions in a critical reactor and plot the results (in all cases, use $\Lambda = 10^{-4}$, 10^{-5}, and 10^{-6} s):
   a. $\rho_1 = 0.75\$$.
   b. $\rho_1 = -1.0\$$ (shut down).
   c. $\rho_1 = -3.0\$$ (shut down).
   d. Estimate first the length of time you want to run the transient; present the rationale of your estimate.
   e. Discuss the transient results, in particular the short-time behavior, asymptotic behavior, and Λ dependence of both.

6a. Find numerically the transient that follows a ρ jump from $\rho_0 = -1\$$ to $\rho_1 = 0$, with a source $s_0 = \beta p_0$ present during the transient. Again use the three Λ values of problem 5.

b. Plot and discuss your results.

7. *Interpretation of the Adjoint Flux.* Prove by application of the point kinetics equations the interpretation of the adjoint flux, $\phi_0^*(\mathbf{r},E)$, as the asymptotic relative flux rise δp_{as} (aside from a proportionality constant) that follows a burst-insertion of independent source neutrons in a critical reactor at location $\mathbf{r}$ with energy E:

$$\delta p_{as} = c\phi_0^*(\mathbf{r},E) \quad .$$

   a. Find the analytical solution for a burst-insertion of independent source neutrons in a critical reactor, $\delta s(t) = s_0\delta(t)$. Use one group of delayed neutrons, $\lambda = \bar{\lambda}$. Find $p(t)$, and especially δp_{as}.
   b. Why is it sufficient to use a single group of delayed neutrons for determining δp_{as}?
   c. Relate s_0 to $\phi_0^*(\mathbf{r},E)$ and present the desired proof.

Review Questions

1. Simplify the point kinetics equations such that they describe a stationary state (critical and subcritical).
2. Give the stationary relation between ζ_k, the reduced precursor integral, and the flux amplitude.
3. Derive the relation between total fission and delayed neutron sources for a stationary system.
4. Give and interpret the flux amplitude equation for a subcritical system.
5. Give and interpret the flux amplitude equation for a critical reactor. (Show the equivalence of the resulting two source multiplication formulas.)
6. Which simplifying assumption may be applied to the delayed neutron source for very short times and why is it justified?
7. How can this most simplifying assumption be improved?
8. Do these two assumptions allow for a combination of the six precursor groups? If yes, which quantity appears instead of the individual decay constants λ_k?
9. Give the definition and typical values for $\overline{\lambda}$.
10. Why can $\overline{\lambda^{in}}$ (definition) not be used in short time kinetics?
11. Explain qualitatively the numerical differences in the values of $\overline{\lambda}$ and $\overline{\lambda^{in}}$.
12. How well does a straightforward Taylor expansion solution of the kinetics equation converge? Explain why!
13. Which conditions have to be fulfilled to yield an asymptotic transient with a "stable" period?
14. Give the general definition of the (instantaneous) inverse period and show in which plot of $p(t)$ it appears as the slope.
15. Describe the essential steps of the derivation of the inhour formula $\rho(\alpha)$ (essentially using formulas).
16. Give the inhour formula; if you do not recall it correctly, the following investigation may help you to correct it.
17. Limits of the inhour formula: (a) exploit the limit $\alpha \ggg \lambda_k$, (b) investigate the limit $\alpha \gg \lambda_k$, and (c) investigate the limit $\alpha \ll \lambda_k$.

18a. In what combination do the precursor decay constants appear in asymptotic transients following very small and very large reactivity insertions?

b. Estimate the stable period for $\rho_1 = 1.25¢$.

19. Give a list of six to seven kinetics equations with decreasing sophistication of the description of delayed neutrons.
20. Describe in words, with graphs, and with formulas the transient

following a step change in reactivity or source: (a) without delayed neutrons, (b) with constant delayed neutron source, and (c) the correct solution (no formula).

21. Describe the two types of transitions that occur in a step-induced transient.
22. Give a sequence of three source multiplication formulas that explain the prompt jump flux as a transitory state between two steady states.
23. Give the formula, $p(t)$, for the transient with the constant delayed neutron source; discuss and interpret both terms (assume a critical system for $t < 0$ and $\rho_1 < \beta$).
24. Describe qualitatively the difference between this approximate solution and the correct one.
25. Derive the formulas for p_{pj}, the prompt jump flux, from source multiplication considerations.
26. Give the formula, $p(t)$, for the transient with one group of delayed neutrons and discuss both terms (with approximate roots).
27. Describe how you remember this formula (you do not have to remember the stable period).
28. What is the difference between this solution and the one with a constant delayed neutron source?
29. Describe the transient that follows a step source insertion into a critical reactor for $\rho_1 > \beta$, $0 < \rho_1 < \beta$, and $\rho_1 < 0$.
30. What are the two basic shortcomings of the solution with one (compared to six) groups of delayed neutrons?
31. Plot qualitatively the characteristic equation $\rho(\alpha)$: (a) the positive branch (on a linear scale), (b) the negative branch (on a logarithmic scale), and (c) the entire formula on an expanded linear scale (without details in the singularity area).
32. Estimate α for $\rho = -\beta$.
33. Give the formula for the analytic solution of this six-delay-group kinetics problem (general formula; no determination of coefficients or roots).
34. Estimate the time it takes to establish the stable asymptotic transient for $\rho_1 < \beta$ for an initially critical reactor.
35. Explain the prompt jump phenomenon, the delayed neutron-induced transition, and the stable period in terms of roots of the characteristic equation.

Seven

MICROKINETICS

The investigations presented in the previous chapters are concerned with the "average" neutron flux. The phenomena discussed included static source multiplication formulas of different types, the prompt jump, delayed transitions into an asymptotic state that may be stationary or consist of a stable exponential rise, and the asymptotic states themselves. The quantitative description of these phenomena was obtained by solving the point kinetics equations, often in a simplified form, in order to clearly identify the cause-and-effect relations for the various phenomena.

A more in-depth understanding can be developed by investigating individual *fission chains,* which—in superposition—make up the average neutron flux. In this alternative view of kinetics phenomena, the nuclear processes in a reactor, which lead to neutron multiplication through fission, producing prompt and delayed neutrons, are examined in detail. The investigation of fission chains in a probabilistic sense is a topic of noise analysis. For the instructional application intended here, a decomposition of the average flux in *average fission chains* suffices. The study of the time dependencies of the average fission chains is called "microkinetics."

An analogous situation exists in thermodynamics where concepts such as temperature, space-dependent temperature distributions, heat transfer, time-dependent temperature transients, etc. are employed to describe basic macroscopic effects. However, the physical nature of the temperature is better understood as a superposition of microscopic atomic or molecular motion. The explicit consideration of macroscopic phenomena as a superposition of the original microscopic effects clearly provides a greater understanding of basic phenomena in thermodynamics as well as kinetics. In addition, qualitative results for various phenomena can be derived more readily from a microscopic understanding than from the corresponding average differential equations.

7-1 Decomposition of Continuous Rates in Fission Chains

The microkinetic understanding of the neutron flux and its transients is independent of the flux level. An experimental investigation of microkinetics, however, is best performed at very low power. Experimental reactors that operate at such a low power that little or no cooling is required are generally called "zero-power" reactors. The typical order of magnitude of the power level is 1 W, with about one to two orders of magnitude variation in either direction. Practically no thermal-neutron zero-power reactors are in operation anymore. Examples of *fast* zero-power reactors are ZPR-3, which was decommissioned in 1970, and the Zero-Power Plutonium Reactor (ZPPR), both at the National Reactor Testing Station in Idaho and ZPR-6 and ZPR-9 at the Argonne National Laboratory in Illinois; the latter two facilities were inactivated early in 1983. For a description of zero-power reactors and their application in fast reactor physics and development, see Ref. 1.

A total power of 1 W corresponds to $\sim 3 \times 10^{10}$ fission/s in the entire reactor. With such a low power level, the statistical fluctuations of the neutron flux become very noticeable and individual *fission chains* can be investigated. If, for example, the reactor power is 1 W in a typical core volume of 1 m^3 with an average neutron lifetime of 3.3×10^{-7} s, a fission occurs in an average cubic centimeter only every 100 neutron lifetimes.

A "fission chain" is normally initiated by the emission of a delayed neutron. The additional neutrons in the chain are *prompt* neutrons from fission events caused by the original neutron and subsequent prompt fission neutrons in the chain. Fission chains may also be initiated by neutrons coming from an independent source, such as spontaneous fission or an (α,n) reaction.

Figure 7-1 illustrates a typical fission chain. Only nonscattering reactions are shown. A delayed neutron is emitted by a precursor, denoted by a double circle. It either causes fission or is captured in the first nonscattering reaction. The average time between two fission events is ν times the generation time; i.e., ~25 microseconds for an LWR. The average life of an entire fission chain comprises many generations; its length is determined by the prompt period as shown below. A fission chain may last for hundreds of generations. One new precursor is created in the fission chain illustrated in Fig. 7-1.

The number of participating neutrons in individual fission chains *may vary considerably*. The major cause of these fluctuations is that the original neutron may either be captured or it may fission a nucleus in its first nonscattering reaction. If capture occurs, the chain is ended, and

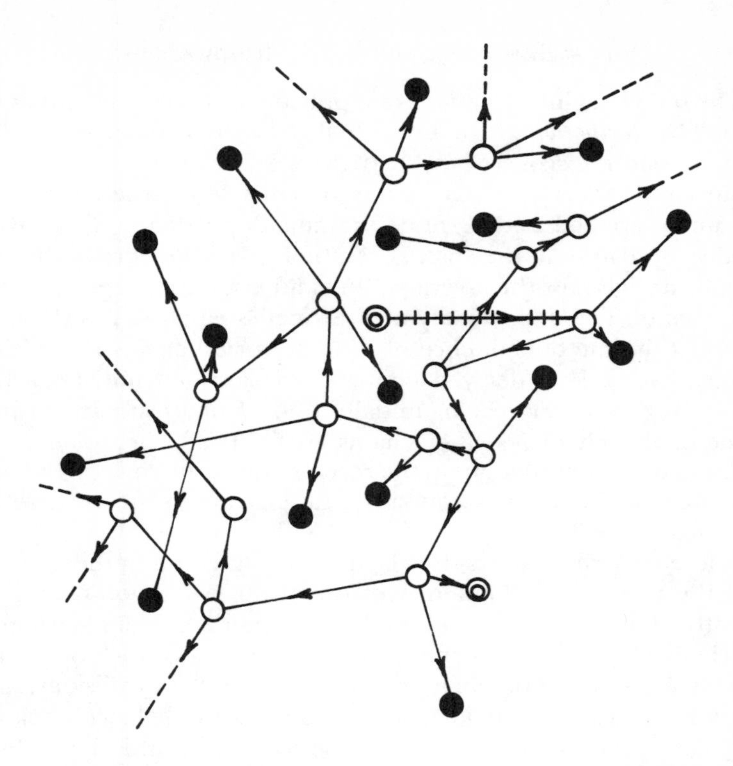

Fig. 7-1. Illustration of a fission chain. A precursor (◎) emits a delayed neutron (—|||||→), which creates prompt neutrons in fission events (○). Capture is indicated by a dot (●).

the total number of participating fission neutrons is zero. In the case of fission, the number of prompt neutrons emitted in an individual fission event is itself subject to considerable fluctuations around its average value ν (see, for example, Ref. 2). The number of emitted neutrons may vary from about one to six, and may even be zero or seven with a small probability. The number of prompt neutrons produced in the first fission obviously will have a significant effect on the subsequent fission chain.

Only *average* chains are considered here, i.e., fluctuations within a chain are neglected. The essential goals in this section are to demonstrate how the "average" flux and fission rate are composed of individual fission chains and to draw conclusions from this demonstration. The term "average" is used on two different levels. Both types of average values yield the same neutron flux, which is *defined* as an average quantity. The flux can be considered either as the sum of "average chains" or as the average of the statistically fluctuating sum of chains because of the linearity of the balance equations:

$$\text{flux} = \Sigma\langle\text{fission chains}\rangle = \langle\Sigma \text{ fission chains}\rangle.$$

The following investigation is simplified by considering only average chains. The *fluctuation* of the individual chains is another subject[1]; it is clearly beyond the scope of this text.

Since only average fission chains are considered, the kinetics equation[a] can be applied to determine the time dependence of the average neutron population in the chain, even though the kinetics equation was originally derived for the (average) flux and not for average (individual) fission chains. The applicability of the kinetics equations to the investigation of the time dependence of average fission chains is easily demonstrated. Suppose a delayed neutron emerges and initiates a fission chain in a zero-power reactor initially void of neutrons. If during the lifetime of this chain no other chains are initiated, then obviously the average neutron populations in the reactor and in the chain are identical. Consequently, the kinetics equation is applicable to the average chain population as well.

If a second chain starts shortly after the initiation of the first fission chain, the reactor will contain neutrons from two fission chains. Consequently, the single chain and total population are no longer identical. Both chains, however, are completely independent of each other due to the practically zero probability for neutron-neutron collisions, and as long as feedback is negligible, i.e., as long as the balance equation is linear. Consequently, each chain can be treated by an independent kinetics equation. The result of both equations may be summed to obtain the flux. Obviously, the same considerations can be extended to many chains with the same conclusions.

The mathematical foundation for the description of the average neutron chain populations can thus be based on the time-dependent diffusion equation or its lumped form,

$$\dot{p} = \frac{\rho - \beta}{\Lambda} p + \frac{1}{\Lambda} s_d + \frac{1}{\Lambda} s \quad . \tag{7.1}$$

In addition to the previous assumptions of constant β and Λ, only step changes of ρ and the independent source are considered in order to avoid obscuring essential features with side effects that are unimportant for this demonstration.

In the application of the kinetics equation, Eq. (7.1), for the calcu-

[a]Examples are the time-dependent diffusion equation or the first of the point kinetics equations (the delayed neutrons have to be treated separately in a manner described in the following).

lation of the (average) flux, the delayed neutron source and the independent source are continuous functions in time (except at the time of a step change of s). These continuous sources, s_d and s, must be decomposed into individual emission events for the investigation of chains. The decomposition of the sources into individual emission events is simpler if the kinetics equation is written in terms of fission and neutron emission rates rather than amplitude functions and reduced sources (see Sec. 3-2B for the derivation of the one-group kinetics equations). The following one-group approximations are substituted in the kinetics equation:

$$p(t) = \frac{\hat{S}_f(t)}{\hat{S}_{f0}} , \tag{7.2a}$$

$$s_d(t) = \frac{\hat{S}_d(t)}{\hat{S}_{f0}} , \tag{7.2b}$$

and

$$s(t) = \frac{\hat{S}(t)}{\hat{S}_{f0}} . \tag{7.2c}$$

Inserting Eqs. (7.2) into Eq. (7.1) yields the balance equation for spatially integrated fission neutron production rates as they are initiated by delayed and independent source neutrons:

$$\frac{d}{dt}\hat{S}_f(t) = \frac{\rho - \beta}{\Lambda}\hat{S}_f(t) + \frac{1}{\Lambda}\hat{S}_d(t) + \frac{1}{\Lambda}\hat{S}(t) . \tag{7.3}$$

A more sophisticated formulation using adjoint weighted reaction rates may be devised along the same line as the derivation presented here. However, its interpretation is more complicated.

The decomposition of $\hat{S}_d(t)$ and $\hat{S}(t)$ into a superposition of the emission of individual, i.e., single, neutrons is simply given by a sum of δ functions:

$$\hat{S}_d(t) = \sum_n \delta(t - t_n) \tag{7.4a}$$

and

$$\hat{S}(t) = \sum_m \delta(t - t_m) , \tag{7.4b}$$

with $\delta(t - t_n)$ denoting Dirac's δ function.[b] The coefficient in front of

---

[b]See Appendix D; note that the δ functions in Eqs. (7.4) have the dimensions of 1/s; they describe neutron production rates per second.

each δ function is 1 since each term describes the emission of one neutron.

Equation (7.3) is first solved without an independent source. The solution of Eq. (7.3), with $\hat{S} = 0$ and $\hat{S}_d(t)$ given by Eq. (7.4a), can be derived readily, giving the fission source that follows a sequence of delayed neutron emissions:

$$\hat{S}_f(t) = \hat{S}_f(0) \exp(\alpha_p t) + \frac{1}{\Lambda} \begin{cases} 0 & \text{for } 0 \le t \le t_1 \\ \sum_n \exp[\alpha_p(t - t_n)] & \text{for } t - t_n > 0 \end{cases} . \tag{7.5}$$

Each exponential function in Eq. (7.5) represents the "die-away" of an average fission chain, which was initiated at time t_n by a delayed neutron, while $\hat{S}_f(0)$ represents the residual of previously initiated chains remaining at $t = 0$, i.e., $\hat{S}_f(0)$ is the microkinetics emission rate of fission neutrons at $t = 0$. The characteristic die-away time is given by the prompt period, τ_p:

$$\tau_p = \frac{1}{|\alpha_p|} = \frac{\Lambda}{\beta - \rho} . \tag{7.6}$$

Figure 7-2 shows a typical solution, Eq. (7.5), for well-separated fission chains in a subprompt-critical reactor ($\alpha_p \simeq -5000/s$).

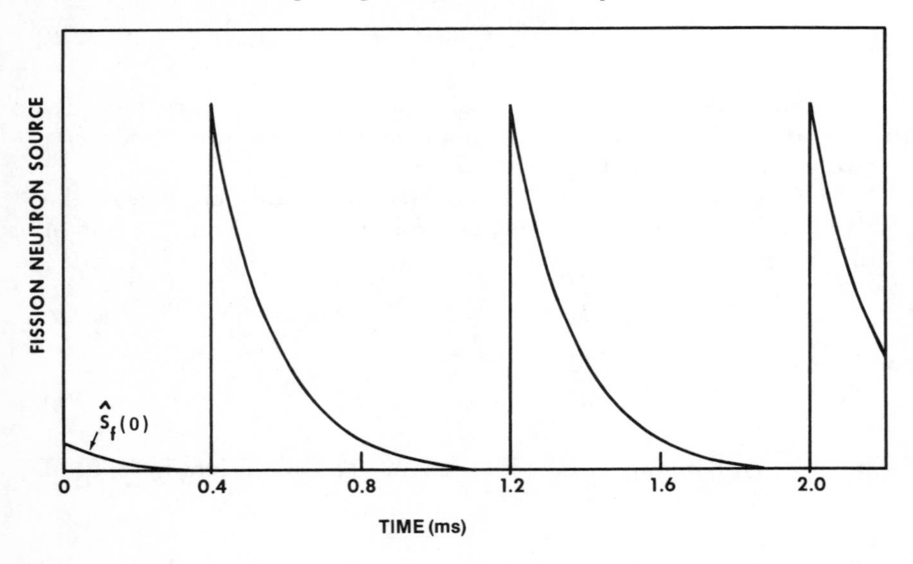

Fig. 7-2. The fission neutron source of average individual fission chains as a function of time for equidistant emission times; $\tau_p = 0.2$ ms.

It should be noted that in the derivation of the kinetics equation, a "quasi-stationary" delayed neutron source was added and subtracted in order to obtain the reactivity definition in full analogy to the static one. Consequently, the fission neutron source $\hat{S}_f(t)$, calculated above, includes the "quasi-stationary" production of delayed neutrons: $\hat{S}_f$ is formed with $\nu\Sigma_f$ rather than with $\nu_p\Sigma_f$. Only prompt neutrons, however, are produced in *actual* fission chains (see Fig. 7-1). Thus, if the number of prompt fission neutrons is desired, $\hat{S}_f(t)$ needs to be modified to reflect prompt fission neutron production rates only, $\hat{S}_{pf}$:

$$\hat{S}_{pf}(t) = \frac{\nu_p}{\nu} \hat{S}_f(t) \quad . \tag{7.7}$$

The slightly fictitious quantity $\hat{S}_f$, however, will continue to be used since its use simplifies the investigation.

7-2 Characteristics of Individual Fission Chains

The average fission chain solution, as given by Eq. (7.5), is investigated in this section. The essential conclusions may either be drawn directly from Eq. (7.5) or derived from it. Differential information is presented and then integral.

7-2A The Lifetime of Average Chains

The three most important conclusions that may be drawn from Eq. (7.5) are:

1. *All* fission chains in a *subprompt*-critical reactor die out. The average decay of the fission chains follows an exponential function with the (negative) *prompt* period determining the average life of the chains.
2. A steady rate of chain-initiating events (e.g., delayed neutron emissions) is obviously required to sustain a finite (average) flux level if $\rho < \beta$.
3. The *average* fission chain in a *superprompt*-critical reactor increases exponentially with the positive prompt period describing the rise of average individual chains.

Note the distinction between conclusions 1 and 3: Individual fission chains in a subprompt-critical reactor may become very long and populous. Eventually, however, *all* chains die out since the losses of prompt neutrons are larger than the gains. If a finite fraction of the individual chains were not to die away, the *average* chain would not die away either. In a superprompt-critical reactor, only the average chain increases. Individual chains may well die out (e.g., if the first source neutron is captured rather than causing a fission).

7-2B The Neutron Density in an Average Chain

All chains have the same initial value of $1/\Lambda$ since all chains start with *one* neutron, which has an average generation time of Λ. A *single* neutron, represented by a neutron flux $1 \cdot \overline{v}$, causes a fission neutron production rate of $1/\Lambda$ (see, for example, the one-group model of Sec. 3-2B):

$$(S_f^{\text{initial}})_{\text{chain}} = \frac{1}{\Lambda} = \overline{v} \cdot \nu\Sigma_f \left[\frac{\text{fission neutrons}}{\text{second}}\right] \quad . \tag{7.8}$$

Note again the use of ν instead of ν_p, which is related to S_f versus S_{pf} as discussed along with Eq. (7.7).

The average chain population decreases with increasing time for $\rho < \beta$. The chain originally consists of *one* neutron; thus, the *average* number of neutrons representing the chain at a given time is <1 even though some individual chains may consist of many neutrons at a certain time.

7-2C The Number of Neutrons in an Average Chain

The total number of fission neutrons participating in an average chain, n_f, can be obtained by integrating the production rate with respect to time:

$$n_f = \frac{1}{\Lambda} \int_{t_n}^{\infty} \exp[\alpha_p(t - t_n) \, dt \quad , \tag{7.9}$$

which gives

$$n_f = \frac{1}{\beta - \rho} \quad . \tag{7.10}$$

For example, in a critical reactor, the number of fission neutrons per average chain is just given by $1/\beta \simeq 300$ or 150 for a fast or thermal reactor, respectively. The number of fission neutrons per chain approaches infinity if ρ approaches $+\beta$. On the other hand, the number of neutrons per chain decreases strongly in subcritical configurations. If a fast (or thermal) reactor is only 1% subcritical, i.e., if $\rho \simeq -3\beta$ ($\rho \simeq -1.5\beta$), then n_f is only 25% (or 40%) of its value at criticality. For strongly subcritical systems, say $k = 0.5$ and $\rho = -1$, there is only about one fission neutron in addition to the original source neutron. Note that n_f does not include the chain-initiating delayed or independent source neutron. It can therefore become <1 as in a subcritical assembly with small k.

7-3 The Fission Chains in Steady-State Reactors

7-3A Comparison with Static Chain Multiplication

In purely static considerations, the delayed neutrons may also be treated as having appeared promptly. The solution of a homogeneous static problem yields an eigenvalue $\lambda = 1/k$ (see Sec. 4-1B). The constant k (or k_{eff}) is called the "multiplication constant." The original introduction of k_{eff} was based on the chain reaction concept. The number of neutrons in a given generation is multiplied by k, which gives the number of neutrons in the subsequent generation of a fission chain. When the chain begins with *one* neutron, the familiar geometrical series for the "total" number of neutrons per chain, n_t^{st}, is obtained (the superscript *st* is attached to indicate the static base of this consideration):

$$n_t^{st} = 1 + k + k^2 + \dots \quad . \tag{7.11a}$$

In a subcritical system, the series in Eq. (7.11a) converges and yields:

$$n_t^{st} = \frac{1}{1 - k} \quad . \tag{7.11b}$$

Equation (7.10), however, yields a different result for the total number of neutrons per chain (i.e., the fission neutrons plus the initiating one):

$$n_t = 1 + n_f = 1 + \frac{1}{\beta - \rho} \quad . \tag{7.12}$$

The results of these two formulas may be very different. For example, n_t^{st} approaches infinity if k approaches 1, whereas Eq. (7.12) yields a finite number of neutrons per chain in a critical reactor. Even in a delayed supercritical reactor where the sequence in Eq. (7.11a) does not converge, the total number of neutrons per chain, as given by Eq. (7.12), remains finite. The reason for this discrepancy stems from very different meanings of the term "chain" in the two equations. Equation (7.12) describes chains that are initiated by a single delayed neutron; n_t contains three contributions—the original chain-initiating neutron, the prompt neutrons produced, and the quasi-stationary delayed neutrons. The latter are actually not yet neutrons; they are still embedded in precursors and do not yet participate in the multiplication. The decay of these precursors will "later" initiate further chains. In Eqs. (7.11), however, all precursors are instantaneously treated as prompt neutrons, and they participate without delay in the multiplication process. Thus, very different chain concepts are applied in Eqs. (7.11) and (7.12). To show the relation of both descriptions, Eqs. (7.11) are derived from Eq. (7.12) in Sec. 7-3C.

7-3B Precursors and Criticality

The number of precursors, n_{pc}, is obtained by multiplication of the number of fission neutrons per average chain with β, the number of precursors per total number of fission neutrons:

$$n_{pc} = \beta n_f = \frac{\beta}{\beta - \rho} \quad , \tag{7.13}$$

where n_{pc} is obtained from n_f by multiplication with β because n_f describes the prompt *and* quasi-stationary delayed neutrons.

Equation (7.13) provides an interpretation of criticality in terms of actual fission chains and the balance of precursors:

$$n_{pc} = \begin{cases} > 1 \text{ for } \beta > \rho > 0 \\ = 1 \text{ for } \qquad \rho = 0 \\ < 1 \text{ for } \qquad \rho < 0 \end{cases} . \tag{7.14}$$

Thus, *the essence of criticality is the conservation of the number of precursors in an average fission chain.*

Even though the individual fission chains die out rapidly, the average flux stays constant in a critical reactor since the precursor population stays constant, and initiates fission chains at a constant (average) rate. In a delayed supercritical reactor, there is more than one precursor produced per average chain, and in a subcritical reactor less than one. In a delayed supercritical reactor, the number of precursors produced and thus the number of chains increases; it decreases in a subcritical reactor without independent source.

For the constant reactivity assumed here, the decreasing precursor population in a subcritical reactor results in a decreasing flux, while the increasing precursor population in a delayed supercritical reactor results in an increasing flux.

7-3C Comparison of Static and Prompt Neutron Chains

The static formula, Eqs. (7.11), includes the delayed neutrons immediately in the multiplication process whereas Eq. (7.12) is based on the realistic situation in which prompt neutrons *and* precursors are produced. The precursors initiate additional chains later in time. If all subsequently initiated chains are added up, the multiplication effect of the delayed neutrons is included and should, therefore, agree with Eqs. (7.11). Let the first chain start with a precursor decay at t_1. Adding the fission neutrons of all chains yields $(n_f)_{\text{total}}$, the total number of fission neutrons in the sequence of chains:

$$(n_f)_{\text{total}} = \frac{1}{\beta - \rho}\left[1 + \left(\frac{\beta}{\beta - \rho}\right) + \left(\frac{\beta}{\beta - \rho}\right)^2 + \ldots\right]$$

$$= \frac{1}{-\rho} \ . \tag{7.15}$$

The terms in the bracket are the number of precursors produced in successive generations that initiate fission chains at later times.

If the reactor is subcritical, the sequence converges, i.e., the rapid die-away of the individual prompt chains is followed by a slow die-away of the precursor population. Adding also the original neutron yields:

$$(n_t)_{\text{total}} = 1 + (n_f)_{\text{total}} = 1 + \frac{1}{-\rho} = \frac{1}{1 - k} \ . \tag{7.16}$$

Thus, the total number of neutrons in the converging series of chains is equal to the number of neutrons in the single "chain" of the static consideration. The latter "chain" is actually a sequence of rapidly disappearing chains of prompt neutrons connected by latent precursor neutrons emitted delayed. Eventually the precursors that carry this sequence of chains will die out or diverge depending on the sign of the static reactivity.

The right sides of Eqs. (7.15) and (7.16) are the static source multiplication factors for the fission and total source, respectively, which were derived in Sec. 6-1A from the balance equation for the average flux. The derivation presented here provides the fission chain or microkinetics interpretation of the average source multiplication that is further explored and summarized in Sec. 7-5.

7-4 Fission Chains in Reactor Transients

7-4A Reactivity-Step-Induced Transients

Consider a reactivity step ($\rho_1 \gtrless 0$) insertion at $t = 0$ in a critical reactor. Figures 7-3 and 7-4 show as dashed lines a selection of chains as they would have occurred if the reactor had stayed critical. But, by the positive reactivity insertion ($\rho_1 < \beta$), the magnitude of the prompt period is suddenly increased [Eq. (7.6)] and the average chain extends out longer in time. Alternatively, the "critical" chains have been shortened in Fig. 7-4 as a result of the negative reactivity insertion.

7-4B Prompt Jump and Delayed Adjustments

The consequence of such a sudden change in the length of the fission chains for the (average) flux, which is composed of many chains,

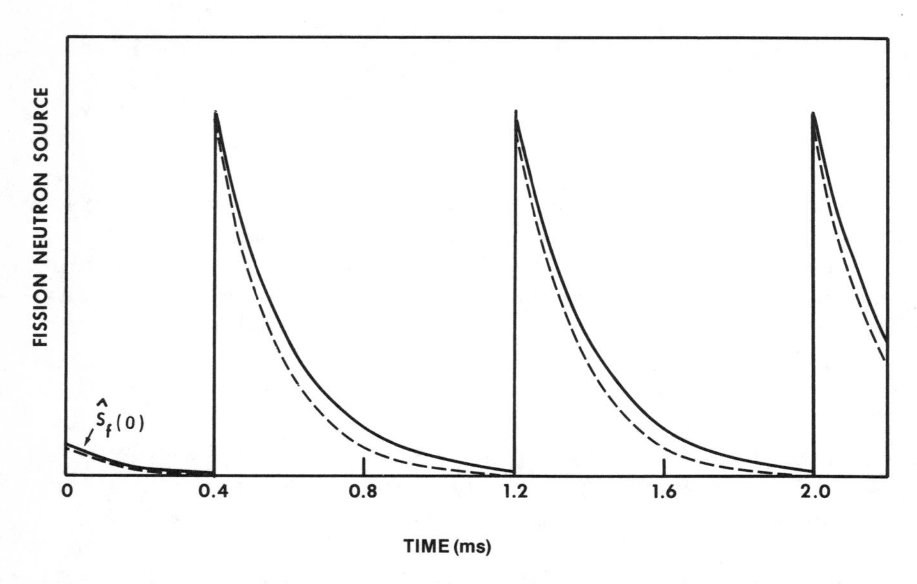

Fig. 7-3. The fission neutron source of average individual fission chains is shown as a function of time for a critical (dashed lines) and a delayed supercritical (solid lines) reactor.

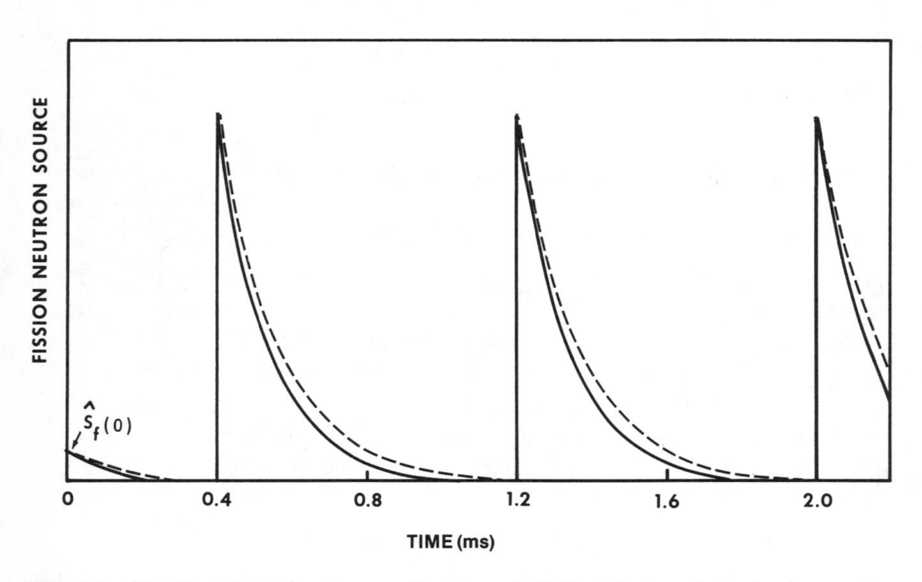

Fig. 7-4. The fission neutron source of average individual fission chains is shown as a function of time for a critical (dashed lines) and a subcritical (solid lines) reactor.

is a prompt jump, up or down, depending on the sign of ρ_1. Since each individual chain starts at the same peak value, some time is required to *gradually realize that the length of a chain has changed.* The time constant for realizing the change in chain length is obviously $|\alpha_p| = 1/\tau_p$, which describes the average chain. This is then also the time constant for the average flux to carry out the prompt jump. This sudden change in the length of the chains and its realization by the flux with τ_p as the period is the *microscopic explanation of the prompt jump phenomenon.*

Another consequence of the change in the chain length is a corresponding change in the number of precursors produced in the chains [see Eq. (7.13)]. After some delay, the number of delayed neutron initiating chains is increased in supercritical transients and decreased in subcritical transients. The subsequent and delayed change in the "density" of chain-initiating events after a preceding change in the length of the chains is the *microscopic explanation of the delayed flux adjustment* following a prompt jump.

7-4C Source-Step-Induced Transients

The sudden insertion of an independent source in a critical reactor can also be analyzed in terms of fission chains. In an example, the source is assumed to have the same strength as the delayed neutron source; possible differences in space and energy distributions of the independent and the delayed neutron sources are disregarded.

The first thing that happens is an increase (e.g., doubling) of the "density" of chain-initiating events. Consequently, the average flux is increased through a prompt jump (e.g., by a factor of 2). The additional chains lead to additional production of precursors. Since the reactor is critical, each chain yields *one* precursor. Each additional new precursor initiates a sequence of chains that—on the average—has an invariable intensity. Thus, each of the independent source neutrons will add one more of these "invariable" chain sequences. Since the independent source is assumed to emit neutrons at a constant rate, these "invariable" chain sequences are added at a constant rate, which in turn leads to the well-known linear increase in the average flux after the prompt jump due to the independent source in a critical reactor.

7-5 Fission Chains and Source Multiplication Formulas

The fission chains are the most direct demonstration of the "source multiplication." One neutron initiates a fission chain in which n_f new neutrons are produced through fission. Thus, multiplication of the source with n_f represents a source multiplication formula. The formulas for the

TABLE 7-I
Types of Source Multiplication Formulas

Number of Fission Chains	Time Dependency	Denominator	Numerator	Applicability
1a. Single chains	Static	$\beta - \rho$	$S_d(+S)$	$\rho \le 0$
b. Single chains	Dynamic	$\beta - \rho$	$S_d(+S)$	$\rho < \beta$
2a. Sequence of chains	Static	$-\rho$ or $1 - k$	S	$\rho < 0$
b. Sequence of chains	Dynamic	$-\rho$ or $1 - k$	S	$\rho < 0$

number of neutrons in chains correspond directly to the source multiplication formulas of the "continuous" treatment, where the flux is not resolved into fission chains. There are two types of source multiplication formulas that can now be related to the physical phenomena of fission chains. Both types can be applied to static and dynamic problems; they are listed in Table 7-I together with their range of applicability.

The formulas of type 1 in Table 7-I describe source multiplication in *single* fission chains [Eq. (7.10)]. Single fission chains contain a finite number of neutrons if $\rho < \beta$. Single chains are initiated by delayed as well as independent source neutrons; thus, both sources have to appear in the numerator. There is no qualitative difference in single fission chains in a supercritical ($\rho < \beta$), critical, or subcritical reactor; there is only the quantitative difference in the length of the chain. Thus, formulas of this type are applicable to stationary or time-dependent neutron fluxes if $\rho \leqslant 0$ or $\rho < \beta$, respectively, as indicated in Table 7-I.

The source multiplication formulas of type 2 in Table 7-I describe the multiplication in *sequences of single chains*. These sequences are "carried" by the precursors produced in the single chains. Obviously, the sequence only converges if the precursors in a sequence eventually die out ($\rho < 0$). The numerator consists only of the independent source neutrons since the multiplication of the precursors is included in the sequence. The denominator is $-\rho$ if the total number of fission neutrons is desired [see Eqs. (6.8a) and (7.15)]. If the neutrons that initiate the sequences are also included, the denominator becomes $1 - k$ [see Eqs. (6.8b) and (7.16)].

Formulas of type 2b in Table 7-I were not discussed in previous sections. They may, for example, be applied when an independent source injects neutrons into a subcritical reactor at a specific time, or as repetitive pulses. The quantitative application of this formula requires the source and the reactivity to change slowly to allow for an approximate convergence of the sequence of chains.

As indicated above, source multiplication formulas that contain the delayed neutron source are not formulas of the same category as the ones based on the independent source only, since the delayed neutron source needs to be precalculated. Therefore, the first source multiplication formula of this type was introduced in Sec. 6-1A as: The flux in a critical reactor may be *viewed* as resulting from the stationary delayed neutron source through source multiplication with $1/\beta$ as the source multiplication factor. The investigation of single fission chains showed that this is exactly the *microscopic physical situation* and not merely a formal relation as it could have appeared in Sec. 6-1A. The fission chains caused by delayed or independent source neutrons are the same. Therefore, the corresponding source multiplication is additive; thus S_d and S appear as a sum in the numerator of the single-chain-type source multiplication formulas. The only difference is that the precursors for the delayed neutron source are built up in thousands of previous fission chains and thus depend on the flux integral, as opposed to the flux-independent source.

An example of the dynamic single-chain source multiplication description of the prompt jump is illustrated in Eqs. (6.92b) and (6.95b). The still-unchanged delayed neutron source, with or without an independent source, initiates the single chains. The chain length has been changed by the reactivity insertion and thus the number of neutrons per chain. The source multiplication formula for the prompt jump, therefore, contains the original delayed neutron source and the n_f value of the lengthened or reduced chain. The understanding of the prompt jump in terms of single fission chains also explains why the simple model with the constant delayed neutron source already yields the correct prompt jump: The alterations of the delayed neutron source, as described by one or six delayed neutron groups, only add further initiating events at much later times.

7-6 Interpretation of the Adjoint Flux

A most revealing interpretation of the adjoint flux, $\phi^*(\mathbf{r},E)$, is obtained by considering the injection of a neutron burst into a critical reactor, at point $\mathbf{r}$ with energy E, and evaluating the subsequent flux rise. This evaluation was formulated as a homework problem in Chapter 6 (problem 7) based on the point kinetics equations. In this section, the solution of the same problem by means of microkinetics is presented.

The corresponding flux rise is first determined using an integrated formulation. The source burst is described in time as a *single* δ function and not as a sequence as in Eq. (7.4b). The integrated description of the injection (at $t = 0$) of a neutron is then given by

$$\hat{S}(t) = \delta(t) \quad . \tag{7.17}$$

The solution for $\hat{S}_f(t)$ follows from Eqs. (7.5) and (7.8) as

$$\hat{S}_f(t) = \frac{1}{\Lambda} \exp(\alpha_p t) \quad , \tag{7.18}$$

where $\alpha_p = -\beta/\Lambda$.

Since the reactor is critical, the proper average precursor decay constant is $\overline{\lambda^{in}}$ as shown in Sec. 6-1B. Thus, after each time interval,

$$\Delta t = 1/(\overline{\lambda^{in}}) \quad ,$$

on the average, a new fission chain appears. With $1/\beta$ neutrons per chain [see Eq. (7.10)], the average fission source addition becomes

$$\Delta \hat{S}_f = \frac{n_f}{\Delta t} = \frac{\overline{\lambda^{in}}}{\beta} \quad . \tag{7.19}$$

With $\overline{\lambda^{in}} = 0.08/\text{s}$ and $\beta = 0.0067$, an integrated fission source addition of ~12 neutrons per second is obtained. This addition is independent of the flux level present in the reactor.

Instead of the simple integrated quantities of Eqs. (7.2) that have been employed for the illustration of microkinetics considerations in this chapter, the adjoint flux weighted quantities are now used in Eq. (7.1) as in standard point kinetics. Thus, the reduced source $s(t)$ is

$$s(t) = \frac{\left(\phi_0^*(\mathbf{r},E), S(\mathbf{r},E,t)\right)}{\left(\phi_0^*(\mathbf{r},E), S_{f0}(\mathbf{r},E,t)\right)} = \frac{(\Phi_0^*,S)}{F_0} \quad . \tag{7.20}$$

The source burst is described by δ functions in $\mathbf{r}$, E, and t:

$$S(\mathbf{r},E,t) = \delta(\mathbf{r})\delta(E)\delta(t) \quad . \tag{7.21}$$

Carrying out the space and energy integrations of the scalar product in Eq. (7.20) singles out the adjoint flux, with the result

$$s(t) = \frac{\phi_0^*(\mathbf{r},E)}{F_0} \delta(t) \quad . \tag{7.22}$$

Using Eq. (7.22) as the source in Eq. (7.1) rather than Eq. (7.17) in Eq. (7.3) yields a flux amplitude addition, Δp_{as}, which is obtained in the same way as $\Delta \hat{S}_f$ of Eq. (7.19), i.e.,

$$\Delta p_{as} = \frac{\phi_0^*(\mathbf{r},E)}{F_0} \cdot \frac{\overline{\lambda^{in}}}{\beta} \quad . \tag{7.23}$$

Multiplying Δp_{as} and F_0 provides the added adjoint flux weighted fission source addition that results from the injection of a single neutron at $(\mathbf{r},E)$:

$$\Delta p_{as}(\Phi_0^*,S_{f0}) = \Delta(\Phi_0^*,S_f) = \phi_0^*(\mathbf{r},E) \cdot \frac{\overline{\lambda^{in}}}{\beta} \quad . \tag{7.24}$$

As shown previously in the simpler analysis, the fission source addition is again independent of the flux level. Its magnitude is proportional to the adjoint flux at the source injection point $\mathbf{r}$ and the source-neutron energy E. As the adjoint flux decreases toward the reactor core boundary, the flux rise caused by the source injection decreases proportionally.

Conversely, the notion of a relative flux rise following a neutron injection may be used to qualitatively infer the relative magnitude of ϕ_0^* as a function of space and energy (see also App. B). For example the *energy* dependence of ϕ_0^* can be qualitatively inferred from the possibility of the source neutron causing fission rather than being captured or leaking out. In a thermal reactor, the chance for causing fission increases with decreasing injection energy since capture and leakage become less and less likely during the remaining slowing down phase. Then $\phi^*(\ ,E)$ increases accordingly. In a fast reactor, the adjoint flux also increases with E moving toward thermal energies. The fact that there is practically no neutron flux at thermal energies in a fast reactor has no bearing on the adjoint flux. The value of $\phi^*(\ ,E)$ is determined by a conjectured neutron injection at energy E.

Homework Problems

1. The evaluation of neutron chains in this chapter is based on the neglect of neutron-neutron (n-n) interaction. Show that this neglect is justified.
   a. Consider two flux values for thermal neutrons, 10^{14} neutron/cm^{2}s and 10^{16} neutron/cm^{2}s (high-flux reactor). Find the n-n collision rate. Assume for simplicity a n-n scattering cross section as for scattering between neutrons and protons (20 barn).
   b. Compare the scattering rate between neutrons with the neutron-proton scattering rate in water.

2. Consider a sequence of equidistant fission chains in a critical reactor; n chains initiated per cubic centimeter and second.
   a. Find the power density, assuming $\beta = 7.5 \times 10^{-3}$, $\nu = 2.5$, and $Q = 200$ MeV per fission.
   b. Find n for a power density of 50 kW/ℓ.

3. Consider a sequence of equidistant fission chains in a critical reactor as in problem 2. Estimate the statistical fluctuations of the fission rate density (per cubic centimeter) by considering the regular variations in the equidistant sequence as a function of n.
   a. Find the maximum and minimum values of the fission rate density across an interval between two chains.
   b. Find the percentage change from the average.

4. In a more sophisticated evaluation of the fluctuations, find the mean-square deviation of the power density about the average power between two chains.

5. Derive a formula for the number of precursors in a converging sequence of fission chains. Plot it as a function of ρ for k between 0.8 and 0.999 (semilog paper); use $\beta = 7.5 \times 10^{-3}$.

6. Consider a reactivity-step-induced transient ($\rho < \beta$) in a critical reactor as composed of fission chains.
   a. "Construct" the power trace from fission chains (using one group of delayed neutrons, $\lambda = \bar{\lambda} = 0.6$/s).
   b. Plot the prompt jump.
   c. Find the asymptotic period from the surplus production of precursors per chain.

7. Find the linear flux rise (slope) from fission source additions following a source drop in a critical reactor. Assume independent source neutrons being injected between delayed neutron emissions: one independent source neutron in every 100th interval between delayed neutrons, in every 10th, and in every interval.

8a. Find the analytic solution of the one-group point kinetics equation for a critical reactor into which an independent source, S, is inserted at $t = 0$. The source is left in the system from $t = 0$ to $t = t_1$, and S is then withdrawn again.
b. Sketch the solution for $S = 10\, S_{d0}$.
c. Discuss the analytic solution and interpret its components.
d. Present an interpretation of the key features of the result in terms of fission chains.

Review Questions

1a. What are "chains of fission neutrons?"
 b. What are the main reasons for fluctuations in fission chains?
 2. Give an approximate balance equation for the fission source $\hat{S}_f(t)$, assuming the delayed neutron source is given $[\hat{S}_d(t)]$.
 3. Give the production rate of delayed neutrons in a form that can be used for the investigation of fission chains.
4a. Sketch the fission source for well-separated fission chains.
 b. Describe the qualitative behavior of average fission chains for sub-prompt- and superprompt-critical reactors.
 5. Give the formula for the number of neutrons per chain (n_{ch}). How many neutrons are contained in an average chain in a thermal reactor ($\beta \simeq 7.5 \times 10^{-3}$) and in a fast reactor ($\beta \simeq 3.3 \times 10^{-3}$) for $\rho_1 = \beta/2$, 0, and $-\beta$?
 6. Assume an average delayed neutron source rate of $1/\Delta t$. Find the corresponding total fission rate and power.
 7. Apply the formula for n_{ch} (fission neutrons per chain) to obtain a formula for the number of precursors produced in a chain. Discuss the result.
8a. Give a source multiplication relation of $\hat{S}_{f0}$ and $\hat{S}_{d0}$ and discuss its physical significance.
 b. What is the essence of criticality in terms of fission chains?
 9. Explain the two types of transitions exhibited in reactivity-step-induced transients in terms of fission chains.

REFERENCES

1. W. G. Davey and W. C. Redman, *Techniques in Fast Reactor Critical Experiments,* an AEC monograph, Gordon and Breach Science Publishers, New York (1970).
2. G. R. Keepin, *Physics of Nuclear Reactors,* p. 60ff, Addison-Wesley Publishing Co., Reading, Massachusetts (1965).

Eight

APPROXIMATE POINT KINETICS

8-1 The Prompt Jump Approximation

In addition to the simplifications of the delayed neutron source (Sec. 6-1), there is another very useful approximation, the so-called "prompt jump approximation" (PJA). It is a method for treating dynamics problems in the subprompt and subcritical reactivity domain. It is devised by exploiting the phenomenon of the rapid prompt jump response to changes in the reactor. This approximate method becomes more accurate for smaller neutron generation times. The PJA results with six delay groups for large classes of fast reactor problems are quite accurate. Therefore, approximate results are sometimes used to replace the results of the complete solution, with substantial savings in computation time. The accuracy of PJA results for thermal power reactors is sufficient for most estimates. The illustrations presented in this chapter are calculated for a fast reactor. Corresponding thermal reactor problems are left to homework assignments.

The prompt jump approximation, together with some other approximations such as the one-delay-group approximation, also simplifies the kinetics equation so that relatively complicated problems can be investigated analytically. This provides solutions that are well suited to describe some of the basic transient phenomena; they can also provide a check on the computer output of complicated dynamic programs.

8-1A Formulation and Implementation of the Prompt Jump Approximation

Investigations in Chapter 6 showed that the neutron flux responds with a prompt jump to a step change of the reactivity or independent source in a subprompt-critical reactor. Step changes are mathematical idealizations of rapid, but actually always gradual, changes. Investigations of hypothetical accidents in reactors have shown that the fastest reactivity insertion rates are for the most part well below 100$ per second. The speed of possible changes in the system that may alter the reactivity is limited by mechanical speeds such as coolant velocity, the motion of boiling fronts or shutdown rods, etc. The prompt jump adjustment time, τ_p, is very small, e.g., around critical ($\rho \simeq 0$):

$$\tau_p \simeq \frac{\Lambda}{\beta - \rho} \simeq \begin{cases} 10^{-4} \text{ s for fast reactors} \\ 10^{-3} \text{ s for light water reactors} \end{cases} . \tag{8.1}$$

Thus, the reactivity insertion *during* the prompt jump period is very small even for the most rapid reactivity insertion rates. Therefore, the idealized prompt jump, i.e., the rapid flux response *after* a finite change in reactivity, should be replaced by *simultaneously* considering reactivity insertions and prompt flux adjustments. The PJA is the method for an approximate treatment of the prompt jump-type flux adjustment *during* a gradual reactivity insertion.

To illustrate the effect of the prompt jump phenomenon during a gradual insertion of reactivity, consider at first the decomposition of a rapid reactivity ramp, 50$/s, into a sequence of small reactivity steps, $\delta\rho = 1¢$. The upper part of Fig. 8-1 shows the reactivity ramp as a dashed line and the staircase-shaped substitute as a solid line. The lower part of Fig. 8-1 shows the PJA flux response to the reactivity insertion as dotted lines and the actual flux increase caused by the staircase reactivity insertion as the solid line. The dashed-dotted lines represent the continuations of the prompt jumps if subsequent stairs were not introduced. The actual sequence of small prompt jumps following the incremental steps is clearly exhibited. The figure also shows that the first prompt jump is practically completed at the time when the third one starts. The same holds for the second and the fourth jump, etc. Consequently, the prompt jump is not yet completed for only the "very recent part" of a gradual reactivity insertion.

The fact that the prompt jump-type flux response occurs so rapidly suggests that a good approximation can be obtained by neglecting the adjustment time completely. This leads to the *prompt jump approximation* (PJA), which consists of *approximating the rapid prompt jump response of the neutron flux to changes in reactivity or independent source by an instantaneous response.*

The qualitative difference between the correct neutron flux amplitude, $p(t)$ (solid line), and the approximate flux as it follows from the PJA, $p_{pj}(t)$ (dotted line),[a] can also be seen from the lower part of Fig.

---

[a]An index pj on the flux calculated by the PJA is used only if it is compared with the correct $p(t)$.

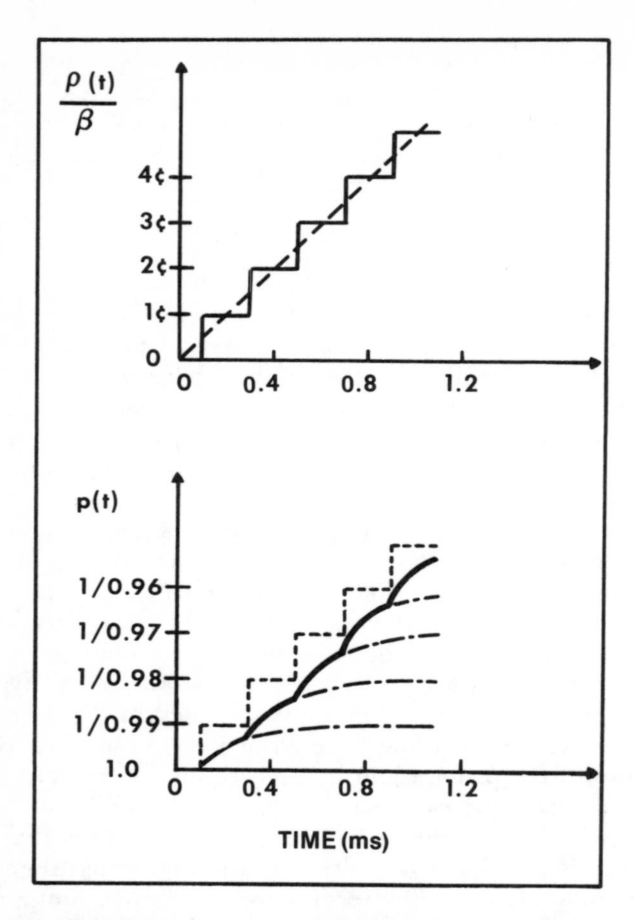

Fig. 8-1. Superposition of the prompt jump transitions following a staircase-type reactivity insertion of 50$/s with τ_p = 0.15 ms.

8-1 (see also Fig. 8-2). Generally, $p_{pj}(t)$ is always somewhat "ahead in time" of the correct flux $p(t)$. This time shift is proportional to the prompt period; it is, therefore, ~10 times larger in light water reactors (LWRs) than in fast reactors. If the flux is rising, $p_{pj}(t)$ is larger than $p(t)$, since it "anticipates" the increase:

$$p_{pj}(t) \geq p(t) \quad , \text{ for increasing } p(t) \quad . \tag{8.2a}$$

If the flux is decreasing, $p_{pj}(t)$ is smaller than $p(t)$ since it anticipates the decrease:

$$p_{pj}(t) \leq p(t) \quad , \text{ for decreasing } p(t) \quad . \tag{8.2b}$$

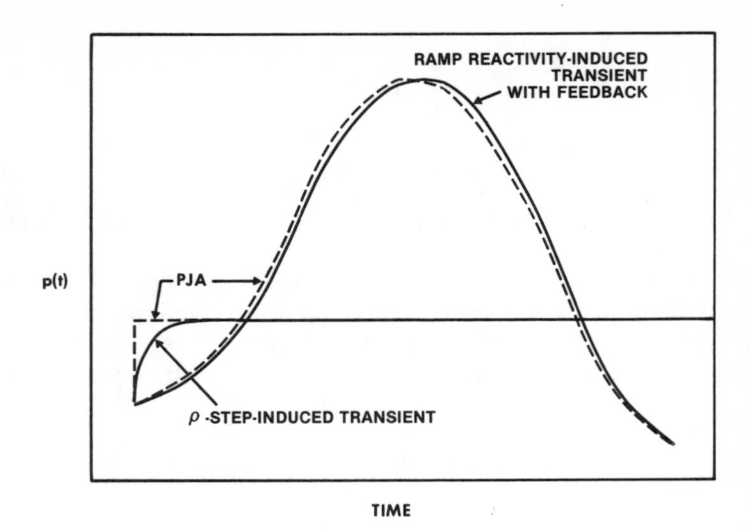

Fig. 8-2. Typical derivations of the flux calculated with the PJA (dashed lines) from reference fluxes (solid lines).

For the case in which the reactivity is constant for a few prompt periods so that the prompt jump can be completed, there is practically no error in p_{pj}. Flux maxima and minima generally occur earlier in $p_{pj}(t)$ than in $p(t)$ (see Fig. 8-2). Equations (8.2) may become invalid in the immediate vicinity of flux extrema, due to the crossover of the two curves.

Figure 8-2 shows a transient in which the flux change is reversed due to reactivity feedback (bell-shaped curves). The solid line represents the actual flux transient and the dashed line represents the PJA. Since the error of the PJA has different signs on most of the up and down phases of the transient, its inaccuracy largely cancels in the flux integral. This compensation is generally important for safety investigations where the integral over the power transient yields a change in temperature or other key variables. The shift of the transient depicted in Fig. 8-2 is of the order of magnitude of τ_p (10^{-3} to 10^{-4} s), which is practically of no importance.

Figure 8-2 also shows the corresponding prompt jump and the PJA solution for a ρ-step-induced transient. The PJA solution is also a step due to the complete neglect of the response period τ_p.

Any practical benefit from the conceptual idea of PJA can only be obtained if its implementation simplifies the kinetics equations. The implementation of the PJA requires that the "rapid prompt jump response," which is described by the period τ_p, is replaced by an instantaneous response:

$$\tau_p = \frac{\Lambda}{\beta - \rho} \rightarrow 0 \text{ in PJA} \tag{8.3a}$$

and thus

$$\Lambda \rightarrow 0 \text{ in PJA} \quad . \tag{8.3b}$$

Therefore, the PJA is often called the "zero lifetime approximation" ($\Lambda \propto \ell$) or "zero Λ approximation."

The form of the point kinetics equations, Eqs. (5.34) or (5.74), readily allows the implementation of the zero Λ approximation: Λ is brought to the left side of Eq. (5.34) and the term $\Lambda\dot{p}(t)$ is neglected since Λ is assumed to be zero. This gives the exact point kinetics equations in the PJA:

$$0 = [\rho(t) - \beta(t)]p(t) + \frac{F_0}{F(t)} \sum_k \lambda_k \zeta_k(t) \tag{8.4a}$$

and

$$\dot{\zeta}_k(t) = -\lambda_k \zeta_k(t) + \frac{F(t)}{F_0} \beta_k(t) p(t) \quad . \tag{8.4b}$$

The PJA of the kinetics equations of the point reactor model, Eqs. (5.74), are given by:

$$0 = [\rho(t) - \beta]p(t) + \sum_k \lambda_k \zeta_k(t) \tag{8.5a}$$

and

$$\dot{\zeta}_k(t) = -\lambda_k \zeta_k(t) + \beta_k p(t) \quad . \tag{8.5b}$$

As indicated above, the accuracy of the point reactor model results may be improved by using a more accurately calculated reactivity than provided by first-order perturbation theory (see also Chapter 11).

Note that the more common forms of kinetics equations, i.e., Eqs. (5.35), etc., do not allow a simple implementation of the zero Λ approximation. Note further that the precursor balance equations, Eqs. (8.4b) or (8.5b), are not affected by the zero Λ approximation since the balance of precursors is physically independent of the neutron generation time.

The neglect of Λ leads to an elimination of $\dot{p}(t)$ in the PJA balance equations, Eqs. (8.4a) and (8.5a). The neglect of $\Lambda\dot{p}$ does not imply that $p(t)$ is constant. It only implies that the time dependence of the flux is directly determined by the time-dependent balance of production of prompt and delayed neutrons without considering $\dot{p}$.

If the reduced delayed neutron source in Eqs. (8.5) is formally retained as a function of time, the following is obtained as the balance equation:

$$0 = [\rho(t) - \beta]p(t) + s_d(t) \quad , \tag{8.6}$$

which can be solved for $p(t)$ resulting in a

time-dependent or instantaneous source multiplication formula:

$$p(t) = \frac{s_d(t)}{\beta - \rho(t)} \quad . \tag{8.7}$$

If an independent source is included, one obtains the very important

time-dependent source multiplication formula:

$$p(t) = \frac{s_d(t) + s(t)}{\beta - \rho(t)} \quad . \tag{8.8}$$

Important conclusions about the transient behavior can be readily drawn from these source multiplication formulas. Application of Eq. (8.7) to a supercritical reactor indicates that the flux will decrease when ρ moves away from β although the reactor is supercritical. Analogously, Eq. (8.8) shows that the flux will increase in a subcritical reactor when ρ moves toward β, provided in both cases that the denominator changes faster than the numerator.

Equations (8.7) and (8.8) are source multiplication formulas that correspond to the multiplication in individual fission chains (see Sec. 7-5). Previous source multiplication formulas of this type, such as Eqs. (6.11), (6.12), and (6.92), can be obtained from Eq. (8.8) by inserting the corresponding values for reactivities and sources. In this sense, Eq. (8.8) may be considered a *general source multiplication formula,* and those previously obtained as special cases. There is, however, an important difference between Eq. (8.8) and the previously derived "special cases." In all previously derived formulas, the reactivity was assumed constant, at least after a step change. Then the average individual fission chains are pure exponential functions [see Eq. (7.5)]. Their integral yields the number of fission neutrons per chain in terms of a source divided by a constant value of $\beta - \rho$. If, however, the reactivity is not sufficiently invariable during a prompt period, the time-dependent denominators of Eqs. (8.7) and (8.8) describe the source multiplication only approximately, since the number of neutrons per chain can only be approximately described by an expression of the form of Eq. (7.10). Thus, the previous formulas with constant reactivity are exact and Eqs. (8.7) and (8.8), even though formally more general, are related only approximately to the individual fission chains.

The type of approximation in Eqs. (8.7) and (8.8) is naturally the same as in the PJA, i.e., $\tau_p \rightarrow 0$. This means that the lifetime of fission chains is reduced to zero, i.e., the multiplication in the chains occurs immediately. If the lifetime of fission chains is assumed to be zero, any continuously varying reactivity may be treated as a constant with the consequence that the same source multiplication formula applies as that for constant reactivities.

A major benefit of the PJA is derived from combination with the one-delay-group kinetics equations. It is then possible to treat analytically the relatively complicated kinetics and dynamics problems (see Chapter 10). This yields a better understanding of the transient behavior than can be provided by numerical results alone.

The one-delay-group kinetics equations corresponding to the prompt jump in the point reactor model are given by:

$$0 = [\rho(t) - \beta]p(t) + \lambda\zeta(t) \tag{8.9a}$$

and

$$\dot{\zeta}(t) = -\lambda\zeta(t) + \beta p(t) \tag{8.9b}$$

with a suitable one-delay-group λ (e.g., $\lambda = \bar{\lambda}$). The calculation of ρ(t) in Eq. (8.9a) may be more sophisticated than in the point reactor model. By eliminating $\zeta(t)$, the kinetic equation in the PJA for one delayed neutron group is obtained in the form:

$$\dot{p}(t) = \frac{\lambda\rho + \dot{\rho}}{\beta - \rho}\, p(t) \quad . \tag{8.10}$$

The PJA thus reduces the one-delay-group model from a second-order to a first-order differential equation. Consequently, problems with the general time dependence of the coefficients (i.e., reactivity) as well as independent source (if included) can be solved analytically. The second-order differential equation for $p(t)$, i.e., the equation without PJA, can be solved in a closed analytic form only for special variations of the coefficients.

If a stationary critical state is assumed for $t \leqslant 0$, the initial condition becomes:

$$p(0) = p_0 \quad . \tag{8.11}$$

A special situation occurs in the PJA when the reactivity variation starts with a step (ρ may then continue with an arbitrary time dependence). Due to the instantaneous response of the flux to the reactivity step, a flux value different from p_0 is obtained at $t = 0+$. It is convenient to use this modified flux as the "initial" value. It is denoted as p^0 and called the "pseudo-initial flux" in order to distinguish it from the actual initial flux p_0.

The pseudo-initial flux may best be derived from the original equations, Eqs. (8.9). An initial stationary state was assumed, with $\lambda\zeta_0 = \beta p_0$. This relation also holds over the reactivity step because the ζ variation starts with a zero slope (see Sec. 6-1B). Inserting $\lambda\zeta_0 = \beta p_0$ into Eq. (8.9a) and solving for $p(0+)$ yields the general initial condition:

$$p(0+) = \begin{cases} p_0 & \text{for gradual reactivity increase } (\rho_1 = 0) \\[2ex] \dfrac{\beta}{\beta - \rho_1} p_0 = p^0 & \text{for initial reactivity step } (\rho_1 \neq 0) \ . \end{cases} \tag{8.12}$$

The PJA solution presented in Fig. 8-2 shows the flux "starting" at the pseudo-initial flux for a step reactivity insertion.

The PJA is a special example of what is mathematically called the "method of singular perturbations." This approximation can be applied if one transition occurs very much faster than the others (in this case, the prompt jump transition compared to the transitions corresponding to the other roots; see Sec. 6-1B). This rapid transition is approximated by an instantaneous transition. If the limit rapid → instantaneous transition in the analytic solution is explicitly carried out, the integrand in the convolution integral corresponding to this transition becomes *singular*. This may be shown readily in the simple example of a transient that follows a reactivity step. Since in the present context interest is focused on the rapid transition only, the other transitions may be eliminated by formally treating $s_d(t)$ as a known function. The solution of the first of the kinetics equations is then given by

$$p(t) = p_0 \exp\left(-\frac{\beta - \rho_1}{\Lambda} t\right) + \int_0^t \frac{s_d(t')}{\Lambda} \exp\left[-\frac{\beta - \rho_1}{\Lambda}(t - t')\right] dt' \ . \tag{8.13}$$

In the limit $\Lambda \to 0$ (i.e., $\tau_p \to 0$), the first term on the right side of Eq. (8.13) disappears for $t > 0$. The integrand, however, becomes singular. Introducing the following as an integration variable,

$$-(\beta - \rho_1)(t - t') = \theta \ , \tag{8.14}$$

gives for the limit $\Lambda \to 0$

$$\lim_{\Lambda\to 0} p(t) = \lim_{\Lambda\to 0}\left\{\frac{1}{\beta - \rho_1}\int_{\theta_0}^{0}\frac{s_d[t'(\theta)]}{\Lambda}\exp(\theta/\Lambda)\,d\theta\right\} \quad . \tag{8.15}$$

The quantities θ_0 and t' depend on t and θ, respectively, as:

$$\theta_0 = -(\beta - \rho_1)t < 0$$

and

$$t' = t + \frac{\theta}{\beta - \rho_1} \quad . \tag{8.16}$$

In the limit $\Lambda \to 0$, the factor $1/\Lambda \cdot \exp(\theta/\Lambda)$ in the integrand becomes infinitely large at $\theta = 0$ (i.e., $t' = t$), and infinitesimally narrow with an integral of 1 (similar to a δ function). The integration over this singular integrand singles out the value of the delayed neutron source at $\theta = 0$, i.e., at $t' = t$. Thus, Eq. (8.15) becomes

$$\lim_{\Lambda\to 0} p(t) = \lim_{\Lambda\to 0}\left[\frac{s_d(t)}{\beta - \rho_1}\int_{\theta_0}^{0}\exp(\theta/\Lambda)\frac{d\theta}{\Lambda}\right] = \frac{s_d(t)}{\beta - \rho_1} \quad , \tag{8.17}$$

since

$$\lim_{\Lambda\to 0}\left[\int_{\theta_0}^{0}\exp(\theta/\Lambda)\frac{d\theta}{\Lambda}\right] = \int_{-\infty}^{0} e^x\,dx = 1 \quad . \tag{8.18}$$

The result of the limit of Eq. (8.17) is analogous to the time-dependent source multiplication formula, Eq. (8.7), which was directly obtained from the kinetics equation in the zero Λ approximation. The approximate character of the source multiplication formulas, Eqs. (8.7) and (8.8), can be viewed as resulting from the application of the singular perturbation method.

8-1B Application of the Prompt Jump Approximation

Some initial applications of the PJA to kinetics problems are discussed in this section. Further applications particularly to problems with reactivity feedback (dynamics problems) are presented in Secs. 10-2 and 10-3.

In the one-delay-group approximation, the general PJA solution can be derived readily for any time variation of the reactivity. From Eq. (8.10) and the initial condition of Eq. (8.11), it follows that:

$$p(t) = p^0 \exp\left[\int_{0+}^{t}\frac{\lambda\rho(t') + \dot\rho(t')}{\beta - \rho(t')}\,dt'\right] \quad , \tag{8.19}$$

where the lower integration limit is 0+, since the instantaneous flux rise caused by a reactivity step is included in p^0.

The general solution, Eq. (8.19), can be substantially simplified by carrying out the integration of the second term in the exponent:

$$\int_0^t \frac{\dot{\rho}(t')\,dt'}{\beta - \rho(t')} = \int_{\rho_1}^{\rho(t)} \frac{d\rho}{\beta - \rho} = -\ln \frac{\beta - \rho(t)}{\beta - \rho_1} \quad . \tag{8.20}$$

Inserting Eq. (8.20) into Eq. (8.19) yields for the general solution:

$$p(t) = \frac{\beta p_0}{\beta - \rho(t)} \exp\left[\int_0^t \frac{\lambda \rho(t')\,dt'}{\beta - \rho(t')}\right] \quad . \tag{8.21}$$

The term in front of the exponent in Eq. (8.21) is the instantaneous source multiplication formula, Eq. (8.7), with $s_d(t)$ replaced by the initial delayed neutron source βp_0. Thus, considering the first factor of Eq. (8.21) alone is equivalent to the combined

prompt jump and constant-delayed neutron source approximation:

$$p(t) = \frac{\beta p_0}{\beta - \rho(t)} \quad . \tag{8.22}$$

The exponential function in Eq. (8.21) describes the effect of the variation of the delayed neutron source in the one-delay-group approximation.

Specific solutions such as the transient that follows a reactivity step, ρ_1, can be obtained directly from the general solution, Eq. (8.21):

$$p(t) = \frac{\beta p_0}{\beta - \rho_1} \exp\left(\frac{\lambda \rho_1}{\beta - \rho_1} t\right) \quad . \tag{8.23}$$

The comparison of Eq. (8.23) with the corresponding solution in Sec. 6-2C, Eq. (6.109), shows that Eq. (8.23) yields the asymptotic transient. The prompt jump transition is reduced to a step (ρ_1 may be positive or negative, but $\rho_1 < \beta$).

Of more interest and importance for the understanding of transients is the application of Eq. (8.21) to problems with time-dependent reactivities. As an example, a ramp-type reactivity insertion followed by a ramp-type counteraction (sawtooth reactivity transients; see Fig. 8-3) is considered:

$$\rho(t) = \begin{cases} at & \text{for } 0 \leqslant t \leqslant t_1 \\ at_1 - a'(t - t_1) & \text{for } t \geqslant t_1 \\ a(2t_1 - t) \text{ if } a' = a & \text{for } t \geqslant t_1 \end{cases} \quad . \tag{8.24}$$

Only the case with $a' = a$ is treated explicitly. The general case $(a' \neq a)$ is discussed only qualitatively.

The integral in Eq. (8.21) is composed of two parts, corresponding to the two different reactivity insertions. Integrating over the first part, $t \leq t_1$, yields

$$\lambda \int_0^t \frac{at'}{\beta - at'}\, dt' = \frac{\lambda}{a}\beta \int_0^{at/\beta} \frac{\tau\, d\tau}{1 - \tau} = \frac{\lambda}{a}\beta h\left(\frac{a}{\beta}t\right) \tag{8.25}$$

with

$$h\left(\frac{a}{\beta}t\right) = -\frac{a}{\beta}t - \ln\left(1 - \frac{a}{\beta}t\right) \quad . \tag{8.26}$$

The second part of the integral for $t \geq t_1$ is basically of the same form:

$$\lambda\int_{t_1}^t \frac{a(2t_1 - t')\, dt'}{\beta - a(2t_1 - t')} = -\frac{\lambda}{a}\beta \int_{at_1/\beta}^{a(2t_1 - t)/\beta} \frac{\tau'\, d\tau'}{1 - \tau'}$$

$$= -\frac{\lambda}{a}\beta\left[h\left(2\frac{a}{\beta}t_1 - \frac{a}{\beta}t\right) - h\left(\frac{a}{\beta}t_1\right)\right] \quad . \tag{8.27}$$

Inserting Eqs. (8.25) to (8.27) into the general solution, Eq. (8.21), gives

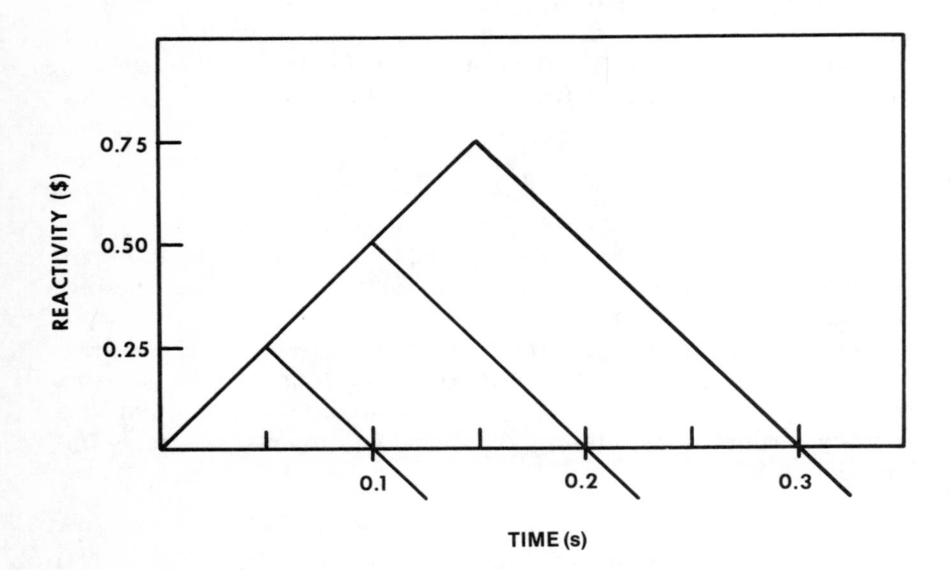

Fig. 8-3. The sawtooth-type time dependence of the reactivity in the transient shown in Fig. 8-4 where a = 5$/s. A 5$/s counteraction may be associated with a reactor scram.

$$p(t) = \frac{p_0\beta}{\beta - at} \exp\left\{-\lambda\left[t + \frac{\beta}{a}\ln\left(1 - \frac{a}{\beta}t\right)\right]\right\} \text{ for } 0 \le t \le t_1$$

and

$$p(t) = \frac{p_0\beta}{\beta - a(2t_1 - t)} \exp\left\{-2\lambda\left[t_1 + \frac{\beta}{a}\ln\left(1 - \frac{a}{\beta}t_1\right)\right]\right. \tag{8.28}$$

$$\left. + \lambda\left[(2t_1 - t) + \frac{\beta}{a}\ln\left(1 - 2\frac{a}{\beta}t_1 + \frac{a}{\beta}t\right)\right]\right\} \text{ for } t \ge t_1 \quad .$$

Of particular interest is the flux at the time ($t = 2t_1$) when ρ is reduced to its initial value:

$$p(2t_1) = p_0 \exp\left\{-2\lambda\left[t_1 + \frac{\beta}{a}\ln\left(1 - \frac{a}{\beta}t_1\right)\right]\right\} \quad . \tag{8.29}$$

The factors in front of the exponential functions in Eq. (8.28) describe, as has already been discussed in the general solution, the instantaneous source multiplication of the initial delayed neutron source. The instantaneous source multiplication yields an increasing flux on the increasing branch of the reactivity transient and yields a decreasing flux when the reactivity is decreasing (in spite of the fact that the reactor is supercritical for $t < 2t_1$). The exponential function superimposes on the basic behavior, which is described by the instantaneous source multiplication, the effect of the increase of the delayed neutron source during the transient.

The effect of the increased delayed neutron source may be estimated by considering the flux at $t = 2t_1$ where the reactivity is back to zero. Table 8-I shows quantitatively the effect of the delayed neutron source, i.e., of the exponential function in Eq. (8.29). The reactivity rates in this example were chosen to be 5\$/s and 50\$/s; three values were used for the maximum reactivity. The PJA with a constant delayed neutron source (CDS) yields 1.000 in all three cases. The results of Table 8-I indicate that the effect of the increased delayed neutron source is relatively small compared to the flux maximum. Therefore, even the very simple model

TABLE 8-I

$p(2t_1)/p_0$ for Symmetrical Sawtooth Reactivity Insertions

Ramp Rate, a	5\$/s	50\$/s	PJA-CDS
$\rho_{max} = 0.25\$$	1.004	1.0004	1.000
$\rho_{max} = 0.50\$$	1.022	1.0021	1.000
$\rho_{max} = 0.75\$$	1.075	1.0072	1.000

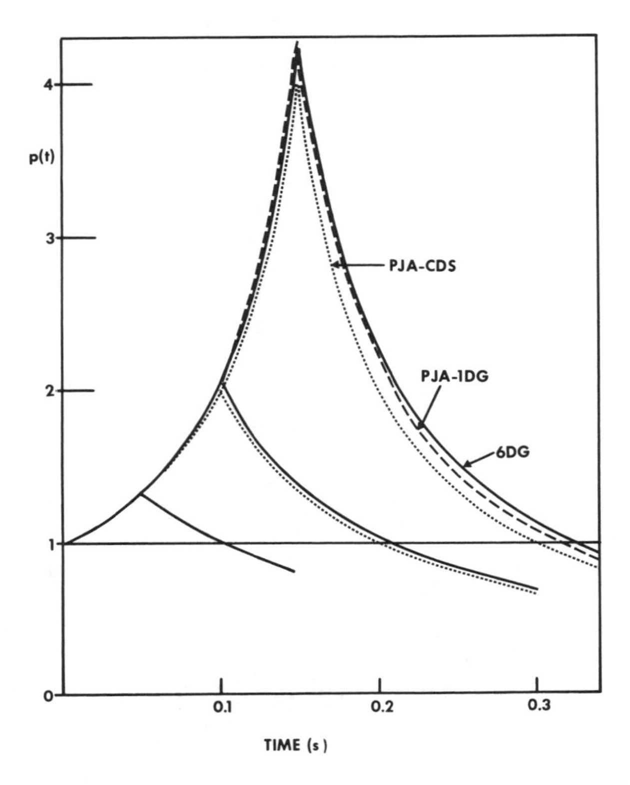

Fig. 8-4. Flux response to a sawtooth reactivity insertion with a ramp rate of 5$/s: six delay groups (6DG) = solid lines; PJA for one delayed neutron group (1DG) = dashed lines; PJA for CDS = dotted lines.

of the instantaneous source multiplication of the initial delayed neutron source gives a good semi-quantitative description of rapid subprompt-critical flux transients in reactors with small generation times (see Fig. 8-4, dotted lines).

Figure 8-4 shows a quantitative comparison of different treatments of the sawtooth reactivity transient depicted in Fig. 8-3. The results of Fig. 8-4 show that the simple source multiplication formula of the PJA-CDS model allows a good semi-quantitative description of rapid transients. The deviations between the PJA results with one-delay-group (dashed lines) and the six-delay-group kinetics are small; they largely result from the one-delay-group model, not from the application of the PJA. The curves in Fig. 8-4 are calculated for a fast breeder reactor. The corresponding curves for an LWR are similar; in fact, the PJA-CDS curves are identical.

The behavior of the flux slope as exhibited in the same case of a sawtooth transient may be formulated more generally by examining the differential equation, Eq. (8.10):

1. During delayed *supercritical* transients, the flux will decrease if the reactivity *decreases* with a rate

$$\dot{\rho} < -\bar{\lambda}\rho \quad . \tag{8.30}$$

2. Similarly, the flux will *increase* when a reactor is *subcritical* provided that ρ increases with a rate

$$\dot{\rho} > -\bar{\lambda}\rho \quad . \tag{8.31}$$

This shows the strong effect of $\dot{\rho}$ on $\dot{p}$ for all rapid subprompt-critical transients. The sign of $\dot{\rho}$ may be more decisive than the sign of ρ itself. In the CDS and PJA approximations, which lead to Eq. (8.22), the flux directly follows the reactivity (thus $\dot{p}$ is proportional to $\dot{\rho}$). By accounting for the variation of the delayed neutron source in the $\bar{\lambda}$ model, the *immediate* dependence of the flux slope on $\dot{\rho}$ is modified by the inequalities (8.30) and (8.31). In the six delay group, the simple inequalities (8.30) and (8.31) are replaced by much more complicated relations with only minor quantitative changes for rapid transients, as becomes evident from inspection of Fig. 8-4.

Flux transients that follow asymmetric reactivity sawtooths may be discussed qualitatively on the basis of Eq. (8.21). Suppose the rising part is a rapid ramp, for example, 5$/s as in Fig. 8-3. The decreasing branch may be variable, such as a' in Eq. (8.24). If $a' > a$, the flux will be turned back even more rapidly than in Fig. 8-4. Generally, the flux decreases after ρ has passed its peak value as long as $a' > \bar{\lambda}\rho_{max}$. If $a' = \bar{\lambda}\rho_{max}$, the flux has a zero slope after the reactivity maximum. For very slow withdrawl rates ($a' < \bar{\lambda}\rho_{max}$), the flux will increase at first but eventually it will decrease when enough reactivity is withdrawn that the inequality (8.31) becomes satisfied. In the limit of $a' = 0$, the flux increase will continue, asymptotically with a stable period.

This quantitative example and this subsequent qualitative discussion clearly demonstrate the importance of simple approximations, such as PJA-one delay group, for achieving an understanding of transient results—an understanding that cannot be derived from numerical results alone. Based on such an understanding, qualitative predictions for flux transients can be made.

8-2 The Prompt Kinetics Approximation

In the summary of delayed neutron source approximations, Eqs. (6.78), the drastic approximation of complete negelect of the delayed neutron source was included. The kinetics based on the resulting differential equation,

$$\dot{p}(t) = \frac{\rho(t) - \beta}{\Lambda} p(t) \quad , \tag{8.32}$$

is often called "prompt" kinetics since only prompt neutron multiplication is treated.

Prompt kinetics may be improved by accounting approximately for the delayed neutrons without adding any complications to Eq. (8.32). Since Eq. (8.32) is not to be changed, the improvement can only appear in the initial condition; the modified initial condition is called the "pseudo-initial condition." In Secs. 8-2A and 8-2B, the pseudo-initial condition is derived for the two basic idealized transients that follow step and ramp reactivity insertions. The term "prompt kinetics approximation" is used here to denote "prompt kinetics with a pseudo-initial condition."

The prompt kinetics approximation (PKA) holds only in the superprompt-critical domain. Applications below prompt critical may lead to physically unrealistic results. For example, an application of Eq. (8.32) to a delayed supercritical transient that starts from a critical reactor yields a prompt *decrease* of the flux whereas the neutron flux actually *increases*. There are, however, important delayed supercritical transients (with nonstationary "initial" conditions) for which the PKA yields not only meaningful but even fairly accurate results in the sub-prompt-critical domain (see Sec. 10-3).

8-2A Superprompt-Critical Transients Following a Step Reactivity Insertion

The application of Eq. (8.32) to a superprompt reactivity step yields a purely exponential flux increase:

$$p(t) = p^0 \exp(\alpha_p t) \quad , \tag{8.33}$$

with

$$\alpha_p = \frac{\rho_1 - \beta}{\Lambda} \quad . \tag{8.34}$$

The factor in front of the exponential function in Eq. (8.33) is the desired pseudo-initial condition.

The straightforward approach to determine the pseudo-initial flux for Eq. (8.33) is to find first an analytical solution in an appropriate model for the step-induced transient and then approximate this solution by the single term, $p^0 \exp(\alpha_p t)$. The kinetics model that provides an appropriate basis for the treatment of the rapid superprompt transients is the $\bar{\lambda}$ kinetics.[b] The solution for step-induced transients was derived earlier in Sec. 6-2C. For ρ_1 that is not too close to β, the following solution was obtained:

$$p(t) = p_0\left[\frac{\rho_1}{\rho_1 - \beta}\exp(\alpha_p t) - \frac{\beta}{\rho_1 - \beta}\exp(\alpha_1 t)\right] , \qquad (8.35)$$

with

$$\alpha_1 = \frac{\rho_1 \lambda}{\beta - \rho_1} . \qquad (8.36)$$

Equation (8.35) could have been applied above to derive the PJA for $\alpha_p < 0$. In terms of the solution (8.35), the PJA consists of the neglect of the first term in the bracket, which disappears rapidly since $\alpha_p < 0$ for $\rho_1 < \beta$. The second term is positive for $\rho_1 < \beta$ and has a positive exponent. It represents the solution (in one-delay-group approximation) after a short transition period. The corresponding pseudo-initial flux is represented by the factor in front of the exponential function [see Eq. (8.12)].

For $\rho_1 > \beta$, the situation is reversed: α_p is positive and the first term increases rapidly while α_1 is negative and the second term becomes negligible compared to the first term after a short transition period, which is essentially given by $1/\alpha_p$. Thus, the factor in front of the first exponential in Eq. (8.35) gives the pseudo-initial flux for the description of step-induced superprompt-critical transients in the prompt kinetics approximation as

$$p^0 = \frac{\rho_1}{\rho_1 - \beta} p_0 . \qquad (8.37)$$

Figure 8-5 shows a comparison of the solution of Eq. (8.33) with the more complete solution (8.35) for $\rho_1 = 1.1\beta$, i.e., $p^0 = 11\ p_0$ in this example.

The most important application of the kinetics approximation is the description of superprompt-critical transients with prompt reactivity feedback. The PKA is particularly adequate if the action of feedback, which requires the accumulation of energy, does not become significant during the prompt jump transition period. In such cases, the description

---

[b] Although the CDS approximation and $\bar{\lambda}$ kinetics yield the same pseudo-initial flux, the latter approximation is applied in order to show the similarity between the PJA and the PKA.

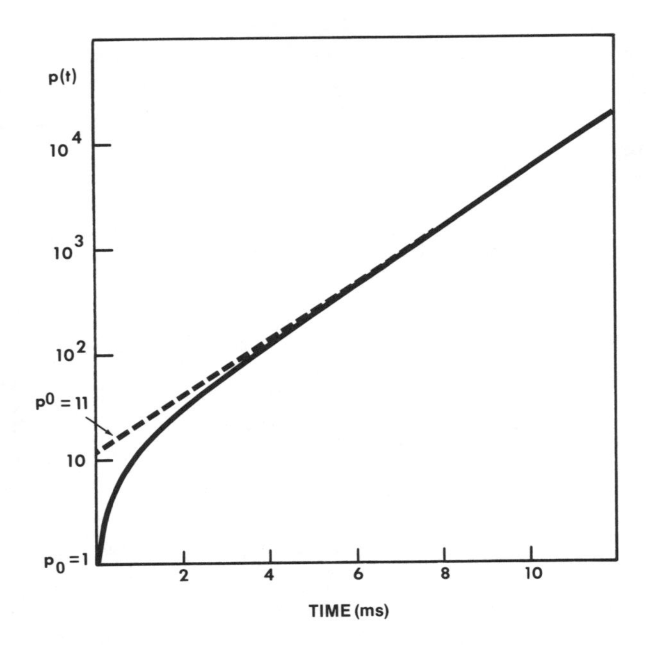

Fig. 8-5. A superprompt-critical transient following a reactivity step, $\rho_1 = 1.1\$$, calculated in $\bar{\lambda}$ kinetics (solid line) and in the PKA (dashed line).

of the transient during the action of feedback is virtually the same in the PKA as in the $\bar{\lambda}$ kinetics mode. The $\bar{\lambda}$ model itself agrees very well with the six-delay-group kinetics for the rapid superprompt-critical transients (six- and one-delay-group results are indistinguishable in Fig. 8-5).

A further discussion of the accuracy of the PKA is presented in Sec. 10-3 together with its application to dynamics problems.

In the CDS approximation, α_1 in Eq. (8.35) is zero. This reduction in accuracy of Eq. (8.35) obviously has no effect on the value of the superprompt pseudo-initial flux. The time required to establish the asymptotic transient is also practically the same since $\alpha_p \gg |\alpha_1|$.

The fact that the CDS approximation yields the same PKA as the $\bar{\lambda}$ kinetics (*and* the six-delay-group kinetics) shows that the time dependence of the precursor population is not very important as long as the reactor is superprompt critical. However, the change in the precursor population during the time the reactor is superprompt critical will be important after the reactivity is reduced below β (see Chapter 10).

The asymptotic flux rise in a superprompt-critical transient is characterized by the prompt period (as is the prompt jump for $\rho < \beta$). However, the rise of the flux to this asymptotic period exponential is faster than the prompt jump. It physically consists of a superprompt-critical source multiplication of the delayed neutron source. This is explored in detail and explained as a composite of the diverging fission chains in homework problems at the end of this chapter.

8-2B The Pseudo-Initial Flux for Transients Following a Ramp Reactivity Insertion

The initial condition for a kinetics model that is applicable in the superprompt-critical reactivity range should, in some way, be based on the flux at the time when $\rho(t)$ exceeds β, i.e., on $p_{pc} = p_{\text{prompt critical}}$. If the reactivity insertion is idealized as a step, the flux p_{pc} is still equal to p_0. For a gradual reactivity insertion, p_{pc} is larger than p_0; it may be many times larger than p_0.

Figure 8-6 shows a ramp reactivity together with the resulting flux transient; t_p denotes the time when the reactivity is just prompt critical [compare Eq. (8.40)]. In the fast reactor example of Fig. 8-6 where a = 10\$/s, then $p_{pc} = p(t_p) \simeq 35\, p_0$. The flux passes through t_p ($t_p = \beta/a$) with a substantial slope.

In the case of gradual reactivity insertion, the PKA is generally applied only for $t \geqslant t_p$. The time scale is thus started with $t = 0$ at t_p; compare Fig. 8-7. The "initial" prompt reactivity, ρ_p, is then zero, i.e., the prompt kinetics equation (8.32) is replaced by

$$\dot{p}(t) = \frac{\rho_p(t)}{\Lambda} p(t) \quad , \tag{8.38}$$

with

$$\rho_p(0) = 0 \quad . \tag{8.39}$$

In the case of a ramp reactivity insertion, the equation,

$$\dot{p} = (at - \beta)\frac{1}{\Lambda}p = (t - t_p)\frac{a}{\Lambda}p \quad , \tag{8.40}$$

with $t_p = \beta/a$, is replaced by:

$$\dot{p}(t) = \frac{at}{\Lambda}p(t) \quad , \tag{8.41}$$

where $t = 0$ corresponds to $t = t_p$ in Eq. (8.40).

The solution of Eq. (8.41) "starts" with a zero slope at the new "initial" time; see the dashed-dotted line in Fig. 8-7. It can therefore be expected that the pseudo-initial flux, p^0, in the case of a ramp reactivity

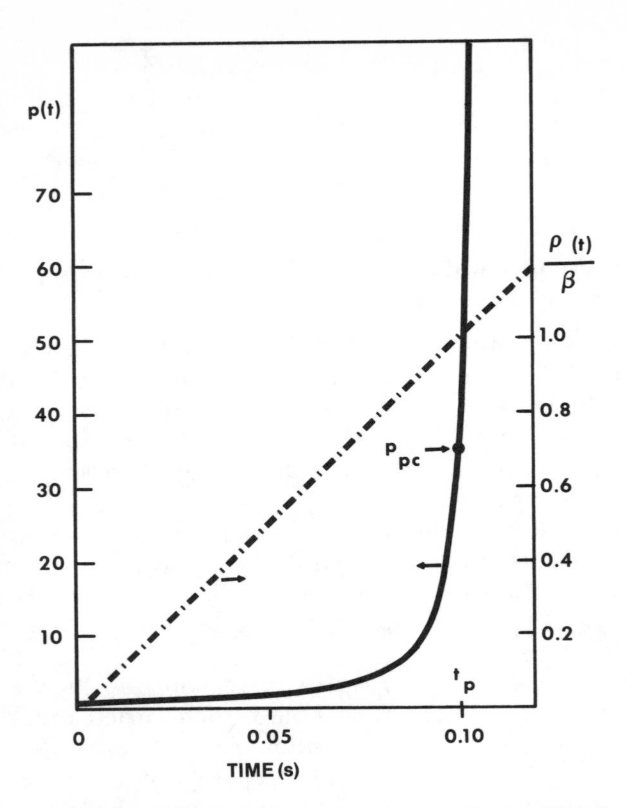

Fig. 8-6. A ramp reactivity insertion where a = 10\$/s and the resulting flux transient is calculated with six delayed neutron groups (solid line). The one-delay-group $\bar{\lambda}$ solution is within the solid line.

insertion, is larger than p_{pc} so that the transient predicted by the PJA can asymptotically agree with the correct transient.

The derivation of a formula for p^0 follows the same approach as in the case of a step-induced transient, i.e., a complete analytical solution in a suitable approximation is derived and matched asymptotically with the solution of the prompt kinetics approximation. It was shown earlier for the case of step-induced transients that the correct pseudo-initial flux is already obtained from the approximation with a constant delayed neutron source. The CDS approximation is thus used for the ramp-induced transients.[c]

---

[c]For a discussion of the accuracy of the CDS approximation in this context, see the last two paragraphs in this chapter.

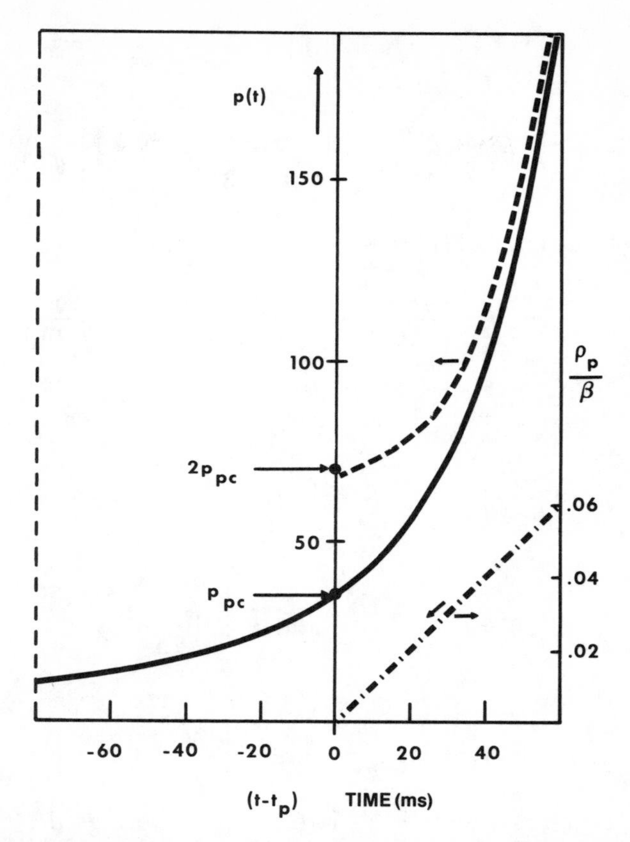

Fig. 8-7. A ramp reactivity insertion where a = 10\$/s and the resulting flux transient on a time scale beginning at t_p is calculated in $\overline{\lambda}$ kinetics (solid line) and in the PKA (dashed line).

The differential equation in the CDS approximation is given by

$$\dot{p} = (at - \beta)\frac{1}{\Lambda}p + \frac{\beta}{\Lambda}p_0 \quad . \tag{8.42}$$

The solution of Eq. (8.42) is to be matched asymptotically with the solution of Eq. (8.40), i.e., with

$$p(t) = p^0 \exp\left[\frac{a}{2\Lambda}(t - t_p)^2\right] \quad . \tag{8.43}$$

The solution of Eq. (8.42) with $p(0) = p_0$ is given by (see Appendix C):

$$p(t) = p_0 \exp\left[\left(\frac{at^2}{2\beta} - t\right)\frac{\beta}{\Lambda}\right]$$
$$\times \left(1 + \frac{\beta}{\Lambda}\int_0^t \exp\left\{-\left[\left(\frac{at'^2}{2\beta} - t'\right)\frac{\beta}{\Lambda}\right]\right\} dt'\right) \quad . \tag{8.44}$$

The integral in Eq. (8.44) leads to error functions,

$$\frac{\beta}{\Lambda}\int_0^t \exp\left\{-\left[\frac{at'^2}{2\Lambda} - \frac{\beta}{\Lambda}t' + \frac{\beta^2}{2\Lambda a}\right]\right\} dt' \, . \, \exp\frac{\beta^2}{2\Lambda a}$$
$$= \exp\frac{\beta^2}{2\Lambda a}\left(\frac{2}{a\Lambda}\right)^{1/2} \beta\int_0^{\theta_p} \exp\left[-(\theta' - \theta_p)^2\right] d\theta'$$
$$= \beta\left(\frac{\pi}{2a\Lambda}\right)^{1/2} \exp\frac{\beta^2}{2\Lambda a}[\operatorname{erf}(\theta - \theta_p) - \operatorname{erf}(-\theta_p)] \quad , \tag{8.45}$$

with

$$\theta' = \left(\frac{a}{2\Lambda}\right)^{1/2} t' \text{ and } \theta = \left(\frac{a}{2\Lambda}\right)^{1/2} t \tag{8.46a}$$

and

$$\theta_p = \theta(t = t_p) = \beta\left(\frac{1}{2\Lambda a}\right)^{1/2} \quad . \tag{8.46b}$$

Inserting Eqs. (8.45) and (8.46) into Eq. (8.44) brings the solution of Eq. (8.42) into the form:

$$p(t) = p_0 \exp(\theta - \theta_p)^2$$
$$\times \{\exp(-\theta_p^2) + \sqrt{\pi}\theta_p[\operatorname{erf}(\theta - \theta_p) + \operatorname{erf}(\theta_p)]\} \quad . \tag{8.47}$$

The flux at prompt critical is obtained from Eq. (8.47) by setting $\theta = \theta_p$:

$$p(t_p) = p_0[\exp(-\theta_p^2) + \sqrt{\pi}\theta_p \operatorname{erf}(\theta_p)] \quad , \tag{8.48}$$

with θ_p given by Eq. (8.46b).

With increasing argument, error functions very rapidly approach their limit value of 1. For $\theta_p \geqslant 2$, the error function in Eq. (8.48) is almost equal to 1. In all practically important transients, i.e., for ramp rates well below 100$/s, θ_p is actually >2. In addition, the exponential function in Eq. (8.48) is negligible compared to the second term if $\theta_p > 2$. For the flux at prompt critical, this simply gives:

$$p_{pc} = p(t_p) = p_0\theta_p\sqrt{\pi} = p_0\beta\left(\frac{\pi}{2\Lambda a}\right)^{1/2} \quad . \tag{8.49}$$

For the calculation of the asymptotic form, the second error function in brackets in Eq. (8.47) can also be replaced by 1. Furthermore, for θ such that $(\theta - \theta_p) \geqslant 2$, the first error function in Eq. (8.47) also assumes its asymptotic value. Thus, the bracket in Eq. (8.47) becomes equal to 2. The exponential function in the braces of Eq. (8.47) may be neglected as in Eq. (8.49). The asymptotic solution, based on the CDS approximation, then becomes

$$p_{as}(t) = 2p_{pc} \exp\left[\frac{a}{2\Lambda}(t - t_p)^2\right] \quad . \tag{8.50}$$

The comparison of this asymptotic solution, Eq. (8.50), and the PKA, Eq. (8.43), provides the desired pseudo-initial condition,

$$p^0 = 2p_{pc} \quad . \tag{8.51}$$

The dashed line presented in Fig. 8-7 represents the solution in the PKA.

The derivations of the pseudo-initial fluxes for step- and ramp-induced superprompt-critical transients are both based on the CDS model. This is fully justified for a step-induced transient for which the asymptotic transient that determines the pseudo-initial flux is established after such a short time that the CDS model is still valid. The same holds for fast ramps. For slower ramps, the delayed neutron source increases during the time interval $0 < t \leqslant t_p$, especially in fast reactors; not as much in thermal reactors. Consequently, the flux at prompt critical will be larger than given by Eq. (8.49). This effect can be largely accounted for by using the correct value of p_{pc} in Eq. (8.50) rather than the approximate value, Eq. (8.49).

The increasing delayed neutron source also has some effect on the factor 2 in front of p_{pc} in Eq. (8.50). However, the application of the prompt PKA in Sec. 10-4 shows that high accuracy is not required for p^0. Therefore, p^0 as given by Eq. (8.51) can also be applied for reactivity ramps slower than 10$/s in fast reactors and for much smaller ramp rates in thermal reactors.

Homework Problems

1. Consider an asymmetric saw-tooth reactivity insertion.
   a. Generalize the analytical solution, given in the text, so that $a' \neq a$.
   b. Tabulate the solution for a set of ramp-rate values on the declining part; use $a = 5\$/s$; $a' = a, 0.75a, 0.5a, 0.25a$, and zero. Specifically, estimate the ramp rate for which $\dot{p} = 0$ at the reactivity top. (Use $\lambda = 0.6/s$ as one delay group λ.)

2a. Use your point kinetics program to find numerically the six-delay-group solutions (^{235}U data of Table 2-III and $\gamma_k = 1.07$), with and without prompt jump approximation, for the reactivity insertions of problem 1.

b. Compare the results with those of problem 1 and discuss the agreement as well as the differences.

3a. Find the point kinetics solution for six delay groups, with and without prompt jump approximation, for the following parabolic reactivity insertions (use the same data as in problem 2):

$$\rho(t) = 0.9\beta - b\beta(t - t_m)^2 \quad \text{for } t \geqslant 0$$

with $bt_m^2 = 0.9$ and $t_m = 0.2s$, $2s$, and $20s$.

b. Plot the results.

c. Estimate the error in the PJA results of the integral

$$J = \int_0^{2t_m} [p(t) - 1] \, dt \quad .$$

4. Consider a ρ step ($\rho_1 > \beta$) in a critical reactor. Figure 8-5 shows a "super-fast" adjustment before $p(t)$ rises exponentially with the prompt period.
   a. Explain, by means of the analytic solution for a constant delayed neutron source, why this super-fast adjustment is due to the delayed neutron source.
   b. Show, by inspection of the differential equation, that the effect of s_{do} on $\frac{d}{dt} \ln p(t)$ becomes smaller with increasing $p(t)$.

5a. Find, by means of the prompt kinetics approximation, the analytical solution to the following reactivity insertion ($\rho_0 = 0$): $\rho_1 = 1.1\beta$:

$$\rho(t) = \rho_1 + at$$

up to t_2 given by $\rho_2 = 1.2\beta$ with $a = 20\$/s$; then

$$\rho(t) = \rho_2 - at \quad \text{for } t \geqslant t_2 \quad .$$

b. Extend your solution into the subprompt-critical domain (to $t_3 = 10t_2$).

c. Discuss the solution, especially the question of whether it can be physically meaningful for $\rho < \beta$. Refer to the corresponding discussion in the text for rising transients. The considerations of fission chains may be helpful in this discussion.

6. Find the PJA solution for a source-drop in a critical reactor for one delay group ($\lambda = 0.6$/s). How does the flux rise depend on the original flux level? Consider S_1 values that correspond to 0.1, 1.0, and 10% additions to the existing fission source. Assume S_1 to have the same space and energy distribution as S_f.

Review Questions

1a. Formulate the prompt jump approximation in words.

b. Name one other area in nuclear engineering where, in principle, a similar approach is applied.

2. How is the idea of the prompt jump approximation implemented in the kinetics equations?
3. Give and interpret the amplitude equation in the prompt jump approximation.
4. Derive the differential equation for $p(t)$ in PJA with one group of delayed neutrons (i.e., eliminate the precursor balance equation).
5. Integrate the one-delay-group kinetics equation in PJA and factorize the solution such that the solution with a constant delayed neutron source appears explicitly.
6. Present as a formula and a graph the PJA solution for the reactivity insertion ($a = 5\$/s$) below (a) with constant delayed neutron source and (b) with t-dependent delayed neutron source (graph only):

$$\rho t = \begin{cases} 0 & \text{for } t < 0 \\ a\beta & \text{for } 0 \leqslant t < t_1 = 0.8\beta/a \\ a\beta(2t_1 - t) & \text{for } t_1 \leqslant t \leqslant 3t_1 \end{cases} .$$

7. Give an example of a transient in which the flux *decreases* while the reactor is *supercritical* (approximate formula and sketch).
8. Give an example of a transient in which the flux *increases* while the reactor is *subcritical* (approximate formula and sketch).
9. Can the flux decrease while the reactor is superprompt-critical? Justify your answer in words.
10. What is prompt kinetics?

11. What is the basic error of prompt kinetics when p_0 is used as the initial flux?
12. How can the straightforward prompt kinetics be augmented to improve its accuracy?
13. Present the mathematical rationale for the definition of a pseudo-initial flux.
14. Find the pseudo-initial flux for a ρ_0-step-induced transient from the one-delay-group solution; also give a sketch.
15. Does this pseudo-initial flux change when you consider six groups of delayed neutrons, and why?
16. Give the pseudo-initial flux from ramp-induced transients (without proof).
17. Present a summarizing sketch for the basic types of transients following a reactivity step.

Nine

MEASUREMENT OF REACTIVITIES

9-1 Survey and Conceptual Problems

9-1A Survey

Reactivities and reactivity increments play an important role in reactor physics, safety, and the fuel cycle since they influence the design, control, and operational schedule of reactors. Most of the practically important reactivity increments can be subdivided into several groups with different applications.

One group of reactivity increments, which make up the basic reactivity balance, affects the fuel enrichment and the design: The reactivity of a "clean and cold" reactor is modified by the reactivity increments resulting from a number of effects such as the temperature increase from "cold" to the operational temperature, the maximum burnup and buildup of new fissile materials, the buildup of fission products during a loading cycle, the shim control required for reactor operation over the full fuel reloading cycle, and the operational control. The reactivity increments in this group generally appear in "complete" combinations, adding up to the reactivity of the critical reactor, that is, to $\rho = 0$. "Incomplete" combinations of $\delta\rho$'s, which do not yield a critical reactor, have merely the character of interim and auxiliary results. The reactor system formally described by such an incomplete combination is not meant to be an operational system.

A second group of reactivity increments is important for the investigation of reactor safety. These reactivities are applied in accident analysis as well as in the design of the safety system. Examples are $\delta\rho$ of individual and combinations of control and shutdown rods, and $\delta\rho$ due to the Doppler effect; expansion of fuel, moderator, control rods, or grid plates; removal of coolant; and the motion of the cladding or fuel.

Such reactivity increments may occur individually or in off-critical combinations; they can represent the driving force and the feedback of reactor transients.

A third group of reactivity increments, which partially overlaps with the first group, appears in fuel cycle problems and in reactor operation. This group includes the time-dependent reactivity increments of the burnup and buildup of fuel isotopes and fission products, the required control reactivities, and the reactivity changes due to reloading operations.

A fourth group includes all kinds of subcritical reactor states. The knowledge of the degree of subcriticality is important during fuel loading or reloading in power reactors, where a sufficient margin of subcriticality must be assured. Subcritical states also play an important role in many experimental investigations of reactor parameters or general reactor physics problems.

The basic methods for measuring reactivities are discussed in this chapter. Major emphasis is on the correct understanding of the principles of the analysis, particularly on the correspondence between experimental results and the consistent theoretical definitions. The discussion of analysis techniques for the various types of experiments is often restricted to their simplest forms. The simple analysis yields only approximate results. More refined analysis techniques for all methods of reactivity measurements have been developed. Their discussion and evaluation is a large field in itself. Extensive presentation is beyond the scope of this book. Most of the analysis techniques are not fully consistent with the well-defined reactivity concepts as they are reviewed in Sec. 9-1B. The reader is referred to the literature for detailed information on analysis methods.[1–7] For the advanced reader, the consistency between measured quantities and the desired reactivity concepts is discussed in Secs. 9-3H and 9-4C.

The discussion of the simple analysis of the reactivity measurements is based on the point reactor model of Sec. 5-2. The reactivity in the point reactor model is found by first-order perturbation theory if the reactor is initially critical [Eq. 5.71)]. For initially subcritical reactors, the reactivity change after the onset of a transient is calculated from Eq. (5.76).

The methods for measuring reactivities can be subdivided into three categories:

1. static measurements
2. dynamic measurements
3. measurements utilizing statistical flux fluctuations (noise).

Only the first two categories are discussed here. The latter group of methods is beyond the scope of this book. The reader is referred to the literature, e.g., Ref. 2, where further information may be found.

The common denominator of all practical reactivity measurements is that they extract information about the reactivity from various kinds of flux transients. In the static methods, only the stationary states prior to and after a transient are exploited. In dynamic methods, the actual transient is analyzed. In statistical methods, time-independent reactor configurations at very low power are normally investigated. Although the *average* power is constant, the individual fission chains have time-dependent neutron populations. The die-away time of fission chains depends on the prompt reactivity of the reactor (see Sec. 7-1). The investigation of individual fission chains provides information on the prompt periods in normally stationary reactors, i.e., Λ/β in critical and $\Lambda/(\beta - \rho_0)$ in subcritical reactors (see Chapter 7).

The analysis of the various reactivity measurements is first discussed in very simple terms in order to illustrate the approaches. This simplified analysis neglects features such as the location of neutron sources and the deformation of the neutron flux, caused by the insertion of reactivities. In addition, the exact theoretical definition of the reactivities, especially the *conventional nature* of these definitions, is disregarded in the simplified analysis. More sophisticated or theoretically consistent analysis approaches are subsequently discussed for some of the measurement methods; these sections (9-1C, 9-3H, as well as parts of 9-2B) may be omitted by less advanced students or readers.

The analysis presented in this chapter is uniformly based on the exact point kinetics equations, which are the theoretical basis for the dynamic reactivity. These equations include the static reactivity in the initial condition for subcritical reactors (see Secs. 5-1B and 5-1C). The use of the exact point kinetics equations as a basis for the consistent interpretation and analysis of *dynamic* reactivity measurements appears to be mandatory. *Static* reactivity measurements could also be analyzed on the basis of the static diffusion equations as such, without converting it to initial or asymptotic states of the point kinetics equations. However, the use of the exact point kinetics equations for the analysis of static reactivity measurements also makes the entire discussion of reactivity measurements more uniform and transparent, without changing any of the essential results.

The problematic nature of reactivity measurements is addressed in the next two sections. Two categories of questions are discussed. The first category of questions centers around dynamic versus static reactivity: Are both types of reactivities needed? If yes, which one is obtained from a given measurement or analysis? Do static and dynamic reactivity measurements yield the static and dynamic reactivities, respectively? If not, how can a measurement or an analysis be devised to yield the desired type of reactivity? The second category of questions centers on the effect of the conventional nature of the reactivity definitions on the exact measurability of reactivities. The two categories of questions are discussed in the next two sections and later in this chapter, especially in Sec. 9-3H.

9-1B Dynamic Versus Static Reactivities

The two basic concepts applied to describe the various reactivity effects are the "static" and the "dynamic" reactivities (see Secs. 6-1A and 5-1B through 5-1D). To provide a concise basis for the discussion of the conceptual problems of reactivity measurements, the features of the static and dynamic reactivity concepts that are pertinent to the subsequent discussion are briefly reviewed.

Static Reactivity Formulas

The definition of the static reactivity, ρ^{st}, is based on the λ mode eigenvalue problem

$$(\mathbf{M} - \lambda \mathbf{F})\Phi_\lambda = 0 \quad , \tag{9.1a}$$

with

$$\rho^{st} = 1 - \lambda \quad . \tag{9.1b}$$

The description of static off-criticality from Eqs. (9.1) is the traditional convention. There is, however, no physical necessity to describe off-criticality from Eqs. (9.1). Other conventions would yield different values for ρ^{st} but would also represent a self-consistent scheme for the description of off-criticality.

The convention, Eqs. (9.1), yields a uniquely defined value of ρ^{st} for each individual state, described by the operators **M** and **F**.

Specific physical changes in the reactor system lead to corresponding changes in the static reactivity given by the static reactivity increment, $\delta\rho^{st}$. This increment naturally depends on both states embracing the reactivity interval described by two sets of operators, **M** and **F**. This distinguishes qualitative reactivity increments from reactivities.

There are several mathematically equivalent formulas that may be used to define or find ρ^{st} and $\delta\rho^{st}$. The exact reactivity increment between states "n" and "$n - 1$" may be obtained as:

$$\delta\rho_n^{st} = \lambda_{n-1} - \lambda_n \quad . \tag{9.2a}$$

This equation becomes the same as Eq. (9.1b) if $\lambda_{n-1} = 1$; the reactivity appears then as the reactivity increment about an arbitrary critical state.

Alternatively, $\delta\rho^{st}$ can be calculated from the so-called "exact perturbation formula" of Sec. 4-3:

$$\delta\rho_n^{st} = \frac{(\Phi^*_{\lambda n-1},[\lambda_{n-1}\mathbf{F}_n - \mathbf{M}_n]\Phi_{\lambda n})}{(\Phi^*_{\lambda n-1}, \mathbf{F}_n\Phi_{\lambda n})} \tag{9.2b}$$

or

$$\delta\rho_n^{st} = \frac{(\Phi^*_{\lambda n-1},[\lambda_{n-1}\delta\mathbf{F} - \delta\mathbf{M}]\Phi_{\lambda n})}{(\Phi^*_{\lambda n-1},\mathbf{F}_n\Phi_{\lambda n})} , \tag{9.2c}$$

where

$$\begin{aligned} \delta\mathbf{F} &= \mathbf{F}_n - \mathbf{F}_{n-1} \\ \delta\mathbf{M} &= \mathbf{M}_n - \mathbf{M}_{n-1} \quad . \end{aligned} \tag{9.2d}$$

In Eq. (9.2b), the dependency on the state "$n-1$" is reflected in $\Phi^*_{\lambda n-1}$ and in λ_{n-1}.

An exact formula for the static reactivity as such can also be devised in analogy to Eq. (9.2b) by setting $\lambda_{n-1} = 1$ and allowing $\Phi^*_{\lambda n-1}$ to become an arbitrary weighting function. The resulting reactivity is determined by the properties of a single state as expected:

$$\rho^{st} = \frac{(\Phi^w,[\mathbf{F} - \mathbf{M}]\Phi_\lambda)}{(\Phi^w,\mathbf{F}\Phi_\lambda)} \quad . \tag{9.3}$$

Dynamic Reactivity Formulas

The definition of the dynamic reactivity is also based on a convention; ρ^{dyn} is conventionally defined by a formula of the same form as Eq. (9.3) with Φ replaced by the actual flux shape function, Ψ at time t, and the weighting function chosen to be the initial adjoint flux, $\Phi^*_{\lambda 0}$, since ρ^{dyn} is related to a given—not an arbitrary—initial state:

$$\rho^{dyn} = \frac{(\Phi^*_{\lambda 0}, [\mathbf{F} - \mathbf{M}]\Psi)}{(\Phi^*_{\lambda 0},\mathbf{F}\Psi)} \quad . \tag{9.4}$$

The dynamic reactivity for an initial stationary state is equal to the corresponding static reactivity: $\rho^{dyn}(0) = \rho_0^{st}$. The appearance of the actual flux shape in Eq. (9.4) reflects the fact that $\rho^{dyn}(t)$ drives the physical transient.

The application of a convention different from Eq. (9.4) yields a different value for the dynamic reactivity.[8,9] This arbitrariness cancels, however, if the kinetics equations are derived and applied consistently, and thus has no effect on the final physical results, just as the conventional nature of the static reactivity definition also has no effect. The dynamic reactivity yields the correct flux transient, and the static reactivity increments add up to the correct critical state. The conventional nature of these reactivity concepts will, however, affect their measurability and the analysis of measurements.

The question may be raised whether *both* reactivity concepts are really required or whether a simpler theory based on only one of these reactivity concepts could be developed. The answer is that both reactivity concepts ρ^{st} and ρ^{dyn} *are* practically required. Neither one can be replaced by the other concept since they serve different purposes. The static reactivity describes the off-criticality of systems in steady, auxiliary states. Actual physical transitions between these auxiliary states are not contemplated because their consideration is conceptually unnecessary. In addition, most of these transitions are practically impossible. In contrast, the dynamic reactivity is introduced to describe physical transitions between actual states of the reactor. The description of these actual transitions requires consideration of additional physical phenomena such as delayed neutrons and feedback effects, which are legitimately disregarded in static reactivity problems.

A time-dependent static reactivity may also be calculated for all physical states appearing in the transient: $\rho^{st}(t)$. The ratio of $\rho^{st}(t)$ and $\rho^{dyn}(t)$ is then available as a computed quantity. Setting ρ^{dyn} equal to ρ^{st} times the computed ratio can sometimes be advantageous; for example, see Eq. (9.20b) and Sec. 9-4C. It may also occasionally be used as an approximation (see Chapter 11). Typical differences in realistic cases can be up to 10 or even 20% of the reactivity (see Sec. 11-4C). In unrealistic cases such as for a state with $k_{eff} = 1.20$ ($\rho^{st} = 0.167$) as it may appear in a reactivity balance of thermal reactors, the deviation could be even larger. But one would not want to calculate the corresponding dynamic reactivity as there is no meaningful time-dependent flux.

9-1C The Exact Measurability of Reactivities

The conventional nature of the definitions of the static and dynamic reactivities poses a principal problem for physically measuring these quantities. The logical solution to this problem is to devise a measurement and an analysis in which the conventional prescription for the *calculation* of the reactivity is exactly *simulated.*

If in the analysis of an experiment the specific conventional prescription of the reactivity definition is not applied, a reactivity quantity is obtained that, in general, corresponds to a different, normally unknown convention. The application of such experimental reactivities and their comparison with theoretical quantities that are defined differently can lead to certain inconsistencies (although the deviation may be small enough to be tolerated). Differences revealed in such comparisons can generally be classed as systematic errors or they should be considered as differences between differently defined quantities.

The definition of ρ^{st}, Eqs. (9.1), can be written in the form

$$(\mathbf{M} - \mathbf{F})\Phi_\lambda = -\rho^{st}\mathbf{F}\Phi_\lambda \quad . \tag{9.5}$$

The right side, containing the factor ρ^{st}, is "conventionally" added in order to obtain a nontrivial solution. The exact *simulation* of this conventional prescription to find ρ^{st} would require the application or withdrawal of a certain amount of the fission neutron source with the same space and energy distribution as $\mathbf{F}\Phi_\lambda$:

$$\delta S(\mathbf{r},E) = -\rho^{st}\mathbf{F}\Phi_\lambda \quad . \tag{9.6}$$

The magnitude of δS and of the calculated $\mathbf{F}\Phi_\lambda$ would determine ρ^{st}. This experiment, in principle, has been discussed in Ref. 3. The source material considered was a spontaneously fissioning material. The measurement of ρ^{st}, however, by simulating the conventional prescription has several serious shortcomings.

If the reactor is subcritical ($\rho^{st} < 0$), the source to be added on the right side of Eq. (9.5) is positive. It may, at least in a "gedanken" experiment (thought experiment), be made up of properly distributed, spontaneously fissioning material. Since the spatial distribution of this independent source is determined by $\mathbf{F}\Phi_\lambda$, the solution of the eigenvalue problem, Eq. (9.5), must be found first. The solution of the theoretical problem, which consists of finding ρ^{st} and Φ_λ simultaneously, must thus be used in setting up the experiment. Other minor shortcomings of this gedanken experiment are that spontaneously fissioning materials also have finite absorption and scattering cross sections, which modify the left side of Eq. (9.5) and thus alter the experiment. If the reactor is supercritical, however, the measurement of static reactivities by adding an independent source does not even work in principle. The source of Eq. (9.6) would have to be negative.

This discussion shows that even in a gedanken experiment it is not possible to measure the static reactivity exactly. This is also true for the λ mode flux, Φ_λ, and the adjoint function, Φ_λ^*, since both functions are only defined simultaneously with ρ^{st}.

The situation with respect to the principle measurability of the dynamic reactivity is different than for the static reactivity, since ρ^{dyn} acts physically by driving flux transients. It could be expected that the proper analysis of transients would yield the dynamic reactivity. However, the arbitrarily introduced denominator $F(t)$ cannot be obtained from transient analysis since $F(t)$ has no effect on the results. Consequently, only ratios of integral kinetics parameters such as ρ/β or ρ/Λ may be inferred from the analysis of transients. The denominator $F(t)$ may be determined separately by a combination of experimental and theoretical information (see Sec. 9-2A). The correct transient analysis for the determination of a dynamic reactivity requires the explicit application of the convention, i.e, the use of the initial adjoint flux as a weighting function in Eq. (9.4).

In summary, dynamic reactivities, divided by one of the other integral kinetics parameters, can be determined from transient analysis. To find ρ^{dyn} consistent with its definition requires the proper use of Φ_0^* or $\Phi_{\lambda 0}^*$ in the analysis.

Since different weighting functions in Eq. (9.4) yield different definitions of dynamic reactivities, it is necessary to single out the specific ρ^{dyn} desired by an appropriate use of the weighting functions Φ_0^* or $\Phi_{\lambda 0}^*$. The use of these theoretically determined weighting functions in the analysis of measurements of ρ^{dyn} does not have the same drawback as the requirement for finding ρ^{st} and Φ_λ from Eq. (9.5). The reason is that the calculation of Φ_0^* and $\Phi_{\lambda 0}^*$ does not yield $\rho^{dyn}(t)$. Thus, the knowledge of the theoretical results, i.e., $\rho^{dyn}(t)$, is not required for performing the experiment or its analysis as in the case of ρ^{st} (compare the discussion in Sec. 9-2A).

9-2 Static Measurements of Reactivities

The two *static* methods most often applied for measuring reactivities infer a reactivity value from the investigation of steady-state neutron fluxes. Most measurements consider two reactor states: one state for which the reactivity is to be determined and another state that provides "calibration reactivity." Transients between the corresponding neutron flux states are not explicitly considered. These two methods are:

1. the source multiplication method
2. the null reactivity method (counter-rod method).

A third method exploits the reactivity dependence of flux shape *deformations*[10]; the shape deformations are introduced by an asymmetrically positioned neutron source:

3. asymmetric source technique.

Only the analyses of the first two methods are discussed in the following. For the analysis of the asymmetric source technique see Refs. 10 and 11.

9-2A Source Multiplication Method for a Single Reactor State

The source multiplication method for a single reactor state has its practical application not actually for the measurement of the reactivity in a given state, but rather for the determination of a large-scale variation of the reactivity, for example, during the first loading of a reactor with fuel. Flux amplitude, source, and reactivity in a subcritical reactor are related by source multiplication formulas, e.g., Eq. (6.8c), which is often written in the following form[a]:

$$p_0 = \frac{s_0}{-\rho_0} = M_s s_0 \quad . \tag{9.7}$$

Here, M_s is called the "source multiplication factor." The source multiplication method is commonly employed in bringing a new reactor system to criticality. If, for example, core loading starts at the center of a planned configuration, the multiplication constant and thus the source multiplication factor M_s increase with increasing radius. When the reactor reaches the critical radius for a cold clean core, the source multiplication factor becomes infinite. This "approach to criticality" is monitored by placing a neutron source in the reactor and measuring the level of the neutron flux for various core sizes (e.g., radii). The counting rate is proportional to M_s. A plot of the inverse counting rate or $1/M_s$, in arbitrary units, as a function of the radius of the loaded configuration becomes equal to zero at the critical radius. Figure 9-1 shows a qualitative $1/M_s$ curve versus relative loading radius. The $1/M_s$ trace is generally curved before reaching the value of zero. The curvature is important for extrapolations to the critical radius based on part of the $1/M_s$ curve.

The counting rate is proportional to the source strength for a given source location. The counting rate also depends on the source location: source neutrons participate more effectively in neutron multiplication when the source is located at the core center rather than at the periphery. However, both ambiguities, source strength and location, have no effect on the $1/M_s$ analysis since only relative values are needed to determine the intersection with zero.

The principal possibility of exploiting the source multiplication formula for a *single* steady state in a more quantitative manner is discussed here as an introduction to the source multiplication methods that are usually applied, which employ a pair of states.

The reactivity ρ_0, which appears in Eq. (9.7), is introduced in the

---

[a]See Sec. 6-1A for other types of source multiplication formulas.

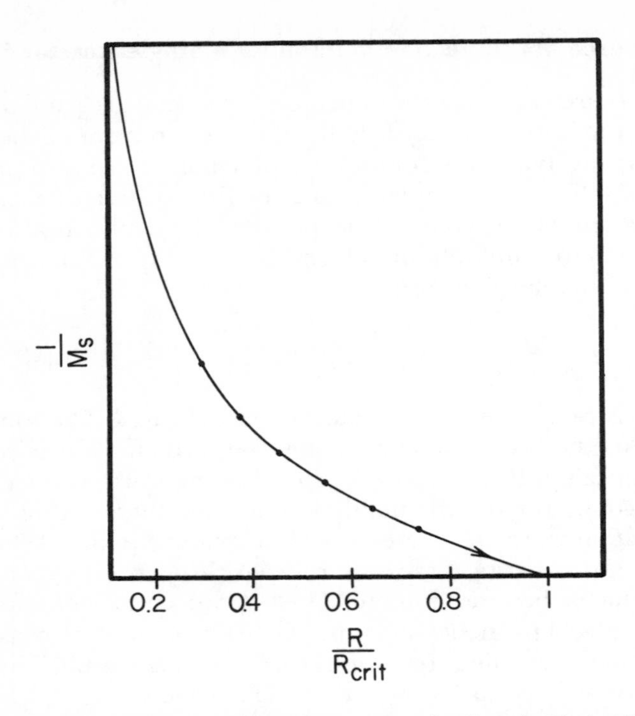

Fig. 9-1. Reciprocal source multiplication factor versus fraction of critical radius loaded.

derivation of this equation as static reactivity:

$$\rho_0 = 1 - \lambda_0 \quad . \tag{9.8}$$

The appearance of $\rho_0 = \rho_0^{st}$ in Eq. (9.7) seems to suggest the possibility for a direct measurement of the static reactivity.

The flux amplitude in the initial state, p_0, can be set equal to one. This leaves s_0, the reduced source, to be determined. According to its definition in Eq. (5.50), s_0, within the framework of the point kinetics equations, is given by:

$$s_0 = \frac{(\Phi_{\lambda 0}^*, S_0)}{(\Phi_{\lambda 0}^*, \mathbf{F}_0 \Phi_0)} \quad . \tag{9.9}$$

The adjoint flux weighting of the source takes into account quantitatively the effect of the source location on source multiplication; the adjoint flux is large at the core center and small at the periphery of the core.

The independent source, S_0, and the fission neutron source $S_{f0} =$

$\mathbf{F}_0\Phi_0$ may in principle be determined experimentally. However, $\phi^*_{\lambda 0}(\mathbf{r},E)$ must be calculated as the solution of the adjoint λ eigenvalue problem. The eigenvalue λ_0 (and thus ρ_0) is simultaneously found with $\Phi^*_{\lambda 0}$. The experimental reactivity is then given by the ratio of the adjoint weighted independent source and the induced fission neutron source:

$$-\rho_0^{exp} = \frac{(\Phi^*_{\lambda 0}, S_0^{exp})}{(\Phi^*_{\lambda 0}, S_{f0}^{exp})} \quad . \tag{9.10}$$

The actual performance of such an experiment would be complicated since it would require a complete mapping of the neutron sources. It may, however, be viewed as a gedanken experiment similar to the one discussed in Sec. 9-1C. Both experiments require the solution of the theoretical problem prior to the analysis. It is not merely a semantic question as to whether such experiments can be called "measurements" of the static reactivity. They are certainly not measurements in the sense of theory-independent determinations of ρ^{st}, since a calculated value of ρ^{st} is required to find the experimental one. The fact that the quantity to be measured has to be known in advance seems to disqualify these "measurements" as "independent determinations."

This raises the question, however, whether Eq. (9.10) may in principle yield an improvement compared to a somewhat inaccurately calculated theoretical value. To investigate this question, a problem that sensitively depends on errors in the reactivity is considered, i.e., the source multiplication in a slightly subcritical reactor. Suppose a stationary reactor contains an independent source that is *much* smaller than the stationary fission neutron source. The reactor is then definitely subcritical but only by a small reactivity. Suppose further that the calculation yields $\Phi^{*\prime}_{\lambda 0}$ and ρ_0', quantities that are somewhat inaccurate, e.g., due to inaccuracies in the group constant set. Since the reactor is only slightly subcritical, a small error in the group constants suffices to change the sign of the reactivity. The corresponding error in the adjoint function is small since $\Phi^*_{\lambda 0}$ does not sensitively depend on the sign of ρ. Even if $\Phi^{*\prime}_{\lambda 0}$ were grossly inaccurate, its use on the right side of Eq. (9.10) gives a reactivity that at least has the correct sign. However, since $\Phi^{*\prime}_{\lambda 0}$ is not grossly inaccurate, its application on the right side of Eq. (9.10) may in principle yield a strong improvement over an inaccurately calculated value, ρ_0'. Thus, the need for prior calculation of $\Phi^*_{\lambda 0}$ and thus ρ_0 does not disqualify this experiment as a measurement of the reactivity since it appears that only good estimates of these quantities are required. In addition, if only a reactivity estimate is desired, an approximate mapping of both neutron sources partially obtained from calculations may suffice.

9-2B Source Multiplication Method Involving Two Reactor States

In actual source multiplication methods, two subcritical states are employed. Either the reactivity in one of the two states or the reactivity increment between the two states is assumed to be known from some other type of measurement such as a dynamic reactivity measurement. This known reactivity is used to "calibrate" the experimental method for this reactor or at least for a certain range of reactivity. An unsophisticated analysis, which is characterized by the disregard of consistent weighting functions and definitions, is discussed first. The formulas used in this simple analysis should be read with this reservation in mind.

In the discussion of the sophisticated analysis, presented subsequent to the simple analysis, two different approaches can be distinguished. The first approach is based on considering the two states linked by a transient, although the transient is not explicitly used. In the second approach, the two states are considered as independent stationary states. Both approaches are discussed in the following.

The pair of source multiplication formulas that corresponds to two states of different reactivities is given by

$$p_0 = \frac{s_0}{-\rho_0} \tag{9.11a}$$

and

$$p_1 = \frac{s_1}{-(\rho_0 + \delta\rho_1)} = \frac{s_1}{-\rho_1} \quad . \tag{9.11b}$$

Here, ρ_0 is the static reactivity of the initial state and ρ_1 is either a dynamic or static reactivity, depending on whether the two states are linked by a transient or not. This difference is mathematically reflected in the weighting function in the reduced source, s_1, as discussed below.

The two source multiplication formulas, Eqs. (9.11), are applied and solved in either one of the following two ways:

1. If the reactivity (ρ_0/β_0) in an initial system is determined by some other method, the values of the reactivity increments $\delta\rho_1$ are determined in terms of ρ_0. Instead of $\delta\rho_1$, one often finds $\rho_0 + \delta\rho_1 = \rho_1$.
2. If a previously determined control rod reactivity $\delta\rho_1$ is inserted into a subcritical reactor with unknown initial reactivities ρ_0, the values of ρ_0 are determined in terms of $\delta\rho_1$.

In the experiment, the flux level is measured with a neutron counter, such as a fission counter. Let CR_0 and CR_1 be the counting rates corresponding to ρ_0 and $\rho_0 + \delta\rho_1$, respectively. In the simplified analysis, counting rates are set directly proportional to the flux level: $p_0 \propto CR_0$

and $p_1 \propto CR_1$; furthermore, s_1 is assumed to be equal to s_0 in this analysis. Solving Eqs. (9.11) for the respective unknown reactivity and dividing by some value of β yields:

$$\frac{\delta\rho_1}{\beta} \simeq \frac{\rho_0}{\beta}\left(\frac{CR_0}{CR_1} - 1\right) , \tag{9.12a}$$

or

$$\frac{\rho_1}{\beta} \simeq \frac{\rho_0}{\beta}\frac{CR_0}{CR_1} \tag{9.12b}$$

and

$$\frac{\rho_0}{\beta} \simeq \frac{\delta\rho_1/\beta}{CR_0/CR_1 - 1} . \tag{9.12c}$$

The reactivities on the right side of Eqs. (9.12) have been calibrated in advance. The "approximately equal to" signs in Eqs. (9.12) are used to indicate that Eqs. (9.12) are based on several simplifications. Dynamic and static reactivities cannot be distinguished at this simple level of the analysis.

The approximations made to obtain the simple formulas of Eqs. (9.12) are now discussed with the states "0" and "1" considered as the initial and final states in a subcritical *transient* (subsequently an analysis for two unrelated states is briefly described). There are three improvements to be considered.

1. The cancellation of the source ratios, s_0/s_1, is an approximation for several reasons. The initial source s_0 is given by Eq. (9.9); after a transient, it is converted into s_1, as given by Eq. (5.50), with $\Phi^*_{\lambda 0}$ from the initial state and Ψ_1 being the flux shape in the final state:

$$s_1 = \frac{(\Phi^*_{\lambda 0}, S_1)}{(\Phi^*_{\lambda 0}, \mathbf{F}_1\Psi_1)} . \tag{9.13}$$

The denominators of Eqs. (9.9) and (9.13) are generally different, since the flux shape Ψ_1 is different from the original shape for almost any reactivity insertion. In addition, $\mathbf{F}_1$ also differs from $\mathbf{F}_0$ if the reactivity change involves movement of fuel. However, if $S_1 = S_0$, there is no change in the numerators of these two equations. In many cases, though, the source changes indirectly; for example, if fuel reloading involves changes of the fuel concentration, the spontaneous fission source and thus the numerators are changed.

2. The terms ρ_0/β_0 and $\delta\rho_1/\beta_1$ are well defined in the framework of the exact point kinetics equations. By introducing a calibration reac-

tivity, ρ_0/β_0 or $\delta\rho_1/\beta_1$, from an experiment that is not consistent with the exact point kinetics equations that describe the two states under consideration, a result is obtained from Eqs. (9.12) that does not appear to have a precisely defined theoretical analog.

3. The correspondence of flux amplitudes and counting rates is not a strict proportionality if the reactivity insertion causes a flux shape deformation. Flux shape corrections are needed as discussed below.

The above approximations can be removed primarily by a sophisticated analysis of the source multiplication method. The improvement requires a combination of experimental and theoretical information (see, for example, Refs. 6 and 12). The proportionality of flux amplitude and counting rate is replaced by the individual relations,

$$p_0 = c_0 \cdot CR_0$$

and

$$p_1 = c_1 \cdot CR_1 \quad . \tag{9.14}$$

where the c's are given in Eq. (9.43). Different β's are introduced for both states, $\beta_0 \neq \beta_1$, and s_1 is distinguished from s_0. This gives, instead of the simple Eqs. (9.12), the following relations for the "measured" reactivity (on the left side) as a function of a calibration reactivity (on the right side):

$$\frac{\delta\rho_1}{\beta_1} = \frac{\rho_0}{\beta_0} \cdot \frac{\beta_0}{\beta_1}\left(\frac{c_0 CR_0}{c_1 CR_1} \cdot \frac{s_1}{s_0} - 1\right) \tag{9.15a}$$

or

$$\frac{\rho_1}{\beta_1} = \frac{\rho_0}{\beta_0} \cdot \frac{\beta_0}{\beta_1} \cdot \frac{c_0 CR_0}{c_1 CR_1} \cdot \frac{s_1}{s_0} \tag{9.15b}$$

and

$$\frac{\rho_0}{\beta_0} = \frac{\delta\rho_1/\beta_1}{\dfrac{c_0 CR_0}{c_1 CR_1} \cdot \dfrac{s_1}{s_0} - 1} \cdot \frac{\beta_1}{\beta_0} \quad . \tag{9.15c}$$

The ratios c_0/c_1, s_1/s_0, and β_1/β_0 are to be estimated theoretically. In practical reactivity measurements, these theoretical correction factors are evaluated with various degrees of sophistication and accuracy.[6,12] Theoretically consistent formulas for these correction factors are discussed in Sec. 9-3H.

The obvious deficiencies of Eqs. (9.12) have been corrected in Eqs. (9.15). But there are still two inconsistencies: first, the corrections use calibration reactivities, which are normally not fully consistent with the theoretical, convention-based definition of ρ; second, the estimation of s_1/s_0 requires the spatial source distributions that are often not well known. Both drawbacks have been largely eliminated in an "inverse spatial kinetics" analysis described in Sec. 9-4C. If the analysis of Eqs. (9.15) were fully consistent with the theory, Eqs. (9.15a) and (9.15b) would yield dynamic reactivities, and Eq. (9.15c), static reactivities.

Instead of two source multiplication formulas resulting from the *same* point kinetics equation, a pair of source multiplication formulas, which represent two independent states, are considered in the following alternative analysis. The adjoint λ mode in state "1" then needs to be applied as a weighting function in the definition of the reduced source:

$$s_1^{(1)} = \frac{(\Phi_{\lambda 1}^*, S_1)}{(\Phi_{\lambda 1}^*, \mathbf{F}_1 \Psi_1^{(1)})} \quad . \tag{9.16}$$

Superscript ones are applied to quantities that are different from the corresponding quantities in Eq. (9.13) due to state "1" being treated as a new initial state: $\Phi_{\lambda 1}^*$ is different from $\Phi_{\lambda 0}^*$ due to the difference in the two systems. But $\Psi_1^{(1)}$ and Ψ_1 are different, although they belong to the same configuration, with $\Phi_1(\mathbf{r},E)$ as the flux. They differ in their normalization:

$$\Phi_1 = p_1 \Psi_1 \text{ versus } \Phi_1 = p_1^{(1)} \Psi_1^{(1)} \quad . \tag{9.17}$$

The second "initial" amplitude $p_1^{(1)}$ may also be set equal to unity, but p_1 is the asymptotic amplitude in the transient-based analysis, i.e., $p_1 \neq 1$. The reactivity in state "1" is also defined differently in the two cases:

$$\rho_1^{(1)} = 1 - \lambda_1 = \rho_0 + \delta\rho_1^{st} \tag{9.18a}$$

is defined as the static reactivity, whereas the dynamic reactivity

$$\rho_1 = \rho_0 + \delta\rho_1 \neq \rho_1^{(1)} \tag{9.18b}$$

has a somewhat different value.

If the two initial fluxes p_0 and $p_1^{(1)}$ are normalized to unity, the ratio of the two "initial" source multiplication formulas yields:

$$\frac{\rho_0}{\rho_1^{(1)}} = \frac{s_0}{s_1^{(1)}} = \frac{(\Phi_{\lambda 1}^*, S_{f1})}{(\Phi_{\lambda 1}^*, S_1)} \frac{(\Phi_{\lambda 0}^*, S_0)}{(\Phi_{\lambda 0}^*, S_{f0})} \quad , \tag{9.19a}$$

where the source ratio may be rewritten by expressing $\Psi_1^{(1)}$ as $p_1\Psi_1$, using Eqs. (9.16) and (9.17), and also reintroducing $p_0 = 1$. This gives

$$\frac{s_0}{s_1^{(1)}} = \frac{s_0}{\tilde{s}_1^{(1)}} \frac{p_1}{p_0} = \frac{\rho_0}{\rho^{(1)}} \quad , \tag{9.19b}$$

with

$$\tilde{s}_1^{(1)} = \frac{(\Phi_{\lambda 1}^*, S_1)}{(\Phi_{\lambda 1}^*, \mathbf{F}_1 \Psi_1)} \quad . \tag{9.19c}$$

Introducing counting rates as in Eq. (9.14) and dividing by $\beta_1^{(1)}$ yields—instead of Eq. (9.15b)—the somewhat different reactivity $\rho^{(1)}$:

$$\frac{\rho^{(1)}}{\beta_1^{(1)}} = \frac{\rho_0}{\beta_0} \cdot \frac{\beta_0}{\beta_1^{(1)}} \cdot \frac{c_0 CR_0}{c_1 CR_1} \cdot \frac{\tilde{s}_1^{(1)}}{s_0} \quad . \tag{9.20a}$$

Note that $\beta_1^{(1)}$ is different from β_1 since it is formed with $\Phi_{\lambda 1}^*$ rather than with $\Phi_{\lambda 0}^*$ as the weighting function.

The difference in Eqs. (9.20a) and (9.15b) is in the adjoint weighting function of the source in state 1, which is $\Phi_{\lambda 0}^*$ in Eq. (9.15b) and $\Phi_{\lambda 1}^*$ in Eq. (9.20a). The analysis represented by Eq. (9.20a) yields the static reactivity $\rho^{(1)}$; an approximate value of $\rho^{(1)}$ had to be calculated along with the determination of $\Phi_{\lambda 1}^*$, consistent with the principal discussion of reactivity determination from the analysis of a single state, presented earlier in this section.

Equation (9.20a) may also be reduced to a simplified form, corresponding to Eq. (9.12b). Thus, the difference in the two analyses' approaches disappears as the analysis is simplified to the level of Eqs. (9.12).

The difference in the results between the two approaches, transient-linked versus independent-states analysis, became apparent when the simple formulas, Eqs. (9.12), were refined with appropriate theoretical correction factors. The method of refining the analysis by means of theoretical corrections may be extended one step further by also linking the dynamic and the static reactivities, ρ_1 and $\rho^{(1)} = \rho_1^{st}$, with a theoretical correction factor derived from exact point kinetics (EPK):

$$\rho_1^{st} = \rho_1 \left(\frac{\rho_1^{st}}{\rho_1} \right)_{\text{EPK}} \quad \text{or} \quad \rho_1 = \rho_1^{st} \left(\frac{\rho_1}{\rho_1^{st}} \right)_{\text{EPK}} \quad . \tag{9.20b}$$

This leads to the conclusion that static reactivity measurements yield static or dynamic reactivities, depending on the type of analysis. The primary results may be converted by completing the analysis with a further correction factor.

The two different procedures for analyzing static source multiplication experiments are illustrated in Fig. 9-2. The first of the two methods described is generally applied. The second method yields a somewhat different reactivity; the difference is explicity related to the use of different adjoint weighting functions, that is, $\Phi_{\lambda 0}^*$ versus $\Phi_{\lambda 1}^*$.

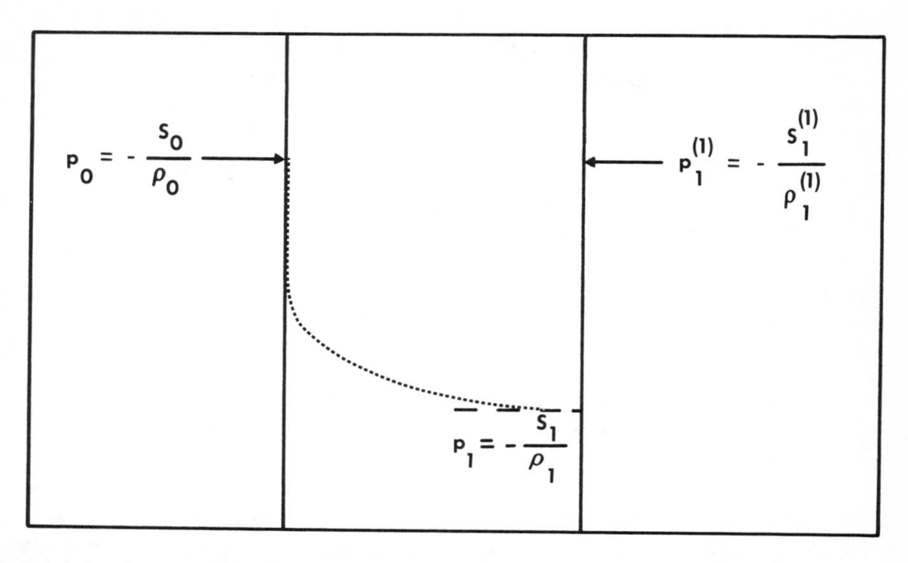

Fig. 9-2. Source multiplication formulas with one initial state (p_0, p_1) and two independent initial states (p_0, $p_1^{(1)}$).

9-2C Null Reactivity Method

The *null reactivity method* is frequently used in critical reactors. The method consists of applying the reactivity to be measured to a critical reactor and restoring (or maintaining) criticality by means of a calibrated control rod (counter rod). The EPK equation for the final critical state is simply given by

$$\dot{p} = 0 = (\delta\rho + \delta\rho_1)p_1 \quad . \tag{9.21}$$

Let $\delta\rho_1$ be the calibrated reactivity and $\delta\rho$ the unknown and desired reactivity. The two reactivities have to cancel, i.e.,

$$\delta\rho = -\delta\rho_1 \quad . \tag{9.22a}$$

Since $\delta\rho_1$ is normally available only in units of β ($= \beta_1$), Eq. (9.22a) is replaced by

$$\frac{\delta\rho}{\beta_1} = -\frac{\delta\rho_1}{\beta_1} \quad . \tag{9.22b}$$

The definitions of these two reactivities, $\delta\rho$ and $\delta\rho_1$, are obtained by splitting the dynamic reactivity as given in the framework of the EPK equation:

$$\delta\rho + \delta\rho_1 = \frac{(\Phi_0^*,[\delta\mathbf{F} + \delta\mathbf{F}_1 - \delta\mathbf{M} - \delta\mathbf{M}_1]\Psi)}{(\Phi_0^*,\mathbf{F}\Psi)} , \tag{9.23a}$$

which yields:

$$\delta\rho = \frac{(\Phi_0^*,[\delta\mathbf{F} - \delta\mathbf{M}]\Psi)}{(\Phi_0^*,\mathbf{F}\Psi)} \tag{9.23b}$$

and

$$\delta\rho_1 = \frac{(\Phi_0^*,[\delta\mathbf{F}_1 - \delta\mathbf{M}_1]\Psi)}{(\Phi_0^*,\mathbf{F}\Psi)} . \tag{9.23c}$$

The fission operator $\mathbf{F}$ and the flux shape Ψ belong to the system after the insertion of *both* reactivity increments. The reactivity increment $\delta\rho_1$ needs to be determined through a prior calibration.

A prior calibration generally involves flux shapes and fission operators that are different from those in Eq. (9.23c). This shows the kind of approximation that is involved in the null reactivity method as well as in other methods that make use of calibrated reactivities. In most cases, the error that results from this inconsistency may be sufficiently small so as to be of no practical importance. If this error is not negligible, approximate corrections may be made in a manner similar to that of the source multiplication method.

If $\delta\rho_1^{cal}/\beta_1$ designates the value found in the calibration, theoretical correction factors can be introduced to obtain, instead of Eqs. (9.22),

$$\frac{\delta\rho}{\beta} = -\frac{\delta\rho_1^{cal}}{\beta_1} \cdot \left(\frac{\delta\rho_1}{\delta\rho_1^{cal}}\right)_{\text{EPK}} \cdot \left(\frac{\beta_1}{\beta}\right)_{\text{EPK}} , \tag{9.24}$$

where β corresponds to the state with only $\delta\rho$ present. The correction factors should be calculated from the EPK formulas.

The null reactivity method can also be applied in subcritical reactors where the calibrated rod ($\delta\rho_1$) is used to maintain the source multiplication rather than criticality. The correction factors shown in Eq. (9.24) may become significant if $\delta\rho_1$ has been calibrated near $\rho = 0$ and is applied to a substantially subcritical reactor.

9-3 Basic Dynamic Reactivity Measurements

9-3A Survey

All dynamics methods are concerned with reactivities that actually cause a transient. The analysis of the transient yields information about

TABLE 9-I
List of Transients for Dynamic Reactivity Measurements

Quantity Changed	Type of Change	Name of Method	Reactivity Range of Applicability
ρ	Step	Asymptotic period	ρ small
ρ	Step	Rod-drop	$\rho \leq 0$
ρ	Step	Rod-jerk	$\rho < \beta$
s	Step	Source-jerk	$\rho < 0$
ρ	Sinusoidal	Pile oscillator	$\delta\rho$ small
s	δ function	Pulsed source	$\rho \leq 0$
ρ	General	Analysis by inverse kinetics	No limitation
s	General		

the dynamic reactivity, normally in the form of ρ/β or $(\beta - \rho)/\Lambda$. The advantage of dynamics methods compared to static methods is that *a calibration is not required.* Dynamics methods can, therefore, be used to calibrate a reactivity for use in static reactivity measurements.

Practical transient analysis can be simplified if the transients are initiated by simple changes. The application of the simplicity suggestion leads to the following three types of changes:

1. step changes
2. sinusoidal change
3. δ function-type change.

The two basic quantities that can be changed are the reactivity and the independent source. Considering both positive and negative step changes and combining the listed types of changes and the basic quantities changed yields the list of easily feasible reactivity measurement methods given in Table 9-I.

9-3B Asymptotic Period Method

The most reliable method for measuring small positive or negative reactivities in a critical reactor is the asymptotic period technique. After the insertion of ρ, the subsequent asymptotic behavior of the neutron population is observed to obtain the stable inverse period α, i.e., the slope of the flux amplitude in a semi-logarithmic plot such as Fig. 6-4. The reactivity is then obtained from an appropriate inhour formula, which relates ρ to the stable inverse period (see Sec. 6-1C for inhour relations). The range of applicability of the asymptotic period method is unfortunately limited to small reactivity changes of $|\rho| \lesssim 30¢$.

9-3C Rod-Drop Method for a Critical Reactor

The reactivity worth of a control rod can be measured by rapidly inserting the rod (fully or in a series of steps) and observing the prompt jump transient. In a simplified analysis, the reactivity insertion is approximated by a step (ρ_1 after the rod drop). Practical rapid rod insertion times are ~0.2 s so that the prompt jump adjustment occurs during the reactivity insertion.

The flux amplitude after the downward prompt jump (sometimes called the "prompt drop") can be found readily if the prompt jump approximation (neglecting $\Lambda\dot{p}$ of Sec. 8-1) is applied with the assumption that the delayed neutron source is still unchanged, which then allows the application of the CDS model of Secs. 6-1B and 6-2B. Both approximations are well justified shortly after the prompt jump. Their implementation in the kinetics equation yields:

$$0 = (\rho_1 - \beta)p_{pj} + \beta p_0 \quad . \tag{9.25}$$

For the desired reactivity ρ_1/β, this gives simply:

$$\frac{\rho_1}{\beta} = \frac{p_{pj} - p_0}{p_{pj}} < 0 \quad . \tag{9.26}$$

After the prompt jump, the flux will decrease further due to the decreasing delayed neutron source (compare Figs. 6-2).

The practical difficulty lies in the determination of the flux level p_{pj} because the actual reactivity insertion is only gradually terminated, i.e., not as a step. Therefore, the prompt flux decrease during the latter phase of the reactivity insertion overlaps with the beginning of the flux decrease due to the decay of the delayed neutron source. Improved accuracy requires the special correction techniques of Sec. 9-3H or an analysis by the inverse kinetics of Sec. 9-4C.

9-3D Subcriticality Measurements by the Rod-Drop Experiment

The rod-drop method may also be used to determine the degree of subcriticality. Since the use of calibrated reactivities is generally avoided in dynamics methods, the subcriticality measurement by the rod-drop method involves two unknown reactivities, say, ρ_0 and $\delta\rho_1$. Two measurements of relative fluxes rather than one as in Eq. (9.26) are therefore needed.

Figure 9-3 shows a typical transient from such an experiment. The insertion of $\delta\rho_1$ causes the flux to jump from its initial level p_0 promptly down to p_{pj}, about 0.5 p_0 in Fig. 9-3; the corresponding prompt transition virtually disappears in the ordinate. After the prompt jump, the flux

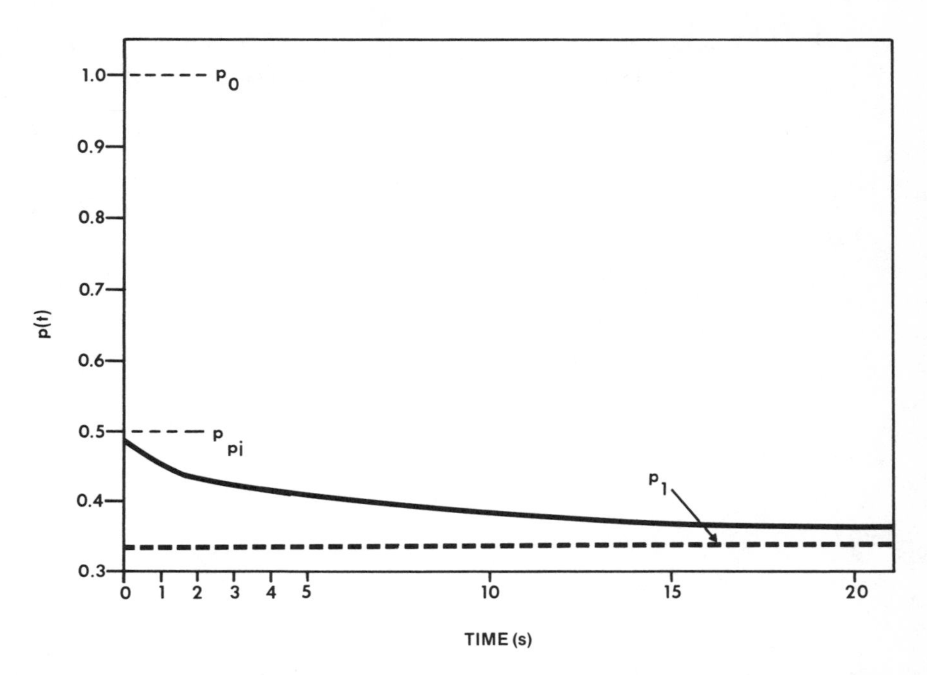

Fig. 9-3. Typical rod-drop transient (solid line) where p_0 is the initial flux level; p_{pj} is the prompt drop; and the asymptotic flux level is represented by the dashed line.

slowly establishes a final asymptotic flux level p_1, about one third of p_0 in Fig. 9-3. These three flux levels are given by the source multiplication formulas of Eqs. (6.92). For p_0, both the single-chain and the chain-sequence formulas are used in the analysis. In a simplified analysis, the changes in β and in the reduced sources are neglected, i.e., $s_1 = s_0$ and $\beta = \beta_0$ are used in all three formulas:

$$p_0 = \frac{s_0}{-\rho_0} = \frac{\beta p_0 + s_0}{\beta - \rho_0} \quad , \tag{9.27a}$$

$$p_{pj} = \frac{\beta p_0 + s_0}{\beta - \rho_1} \quad , \tag{9.27b}$$

and

$$p_1 = \frac{s_0}{-\rho_1} = \frac{\beta p_1 + s_0}{\beta - \rho_1} \quad , \tag{9.27c}$$

with

$$\rho_1 = \rho_0 + \delta\rho_1 \quad . \tag{9.27d}$$

Dividing the single-chain formulas for p_{pj} and p_0 and the chain-sequence formulas for p_1 and p_0 allows a ready elimination of the source:

$$\frac{p_{pj}}{p_0} = \frac{\beta - \rho_0}{\beta - \rho_1} \text{ and } \frac{p_1}{p_0} = \frac{\rho_0}{\rho_1} \quad . \tag{9.28a}$$

Also eliminating ρ_1 yields the so-called *"three-point-formula for the initial subcriticality"*:

$$\frac{\rho_0}{\beta} = \frac{\left(\frac{1}{p_0} - \frac{1}{p_{pj}}\right)}{\left(\frac{1}{p_1} - \frac{1}{p_{pj}}\right)} \quad . \tag{9.28b}$$

From Eq. (9.28a), one readily obtains ρ_1 as:

$$\frac{\rho_1}{\beta} = \frac{p_0}{p_1}\frac{\rho_0}{\beta} \quad . \tag{9.28c}$$

Application of Eqs. (9.28) to the transient depicted in Fig. 9-3, with $p_0 = 1$, $p_{pj} = 0.5$, and $p_1 = 1/3$, yields $\rho_0 = -\beta$; also $\rho_1 = -3\beta$ and thus $\delta\rho_1 = -2\beta$. These are the values that have been used to generate the transient of Fig. 9-3.

The problem in this kind of an analysis is again the determination of the transitory flux value p_{pj}. The inaccuracy of the experimental "value" of p_{pj} limits the accuracy of the entire method. Apparently, this problem does not occur in the asymptotic period method. But since both methods are for the most part applied to different configurations, they normally cannot be substituted for each other. For a refined analysis of this experiment, see Ref. 5 and Sec. 9-4C.

9-3E Source- and Rod-Jerk Methods

In the source-jerk method, the same type of reactor configuration is established as with the rod-drop into a critical reactor, i.e., a subcritical source-free reactor. In both cases, the flux responds with a prompt jump from p_0 down to p_{pj}. The analysis is the same as in Eqs. (9.25) and (9.26); for the initial reactivity, it yields:

$$\frac{\rho_0}{\beta} = \frac{p_{pj} - p_0}{p_{pj}} < 0 \quad . \tag{9.29a}$$

A simple prompt jump-type of analysis for the rod-jerk out of a critical or a subcritical reactor yields the same formulas as for the rod-drop. The only difference is in the sign of the flux change and thus in the resulting reactivity; for example, for a rod-jerk out of a critical reactor, one obtains instead of Eq. (9.26)

$$\frac{\rho_1}{\beta} = \frac{p_{pj} - p_0}{p_{pj}} > 0 \quad . \tag{9.29b}$$

9-3F The Pile Oscillator Method

The pile oscillator method consists of moving a control rod or a sample of material sinusoidally up and down, preferably in a range in which the reactivity depends linearly on the incremental change in control rod or sample position. This leads to a sinusoidal variation of the reactivity.

The basics of the analysis may be understood by applying the simple formulas of the prompt jump approximation. Neglecting or formally including the variation in the delayed neutron source yields the following two equations for a near-critical reactor [compare Eq. (8.22)]:

$$p(t) = \frac{\beta p_0}{\beta - \rho(t)} \text{ (with constant } s_{d0}\text{)} \tag{9.30a}$$

and

$$p(t) = \frac{s_d(t)}{\beta - \rho(t)} \text{ (with variable } s_d\text{)} \quad . \tag{9.30b}$$

Equation (9.30a) shows that a reactivity oscillation causes a flux oscillation. The analysis of the resulting flux transient provides information about $\rho(t)$, specifically ρ_{max}. Considering Eq. (9.30a) shows that due to its nonlinear dependence on ρ, a reactivity increase (ρ_{max}) leads to a flux increase ($p_{max} - p_0$) that is larger than the flux decrease ($p_0 - p_{min}$) caused by the same reactivity decrease ($-\rho_{max}$). This has a direct effect on the change of the delayed neutron source over a full period of the reactivity variation. Compared to the stationary production rate, the overproduction of precursors that occurs when $p(t) > p_0$ exceeds the underproduction for $p(t) < p_0$ since $(p_{max} - p_0) > (p_0 - p_{min})$. Therefore, there is a net increase of the precursor population over a full period. This leads to an increase in the flux over a period that affects $p(t)$ in Eq. (9.30b). The flux thus oscillates around a slightly increasing average value as shown in Fig. 9-4. Here, ρ_{max} = 25¢; then, Eq. (9.30a) would yield a periodic flux variation with maxima at 4/3 and minima at 0.8. Figure 9-4 shows the flux as it is affected by the increase of the delayed neutron source; the $s_d(t)$ increase amounts to ~23% over the depicted 10-s time span.

A semi-quantitative evaluation of the flux response to a reactivity

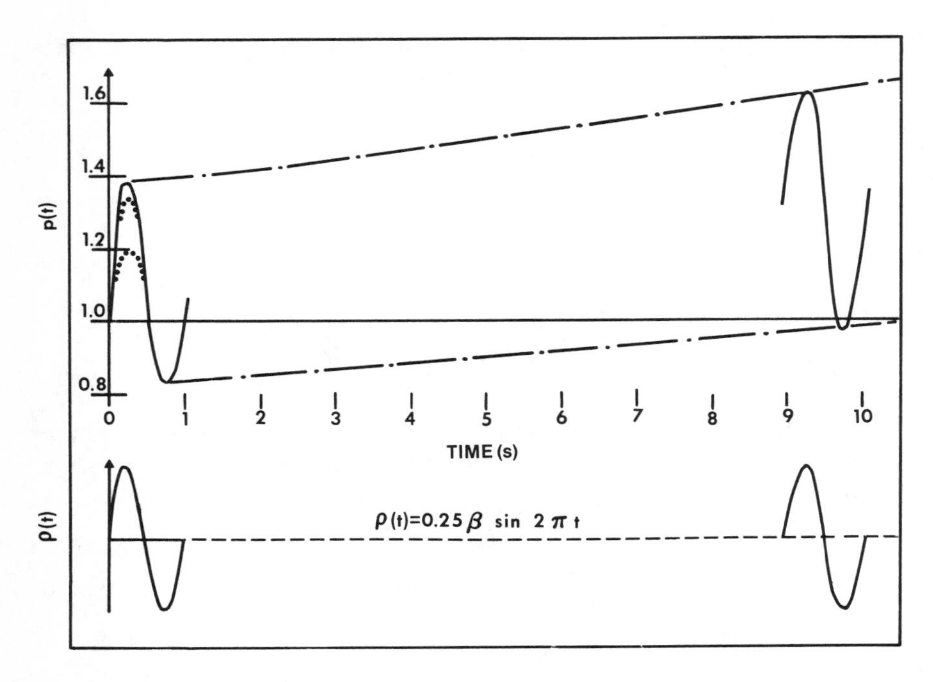

Fig. 9-4. Flux response (upper curve) to a sinusoidal reactivity variation (lower curve).

oscillation, of which the first and the tenth wave are depicted in Fig. 9-4, can serve as an example for evaluating a computer printout of a transient calculation. To this end, the first two half-waves of $p(t)$ for constant $s_d = s_{d0}$ are shown in Fig. 9-4 as dotted lines, the positive half-wave—with its maximum at 4/3—and the negative half-wave, reflected and drawn under the positive one. Its maximum is then at 1.20. The difference between the two maxima is 0.133; the corresponding integral, assuming a parabolic shape, is about 1/3 of the difference of the maxima, i.e., $0.133/3 \simeq 0.044$. Stretching this difference over a full wavelength yields a 2.2% overproduction per wavelength of delayed neutrons compared to the constant rate of s_{d0}. Over a small number of waves, the rate increase can be assumed to be a multiple of the change of s_d per wavelength times the number of waves. This then gives the 22 to 24% increase over the 10 waves shown in Fig. 9-4, in good agreement with the six-delay-group results.

Apparently, such a semi-quantitative understanding of the output of computer calculations can provide the confidence needed to consider the computer output to be the results of the stated problem. A reversal of this semi-quantitative evaluation allows us then to infer the maxima of the reactivity oscillation, which can then be more quantitatively confirmed by computer analysis.

The sinusoidal character of the reactivity oscillation needs to be confirmed independently: for example, if the oscillation is induced by a harmonic variation of a control rod insertion, the variation would have to be in an area where the axial dependency of the adjoint flux is practically linear.

9-3G Pulsed Source Method

In the pulsed source method, a neutron beam is injected into a critical or subcritical reactor core. The injected neutrons initialize fission chains (see Sec. 7-1). The neutron flux is suddenly[b] increased by the appearance of many additional fission chains. The prompt die-away of the flux that results from these fission chains can be used to obtain information about the prompt period.

The space, energy, and angular distribution of the source neutrons is very different from the equilibrium flux of the residing neutron population. After a sufficiently large number of collisions, the neutrons of the additional fission chains represent an additional flux with a distribution close to the flux of the already existing neutron population. Subsequent to this adjustment period, the result for the average fission chains derived in Sec. 7-1 can be applied. The die-away of the additional flux is then described by:

$$\delta p(t) = \delta p_0 \exp\left(-\frac{\beta - \rho_0}{\Lambda} t\right) \quad . \tag{9.31}$$

Figure 9-5 shows a qualitative sketch of a typical transient caused by a pulse of neutrons. The flux increases during the injection period ("*in*") and adjusts its shape to the residing flux during an adjustment period ("*ad*"). The additional flux subsequently dies away with the prompt period describing the decrease. The slope of this decrease on a logarithmic plot gives the prompt periods:

prompt period for a critical reactor:

$$\alpha_c = -\frac{\beta}{\Lambda} \tag{9.32}$$

and

[b]The source is nearly a δ function in time: $s(t) = s_{\text{total}} \cdot \delta(t)$.

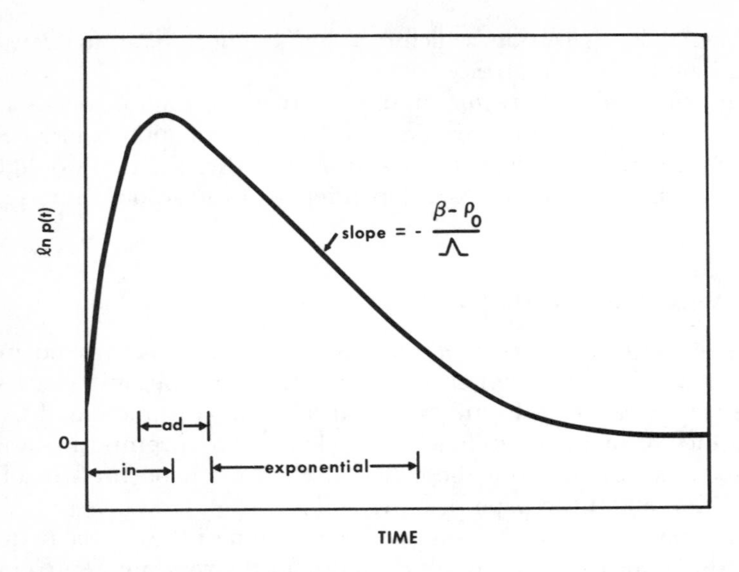

Fig. 9-5. Typical flux transient caused by a pulse of neutrons; "in" is the injection period and "ad" is the adjustment period.

prompt period for a subcritical reactor:

$$\alpha = -\frac{\beta - \rho_0}{\Lambda} = \alpha_c + \frac{\rho_0}{\Lambda} \quad . \tag{9.33}$$

The combination of measured periods in a critical and subcritical configuration yields the reactivity of the latter:

$$\frac{\rho_0}{\beta} = 1 - \frac{\alpha}{\alpha_c} \quad . \tag{9.34}$$

This type of evaluation of the pulsed neutron method was first introduced by Simmons and King.[13] Other types of evaluations were introduced by Gozani[14] and Garelis and Russell.[15] For a comparison of these different techniques, see, for example, Ref. 2.

9-3H Theoretically Consistent Flux Shape Corrections[c]

The analysis of *static* reactivity measurements may yield dynamic or static reactivities as primary results, depending on the type of analysis. The normally applied analysis yields the dynamic reactivity. Both results

---

[c]This section should not be included in an introductory course.

can be converted into each other by a theoretical conversion factor, Eq. (9.20b). *Dynamic* reactivity measurements yield as the primary result a dynamic reactivity; if a static reactivity is desired, the analogous conversion factor as in Eq. (9.20b) can be applied.

The analysis of reactivity measurement techniques was first presented above in a simple form in order to illustrate the basic approaches. Subsequently, theoretical refinements have been introduced in the form of correction factors and weighting factors. Most of these correction factors have been introduced during the historical development of reactivity measurement methods. However, the degree of sophistication of their calculation varied. The major difference occurs in the calculation of the flux-shape correction factors. The λ mode fluxes have been employed in part of the earlier literature rather than source-driven fluxes (solutions of inhomogeneous problems), which are now quite commonly used. Both fluxes may be quite different if the reactor is far subcritical or if the independent source is not proportional to the fission source.

The basic quantity derived from the analysis is ρ/β; it is defined as [see also Eqs. (5.26)]:

$$\frac{\rho(t)}{\beta(t)} = \frac{(\Phi^*_{\lambda 0},[\mathbf{F} - \mathbf{M}]\Psi)}{(\Phi^*_{\lambda 0},\mathbf{F}_d\Psi)} \quad . \tag{9.35}$$

The flux *shape*, $\Psi = \psi(\mathbf{r},E,t)$, is obtained from the flux by factorization:

$$\phi(\mathbf{r},E,t) = p(t)\psi(\mathbf{r},E,t) \quad . \tag{9.36a}$$

The factorization is made unique by means of the conventional constraint condition (compare Sec. 5-1A):

$$\int_V\int_E \frac{1}{v(E)}\,\phi^*_{\lambda 0}(\mathbf{r},E)\psi(\mathbf{r},E,t)\,dE\,dV = K_0 \quad . \tag{9.36b}$$

The conventional basis of the definition of the dynamic reactivity must appear in a consistent analysis, as has been indicated above. This becomes evident if one realizes that the exact time-dependent neutron balance equations as well as their solution do *not* involve the concept of reactivity; one can in principle solve the balance equation directly, such as in its multigroup diffusion form (see Chapter 11). The dynamic reactivity appears conceptually only if one condenses the detailed balance equations, e.g., in the form of the exact point kinetics equations. It is obvious that the way in which this condensation is performed affects the resulting reactivity value.

Furthermore, the experiments are analyzed by solving certain expressions, obtained from the point kinetics equations, for ρ. The result obviously depends on the form of the kinetics equations, which is in part determined by the constraint condition of Eq. (9.36b). A different normalization, e.g., of the power shape, would lead to a different form of the kinetics equation (see Sec. 5-1A) and thus leads, precisely speaking, to a different definition of the dynamic reactivity. Therefore, both conventions, the use of $\Phi^*_{\lambda 0}$ in Eqs. (9.35) and (9.36b) *and* the constancy of the integral of Eq. (9.36b), should appear explicitly in a consistent dynamic reactivity analysis. These conventions have to be applied in the calculation of the appropriate correction factors.

The principal relation between the theoretical description and the measurement of dynamic reactivities is essentially the same for all dynamics methods discussed above. It is, therefore, only discussed for one of the experimental analyses, for the simple analysis[d] of the rod-drop into a *critical* reactor, as presented in Sec. 9-3C. The discussion applies *mutatis mutandis* to the other experiments.

The behavior of the flux shape in the rod-drop experiment consists of a rapid and a slow transition:

$$\psi_0(\mathbf{r},E) \xrightarrow{\text{rapid}} \psi_{pj}(\mathbf{r},E,t) \xrightarrow{\text{slow}} \psi_{as}(\mathbf{r},E) \quad . \tag{9.37}$$

The initial flux shape promptly adjusts, essentially *during* the rod-drop, to the changed reactivity condition. Subsequent adjustment results from the variation of the delayed neutron source; particularly, the spatial distribution of the delayed neutron source adjusts itself to the change in ψ (see also Chapter 11).

The desired quantity to be measured in the rod-drop experiment is the reactivity formed with the asymptotic flux distribution ψ_{as}; however, from the analysis based on the amplitude change during the prompt jump [e.g., Eq. (9.26)], a reactivity formed with ψ_{pj} rather than ψ_{as} is obtained. (For a discussion of the difference of these two reactivities, compare Sec. 11-4A.)

The flux *shape* appears in the correction factors introduced above. The primary quantity in the analysis of the rod-drop reactivity measurement is the flux level or amplitude, e.g., as given in Eq. (9.26):

$$\frac{\rho_1}{\beta} = \frac{p_{pj} - p_0}{p_{pj}} \quad . \tag{9.38}$$

The flux level is normally measured by means of a neutron counter located in a small volume within or outside of the reactor core. Let

---

[d]A more general discussion of a consistent rod-drop experiment analysis is presented in Sec. 9-4C.

$\Sigma^C(\mathbf{r},E)$ describe the space and energy dependence of the counter sensitivity such that

$$CR(t) = \int_{\text{counter}}\int_E \Sigma^C(\mathbf{r},E)\phi(\mathbf{r},E,t)\ dE\ dV \tag{9.39}$$

gives the counting rate. In exact point kinetics, the flux is factored into an amplitude and shape function as given by Eq. (9.36a). This is inserted into Eq. (9.39). The ratio of the counting rates is thus related to the ratio of the flux amplitudes and the reaction rates of the shape functions, R_ψ^C, as follows:

$$\frac{CR(t)}{CR_0} = \frac{p(t)}{p_0}\frac{\int_{\text{counter}}\int_E \Sigma^C(\mathbf{r},E)\psi(\mathbf{r},E,t)\ dE\ dV}{\int_{\text{counter}}\int_E \Sigma^C(\mathbf{r},E)\psi(\mathbf{r},E,0)\ dE\ dV}$$

$$= \frac{p(t)}{p_0}\frac{R_\psi^C(t)}{R_\psi^C(0)}\quad . \tag{9.40}$$

The ratio of the integrals is to be calculated from the exact point kinetics (EPK), by employing the conventions of Eqs. (9.36):

$$\frac{CR(t)}{CR_0} = \frac{p(t)}{p_0}\left[\frac{\int_{\text{counter}}\int_E \Sigma^C(\mathbf{r},E)\psi(\mathbf{r},E,t)\ dE\ dV}{\int_{\text{counter}}\int_E \Sigma^C(\mathbf{r},E)\psi(\mathbf{r},E,0)\ dE\ dV}\right]_{\text{EPK}}\quad . \tag{9.41}$$

Solving Eq. (9.41) for p/p_0 gives the consistent evaluation of p/p_0 in terms of counting rates:

$$\frac{p(t)}{p_0} = \frac{CR(t)}{CR_0}\left[\frac{R_\psi^C(0)}{R_\psi^C(t)}\right]_{\text{EPK}}\quad . \tag{9.42}$$

The c's of Eqs. (9.14) are then given by the EPK estimates of the shape integrals:

$$c(t) = \frac{1}{[R_\psi^C(t)]_{\text{EPK}}} = \frac{1}{\left[\int_{\text{counter}}\int_E \Sigma^C(\mathbf{r},E)\psi(\mathbf{r},E,t)\ dE\ dV\right]_{\text{EPK}}}\quad . \tag{9.43}$$

Thus, the consistent amplitude function is obtained as a ratio of the measured and a specially calculated reaction rate:

$$p(t) = \frac{CR(t)}{R_\psi^C(t)} = c(t)CR(t)\quad . \tag{9.44}$$

A possible inaccuracy in the calculated $R_\psi^C(t)$ largely cancels since only ratios are needed, as in Eq. (9.42).

9-4 Inverse Kinetics

9-4A Inverse Point Kinetics

The most general method for measuring the dynamic reactivity is the so-called "inverse kinetics" method. The method basically consists of measuring $p(t)$, calculating β_k and Λ, and solving the kinetics equations for the reactivity. It was first proposed and investigated in 1959 by Corben.[16] An inverse kinetics procedure for reactivity coefficients was independently developed in 1963 by Ott.[17] The first application of inverse kinetics for determining various reactivities in an intermediate and fast neutron spectrum facility was presented by Carpenter[18,19] in 1965.

The advantage of the inverse kinetics method of determining reactivities is that it works, in principle, for any kind of reactivity insertion and not just for special transients. The method can be used, therefore, to find time-dependent reactivities with on-line computers.[20,21] Murray et al.[22] applied inverse kinetics in a different way. They determined by this inverse method the time-dependent reactivity insertions needed to produce desired power variations.

In practically all previous applications, the inverse kinetics method was derived from the kinetics equations of the point reactor model (PRM) (see Sec. 5-2). Therefore, what is usually referred to as "inverse kinetics" can be called more precisely "inverse point kinetics." It is also known as the "power history method."

In the PRM, β_k and Λ do not depend on time. Thus, β_k, Λ, and counter efficiency have to be calculated only once for the transient (or reactor).

The PRM kinetics equation,

$$\Lambda \dot{p}(t) = [\rho(t) - \beta]p(t) + s_d(t) + s(t) \quad , \tag{9.45}$$

is solved for $\rho(t)$. This gives:

$$\frac{\rho(t)}{\beta} = 1 - \frac{s_d(t)}{\beta p(t)} + \frac{\Lambda \dot{p}(t)}{\beta p(t)} - \frac{s(t)}{\beta p(t)} \quad . \tag{9.46a}$$

The delayed neutron source is determined from the flux amplitude through convolution integrals (see, for example, Sec. 6-1C):

$$s_d(t) = \sum_k \lambda_k \beta_k \int_{-\infty}^{t} p(t') \exp[-\lambda_k(t - t')]\, dt' \quad , \tag{9.46b}$$

where $p(t)$ is determined experimentally.

The difference between the first two terms on the right side of Eq. (9.46a) can be written in the form

$$r_p(t) = 1 - \frac{s_d(t)}{\beta p(t)}$$

$$= \frac{1}{p(t)} \sum_k \lambda_k \frac{\beta_k}{\beta} \int_{-\infty}^{t} [p(t) - p(t')] \exp[-\lambda_k(t - t')]\, dt' \quad . \qquad (9.46c)$$

Thus, $r_p(t)$ is the basic component for the description of the transition of $p(t)$; it is zero initially and it will be zero again in an asymptotic stationary state.

A practical problem in applying the inverse kinetics equation (9.46a) arises from statistical fluctuations in counting rates, which form the basis for the measurement of $p(t)$. Inserting fluctuating or only partially smoothed values of $p(t)$ in the denominators[e] of Eq. (9.46a) yields a bias in the results since the actually occurring inverse of the average of p over a time interval Δt is different from the formally appearing average of the inverse of p:

$$\frac{1}{\langle p \rangle_{\Delta t}} \neq \left\langle \frac{1}{p} \right\rangle_{\Delta t} \quad . \qquad (9.47)$$

Various techniques for smoothing the statistical fluctuations in the inverse kinetics analysis have been developed in recent years.[19–23] Most procedures have been developed primarily for the analysis of the rod-drop transient between two subcritical states, i.e., for the determination of initial subcriticality (see the following paragraph). In a method devised by Yang and Albrecht,[23] the flux approximation employed to smooth $p(t)$ depends on $\delta\rho_1$, via the roots of the inhour equation. This permits the elimination of $\delta\rho_1$ by treating it as a search variable, which is determined by the conditon of yielding the same value ρ_0 for all times.

As indicated previously, the inverse kinetics can be applied to improve the simplified analysis of reactivity measurements. The simplified analysis of the rod-drop into a subcritical reactor, for example, is based on Eqs. (9.27); these equations also follow from Eq. (9.46a) if the same assumptions are introduced. In both cases, the equation for the initial state yields:

$$\frac{\rho_0}{\beta} = -\frac{s_0}{\beta p_0} \quad . \qquad (9.48a)$$

---

[e]For the effect of $p(t)$ errors on the $\dot{p}$ term, see the last paragraph of Sec. 9-4B.

If $\Lambda\dot{p}$ and the small changes of $s_d(t)$ and $s(t)$ during the prompt jump are neglected, Eq. (9.46a) yields, after the prompt jump,

$$\frac{\rho_0 + \delta\rho_1}{\beta} = 1 - \frac{\beta p_0}{\beta p_{pj}} - \frac{s_0}{\beta p_{pj}} , \tag{9.48b}$$

which is equivalent to Eq. (9.27b) where p_{pj} was assumed to be determined experimentally. The inverse point reactor kinetics gives, for the final stationary state,

$$\frac{\rho_0 + \delta\rho_1}{\beta} = - \frac{s_0}{\beta p_1} , \tag{9.48c}$$

which is equivalent to Eq. (9.27c).

The use of the complete time dependence of $p(t)$ and $s_d(t)$ obviously provides an improvement (see, for example, Refs. 19 to 21) over the three point method. This improvement is particularly important for corrections of the deviation of the reactivity insertion from a step function.

If there is no independent source in the reactor, either initially or during the transient, Eq. (9.46a) can be simplified by dropping the source term. This gives the inverse point kinetics equation for a source-free reactor (e.g., an initially critical reactor):

$$\frac{\rho(t)}{\beta} = 1 - \frac{s_d(t)}{\beta p(t)} + \frac{\Lambda\dot{p}(t)}{\beta p(t)} . \tag{9.49}$$

Most applications of inverse kinetics are concerned with subcritical systems with an independent source present. Equation (9.46a) then needs to be either completed by information on the source, or have the source eliminated through an expanded inverse kinetics analysis. The source elimination problem is addressed in Sec. 9-4C.

9-4B Inverse Kinetics for Reactivity Coefficients

Inverse kinetics analysis can also be applied to determine reactivity increments, and thus reactivity coefficients. Both ρ_0 and $\delta\rho_1$ appear in Eqs. (9.48) and consequently either one may be eliminated to obtain the desired reactivity or reactivity increment by proper use of the time-dependent flux transient.

A special inverse kinetics procedure can be devised for reactivity increments that are due to the action of a feedback effect. Such a procedure was proposed by one of the authors[17] for the determination of the Doppler reactivity coefficient from flux transients in the SEFOR reactor[24] see also Sec. 10-2.

The reactivity increment that results from Doppler feedback is given approximately by (see Sec. 10-1B):

$$\delta\rho(t) = -\gamma \int_0^t [p(t') - p_0]\, dt' \quad . \tag{9.50}$$

If a positive reactivity ρ_1 is quickly inserted at $t = 0$, for example, by rapid rod-jerk, the flux will rise promptly. This leads to a corresponding temperature increase, which will, through Doppler feedback, counteract the positive transient-initiating reactivity:

$$\rho(t) = \rho_1 - \gamma \int_0^t [p(t') - p_0]\, dt' = \rho_1 + \delta\rho(t) \quad . \tag{9.51}$$

There are basically two procedures for finding γ from an inverse kinetics analysis: The inverse kinetics can be applied in a straightforward manner to find $\rho(t)$. Since ρ_1 is constant, after it has been completely inserted, the time-dependent part can be found by subtracting an approximate value for ρ_1. Dividing the time-dependent part, $\delta\rho(t)$, by the integral in Eq. (9.51) yields γ. The disadvantage of this procedure is that the error of the subtracted value of ρ_1 may be magnified in the small quantity $\delta\rho(t)$.

The kinetics equation can also be solved[17] directly for γ:

$$\frac{\gamma}{\beta} = \frac{\dfrac{\rho_1}{\beta} - r_p(t) - \dfrac{\Lambda}{\beta}\dfrac{\dot{p}(t)}{p(t)}}{\displaystyle\int_0^t [p(t') - p_0]\, dt'} \quad , \tag{9.52}$$

where r_p is defined by Eq. (9.46c) and ρ_1 can be eliminated by a search. Its value is determined by the condition that the right side of Eq. (9.52) is constant. The value of this constant is γ/β. The right side of Eq. (9.52) is sensitively dependent on the correct value of ρ_1 being inserted. The reason for this is that the denominator of Eq. (9.52) is initially zero and then increases nearly linearly with time for small t after the prompt jump:

$$\int_0^t [p(t') - p_0]\, dt' \simeq (p_{pj} - p_0)t \quad \left.\begin{matrix}\text{for small } t,\\ \text{after the prompt jump}\end{matrix}\right. \quad . \tag{9.53}$$

In a short time interval after the rod-jerk, $p(t')$ has approximately its prompt jump value. The proportionality of Eq. (9.53) to t causes the right side to exhibit a hyperbolic behavior in time for a given trial value of ρ_1. A computer simulation of this analysis[17] is shown in Fig. 9-6, in which the right side of Eq. (9.52) is plotted for three values of ρ_1/β

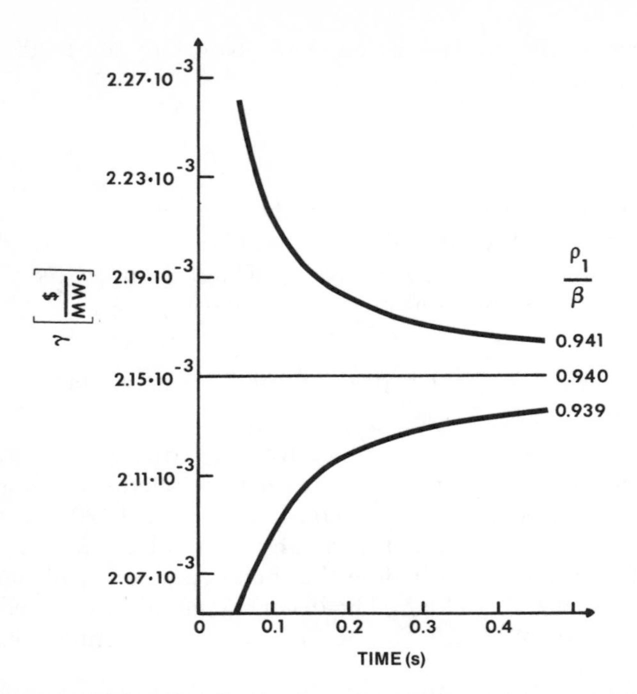

Fig. 9-6. The quantity $\gamma(t)$ of Eq. (9.52), with ρ_1 as a parameter for an excursion following a step insertion of $\rho_1 = 0.940\$$ (after Ref. 17).

inserted during the search. The sensitivity of this procedure is exhibited by the fact that even a deviation by only 0.001$ from the correct ρ_1/β causes a very significant hyperbolic time dependence, which indicates an error in the inserted trial value of ρ_1.

The accuracy of the inverse kinetics analysis may be practically limited not only by the errors in measuring and smoothing $p(t)$ but also by the calculation of the delayed neutron source from $p(t)$ [compare Eq. (9.46b)]. The latter inaccuracy is primarily due to uncertainties in the delayed neutron data. The resulting error can be corrected in a first-order approximation by including an $s_d(t)$ correction in the search.[17]

The $\Lambda\dot{p}$ term is small in the entire subprompt-critical domain. The reason is the same as the one that justifies the applicability of the prompt jump approximation (Sec. 8-1), i.e., the small effect of the $\Lambda\dot{p}$ term on the kinetics results due to the smallness of the neutron generation time. The fact that the contribution of the $\Lambda\dot{p}$ term to the inverse kinetics result is small also implies that the resulting error is small, provided that the value of $\dot{p}$ is not accidentally increased by improper smoothing of the statistical fluctuations of the measured $p(t)$ values.

9-4C Inverse Spatial Kinetics[f]

The inverse point kinetics-type of analysis of reactivity measurements can be augmented such that it becomes consistent with the exact kinetics equations. The resulting approach is called "inverse spatial kinetics," where the term "spatial" refers to both space and energy dependence in the same way as the term "point" in point kinetics refers to the reduction of the space *and* energy dependence.

The main difference between inverse point and inverse spatial kinetics is that the flux shape is assumed to be independent of time in the former, whereas the time dependency of the flux shape is considered in the latter. The neglect of the flux shape variation can cause significant errors in the inverse kinetics analysis of reactivity measurements; the corresponding error in a control rod worth measurement can easily exceed 10%.

If the exact kinetics equations are formally solved for the reactivity, the same form of inverse kinetics equation as Eq. (9.46a) is obtained. The differences are that β_k and Λ depend on time and $p(t)$ is to be obtained with explicit use of the conventional constraint condition in the calculation of the time-dependent flux shape. Furthermore, $s_d(t)$ and $s(t)$ are determined from their exact definitions in Secs. 5-1B and 5-1C. The "exact inverse kinetics equation" is given by:

$$\frac{\rho(t)}{\beta(t)} = r_p(t) + \frac{\Lambda(t)}{\beta(t)}\frac{\dot{p}(t)}{p(t)} - \frac{s(t)}{\beta(t)p(t)} \quad , \tag{9.54}$$

with

$$r_p(t) = \frac{1}{p(t)}\sum_k \lambda_k \int_{-\infty}^{t} \left[\frac{\beta_k(t)}{\beta(t)}\,p(t) - \frac{\beta_k(t')F(t')}{\beta(t)F(t)}\,p(t')\right] \times \exp[-\lambda_k(t - t')]\,dt' \quad , \tag{9.55}$$

where $\beta_k(t)$, $\Lambda(t)$, $F(t)$, and $s(t)$ are given by Eqs. (5.23), (5.26), and (5.50). The time-dependent flux shape function, $\psi(\mathbf{r},E,t)$, is explicitly needed for inverse spatial kinetics. Methods for calculating this shape function are the subject of spatial dynamics, which is discussed in Chapter 11.

An inverse spatial kinetics method for the analysis of rod-drop measurements has been developed by Luck et al.[25] The basic approach consists of finding $\rho(t)/\beta(t)$ from Eq. (9.54). The three different types of quantities in Eq. (9.54) have to be treated in a different manner.

---

[f]This section should not be included in an introductory course.

First, there is the amplitude function, which must reflect, consistently, the experimental transient information, available in the form of a counting rate, $CR(t)$.

Second, there are several ratios [Λ/β, β_k/β, and $F(t')/F(t)$] that are constant in the inverse point kinetics analysis; their time dependence exhibited in the inverse spatial kinetics is estimated theoretically, since it is not available experimentally.

Third, the last term in Eq. (9.54) contains the independent source, which needs to be eliminated. Equations (5.26) and (5.50) give for s/β:

$$\frac{s(t)}{\beta(t)} = \frac{(\Phi^*_{\lambda 0},S)}{(\Phi^*_{\lambda 0},\mathbf{F}\Psi)} \frac{(\Phi^*_{\lambda 0},\mathbf{F}\Psi)}{(\Phi^*_{\lambda 0},\mathbf{F}_d\Psi)} = \frac{(\Phi^*_{\lambda 0},S)}{F(t)} \frac{F(t)}{(\Phi^*_{\lambda 0},\mathbf{F}_d\Psi)} \quad . \tag{9.56}$$

In analyzing a reactivity measurement, one normally does not want to assume that the independent source is known. Dominating contributors to the independent source are the spontaneously fissioning nuclei, such as ^{240}Pu and higher actinide isotopes. The concentration and space distribution of these isotopes may not be known accurately enough for practical purposes. One therefore has to provide the additional information that will allow elimination of the source (see below).

If the independent source is constant in time ($S = S_0$), one can write s/β as a ratio with a time-independent numerator:

$$\frac{s(t)}{\beta(t)} = \frac{s_0 F_0}{\beta(t)F(t)} \quad . \tag{9.57}$$

The method developed in Ref. 25 deals with either a single transient or with a sequence of two or more transients. In this text, the analysis of a two-transient sequence is briefly described. Since the inverse spatial kinetics is the most general method, its application to the analysis of a sequence of two transients illustrates particularly well the basic features of dynamic as well as static reactivity analysis.

The source elimination procedure exploits the fact that the flux transient, which follows a sufficiently rapid change in the reactor, is still proceeding after the physical change in the reactor is completed; the delayed neutron population, which affects the flux, needs many seconds to several minutes to complete its adjustment to the changed fluxes (see Fig. 9-3). To conceptually and explicitly account for variation in the physical state of the reactor, as the basis for the source elimination procedure, the dynamic reactivity in Eq. (9.54) is converted into a static reactivity by means of a theoretical conversion factor, as in Eq. (9.20b):

$$\rho^{dyn}(t) = h(t)\rho^{st}(t) \quad , \tag{9.58}$$

where ρ^{st} is independent of time after the physical change in the system is completed; ρ^{dyn} still depends on time while it adjusts to its asymptotic value (see Sec. 11-4). Thus, during the transition into the asymptotic state that follows, the completion of a change in the reactor holds

$$\rho^{dyn}(t) = h(t)\rho^{st} \quad . \tag{9.59}$$

During the first of the transients in the sequence, the adjoint weighted source, s_0, is to be determined. The static reactivity, $\rho^{st}(t)$, initiates the first transient. For simplicity, $\rho^{st}(t)$ is measured in initial dollars (β_0). This gives instead of Eq. (9.54):

$$\frac{\rho^{st}(t)}{\beta_0} = -\frac{a(t)}{\beta_0} s_0 + b(t) \quad , \tag{9.60}$$

with

$$a(t) = \frac{F_0}{F(t)h(t)p(t)} \quad , \tag{9.61a}$$

$$F(t) = (\Phi^*_{\lambda 0}, \mathbf{F}\Psi) \quad , \tag{9.61b}$$

and

$$b(t) = \left[\frac{\beta(t)}{\beta_0} r_p(t) + \frac{\Lambda(t)}{\beta_0}\frac{\dot{p}(t)}{p(t)}\right]\frac{1}{h(t)} \quad , \tag{9.61c}$$

where $r_p(t)$ is given in Eq. (9.55) and $p(t)$ is related to the counting rate by Eq. (9.44).

For each transient, especially for the first one, one can identify four phases: the time-independent state prior to the transient, the static reactivity insertion (terminated at $t < t_1$), the adjustment of the delayed neutrons, and the final asymptotic state. One obtains four corresponding source multiplication formulas, which are solved for the reactivity as in Eq. (9.60):

$$\frac{\rho_0^{st}}{\beta_0} = -\frac{s_0}{\beta_0 p_0} \quad , \text{ for } t \le 0 \quad , \tag{9.62}$$

$$\frac{\rho^{st}(t)}{\beta_0} = -\frac{a(t)}{\beta_0} s_0 + b(t), \text{ for } 0 \le t \le t_1 \quad , \tag{9.63}$$

$$\frac{\rho_1^{st}}{\beta_0} = -\frac{a(t)}{\beta_0} s_0 + b(t) \quad , \text{ for } t_1 \le t \quad , \tag{9.64}$$

and

$$\frac{\rho_1^{st}}{\beta_0} = -\frac{F_0 \, s_0}{F_1 \, \beta_0 \, p_1 h_1} \quad , \text{ asymptotically} \quad . \tag{9.65}$$

Note that $b(t)$ is zero in stationary states since β, F, and p are then independent of time.

The two stationary source multiplication formulas, Eqs. (9.62) and (9.65), contain three unknowns, $\overset{st}{\rho_0}$, $\overset{st}{\rho_1}$, and s_0. Thus, one additional relation between these three quantities must be extracted from Eqs. (9.63) or (9.64). Equation (9.63), however, contains additional unknowns, in the form of the function $\rho^{st}(t)$; the detailed time dependence of the reactivity insertion cannot normally be assumed to be known. Thus, only Eq. (9.64) can be used to provide the needed additional relation.

The quantities $a(t)$ and $b(t)$ in Eq. (9.64) are known, based on a combination of experimental (counting rate) and theoretical information. Subtraction of Eq. (9.65) from (9.64) yields an equation for s_0:

$$0 = -\frac{\Delta a(t)}{\beta_0} s_0 + b(t) \quad , \tag{9.66}$$

where $\Delta a(t)$ is given by

$$\Delta a(t) = a(t) - a_1 \quad , \tag{9.67}$$

with

$$a_1 = \frac{F_0}{F_1 p_1 h_1} \quad . \tag{9.68}$$

Equation (9.66) can be solved readily for s_0:

$$s_0 = \frac{b(t)}{\Delta a(t)} \beta_0 \quad . \tag{9.69}$$

Apparently, the time dependency in the numerator and denominator of Eq. (9.69) must be the same, so that it cancels and can yield a time-independent result. Practically, however, both numerator and denominator will exhibit some fluctuations, due to fluctuations in the counting rates. Thus, properly averaged quantities should be used in lieu of fluctuating values. This yields the "calibration" of the *source* as:

$$s_0 = \frac{\overline{b(t)}}{\overline{\Delta a(t)}} \beta_0 \quad . \tag{9.70}$$

The experimentally determined source, s_0 per Eq. (9.70), can be inserted in Eq. (9.62), which provides the "calibration" of the *initial reactivity:*

$$\frac{\overset{st}{\rho_0}}{\beta_0} = -\frac{1}{\overline{p}_0} \frac{\overline{b(t)}}{\overline{\Delta a(t)}} \quad . \tag{9.71}$$

Furthermore, the calibrated source and the experimentally determined asymptotic flux amplitude $\bar{p}_1$ yield, through Eq. (9.65), the desired final reactivity ρ_1:

$$\frac{\overset{st}{\rho_1}}{\beta_0} = -\frac{s_0}{\bar{p}_1}\frac{F_0}{F_1\beta_0 h_1} \quad . \tag{9.72}$$

The second factor contains the theoretically determined quantities F_0, F_1, β_0, and h_1.

The determination of the initial values of source and reactivity provides a calibration that may also be applied in subsequent transients. Let the reactivity during the second transient be changed from ρ_1 to ρ_2. Again, one can distinguish four phases of this transient with the first one being the asymptotic phase of the first transient. The four corresponding reactivity formulas are given by:

$$\frac{\overset{st}{\rho_1}}{\beta_0} = -\frac{F_0 s_0}{\beta_0 F_1 \bar{p}_1 h_1} \text{ (prior to second transient)} \quad , \tag{9.73}$$

$$\frac{\rho^{st}(t)}{\beta_0} = -\frac{a(t)}{\beta_0} s_0 + b(t) \begin{pmatrix}\text{during physical change} \\ \text{of reactor}\end{pmatrix} \quad , \tag{9.74}$$

$$\frac{\overset{st}{\rho_2}}{\beta_0} = -\frac{a(t)}{\beta_0} s_0 + b(t) \begin{pmatrix}\text{after physical change of} \\ \text{reactor, during delayed} \\ \text{transition}\end{pmatrix} \quad , \tag{9.75}$$

and

$$\frac{\overset{st}{\rho_2}}{\beta_0} = -\frac{F_0 s_0}{\beta_0 F_2 \bar{p}_2 h_2} \text{ (after delayed transition)} \quad . \tag{9.76}$$

With s_0 being known from a prior calibration, the reactivity can now be determined even during the physical change of the reactor, i.e., $\rho^{st}(t)$ may be found from Eq. (9.74), provided that the counting rate is sufficiently large to determine a $p(t)$ trace during the $\rho^{st}(t)$ insertion.

If the third phase contains sufficient information, i.e., if the $\overset{st}{\rho_2}$ insertion is completed rapidly so that usable values for $b(t)$ and $\Delta a(t) = a(t) - a_2$ can be determined [note that $b(t)$ and $\Delta a(t)$ are zero asymptotically], the source can be newly calibrated. This is particularly important when the source has changed significantly during the reactivity insertion.

In many practical cases, the insertion of the reactivity may be so slow (e.g., unloading of fuel subassemblies) that the delayed transition is practically completed during the reactivity insertion. Then, $b(t)$ is negligibly small:

$$b(t) \simeq 0 \begin{pmatrix}\text{during very slow} \\ \text{reactivity insertions}\end{pmatrix} \quad . \tag{9.77}$$

The third transient phase merges then with the second one and the reactivity is obtained from the residual formula:

$$\frac{\rho^{st}(t)}{\beta_0} = -\frac{F_0 s_0}{\beta_0 F(t) p(t) h(t)} \quad . \tag{9.78}$$

In Equation (9.78), s_0 can be replaced by the initial reactivity, calculated from Eq. (9.70). This gives the *consistent source multiplication formula:*

$$\frac{\rho^{st}(t)}{\beta_0} = \left(\frac{\rho_0^{st}}{\beta_0}\right)^{cb} \cdot \frac{p_0}{p(t)} \cdot \frac{F_0}{F(t)} \cdot \frac{h_0}{h(t)} \quad . \tag{9.79}$$

This equation contains a calibrated reactivity that is obtained within the same sequence of transients and is "theoretically consistent" with its application. This distinguishes Eq. (9.79) from earlier source multiplication formulas (Sec. 9-2b) where a calibration reactivity is used, which is normally unrelated to its subsequent application.

In Equation (9.79), h_0 is used instead of unity to show the symmetry of the consistent source multiplication formula. The first factor is obtained through the described calibration; the second factor contains the counting rates and the corresponding correction factor due to flux shape deformation of Eq. (9.44). The third reflects the conventional choice of the denominator of the lumped kinetics parameters. It appears in Eq. (9.79) since the reactivity is measured here in initial rather than time-dependent dollars. The fourth factor converts the dynamic reactivity, which would be the *direct* result, into a static reactivity. It is given by

$$\frac{h_0}{h(t)} = \frac{1}{h(t)} = \frac{\rho^{st}(t)}{\rho^{dyn}(t)} = \frac{(\Phi_{\lambda 0}^*, \mathbf{F}\Psi)}{(\Phi_{\lambda 0}^*, [\mathbf{F} - \mathbf{M}]\Psi)} \frac{(\Phi_{\lambda 0}^*, [\mathbf{F} - \mathbf{M}]\Phi_\lambda)}{(\Phi_{\lambda 0}^*, \mathbf{F}\Phi_\lambda)} \quad . \tag{9.80}$$

In many practical cases, such as fuel reloading, the source is changed along with the change of the reactivity. Suppose the source stayed constant during the first transient; then the calibration of the original source and reactivity can be performed as described above. The source may change in subsequent transients. If the physical reason for the source change is known, one may approximately correct for the source variation. Equation (9.74) is then replaced by

$$\frac{\rho^{st}(t)}{\beta_0} = -\frac{a(t)}{\beta_0} \cdot \frac{s(t)}{s_0} \cdot s_0 + b(t) \quad . \tag{9.81}$$

For a slow transient one obtains, instead of Eq. (9.79), the *consistent source multiplication formula with source correction:*

$$\frac{\rho^{st}(t)}{\beta_0} = \left(\frac{\rho_0^{st}}{\beta_0}\right)^{cb} \cdot \frac{p_0}{p(t)} \cdot \frac{F_0}{F(t)} \cdot \frac{h_0}{h(t)} \cdot \frac{s(t)}{s_0} \quad . \tag{9.82}$$

The consistent source multiplication formulas, Eqs. (9.79) and (9.82), represent the limit of the inverse spatial kinetics analysis for very slow reactivity insertions. There is a smooth transition from inverse spatial kinetics into consistent source multiplication analysis. As long as the flux varies with time and as the delayed neutron transition is not completed, inverse spatial kinetics gives a more accurate result than the consistent source multiplication formula. Equations (9.79) and (9.82) represent the limit of Eq. (9.74) for $b(t) \rightarrow 0$. As long as $b(t)$ is still noticeable, one has, instead of Eq. (9.82),

$$\frac{\rho^{st}(t)}{\beta_0} = \left(\frac{\rho_0^{st}}{\beta_0}\right)^{cb} \cdot \frac{p_0}{p(t)} \cdot \frac{F_0}{F(t)} \cdot \frac{h_0}{h(t)} \cdot \frac{s(t)}{s_0} + b(t) \quad , \tag{9.83}$$

with $b(t)$ from Eq. (9.61). Asymptotically, $b(t)$ vanishes and the other quantities become independent of time.

Homework Problems

1. Devise a static reactivity measurement method involving two states (p_0 and p_1) with the same reactivity, ρ_0, using a precalibrated source change (s_1/s_0). Find ρ_0 and state the main approximation in this analysis.

2. *Rod-Drop Analysis.* In an experimental project, a rod drop from a negative reactivity ρ_0 down to ρ_1 is carried out. This assignment prepares the subsequent analysis, by calculating a typical rod-drop transient and analyzing the calculated transient in a "dry run." Use the following data for PUR-1: $\Lambda = 10^{-5}$ s and $\beta_k = \gamma\beta_k^{phys}$ with $\gamma = 1.08$. Take the β_k^{phys} from the data presented in Chapter 2 for ^{235}U.
   a. Find $p(t)$ numerically from a point kinetics program for the transient $\rho_0 = -\beta \rightarrow \rho_1 = -2\beta$ in a step change.
   b. Plot $p(t)$ on a proper scale (or on two different scales).
   c. From the two stationary flux levels, find a formula for the ratio of the two reactivities.
   d. Take your numerical p values and find (check) the reactivity ratio.
   e. Find from the prompt jump formula another equation that then allows you to determine the reactivities individually.
   f. Take this formula and find (check) ρ_0 and ρ_1.
   g. Suppose p_0 is established by applying the source at t_{-1}. How long would you have to wait until p_0 is established with a 0.1% accuracy?

3. *Pulsed Source Experiment.*
   a. Consider a critical experimental reactor with a fission neutron source of 7.5×10^{14} neutron/s. Suppose $\nu = 2.5$. What is the reactor power?
   b. Suppose β/Λ is measured with the pulsed source method. A pulse with 10^5 neutron/pulse is applied. To obtain good statistics, the pulsed source is injected 10 times per second for 90 min.
      - Is the pulse rate small enough to allow good separation of the individual pulses?
      - What is the increase in the reactor power after 90 min?
      - Is there a further change in power after the experiment is completed? (Use $\beta = 0.0075$ and $\Lambda = 5 \times 10^{-6}$ s).
4. Estimate the increase of the flux average per period in a pile oscillator experiment with $\rho_{max} = 1¢$ and $10¢$ for 1000 oscillations by evaluating the under and overproduction of precursors (one delay group) during the first oscillation.
5. Apply the inverse kinetics formula for the reactivity to a transient

$$p(t) = p^0 \exp(\alpha_1 t) \quad .$$

Use a one-delay-group approximation. Discuss your result.

Review Questions

1. Name the three groups of methods for the measurement of reactivities.
2. Describe briefly the principal situation of the measurability of static dynamic reactivities.

Note: In the following, present only the simple point kinetics analyses of the individual methods:

3. Present the analysis of the "source multiplication" method using a calibrated control rod.
4. Present the analysis of the "null reactivity" method.
5. What is the common feature of all dynamic methods of reactivity measurements?
6. Classify and list the dynamic methods.
7. Present the analysis of the rod-drop method.
8. Present the analysis of the rod-jerk method.
9. Present the analysis of the source-jerk method.

10. Describe the transient caused by a reactivity oscillation.
11. Give the idea and the formula of the inverse point kinetics method.
12. Give the idea and formula of the inverse point kinetics method for reactivity increments.

REFERENCES

1. T. Gozani, "Consistent Subcritical Fast Reactor Kinetics," p. 109 in *Dynamics of Nuclear Systems,* D. L. Hetrick, Ed., The University of Arizona Press, Tucson (1972).
2. W. G. Davey and W. C. Redman, *Techniques in Fast Reactor Critical Experiments,* an AEC monograph, Gordon and Breach Science Publishers, New York (1970).
3. "Reactivity Measurements," technical report IAEA-108, p. 259, International Atomic Energy Agency, Vienna (1969).
4. D. H. Shaftman, "Estimation of Degree of Subcriticality of ZPR Fast-Criticals Configurations by Methods of Neutron Source Multiplication," ZPR-TM-102, Argonne National Laboratory (Apr. 1972).
5. E. F. Bennett, "Methods and Errors in Subcriticality Measurements by Rod-Drop-Flux-Profile Analysis," ZPR-TM-139, Argonne National Laboratory (Apr. 1972).
6. A. R. Buhl, J. C. Robinson, and E. T. Tomlinson, "Intercomparison of Nonperturbing Techniques for Inferring the Reactivity of Fast Reactors," *Nucl. Technol.,* **21,** 67 (Jan. 1974).
7. E. F. Bennett, S. G. Carpenter, C. E. Cohn, and D. H. Shaftman, "Argonne Experience in Measurement of Reactivity of Subcritical Fast Reactor System," *Trans. Am. Nucl. Soc.,* **16,** 290 (1973).
8. E. P. Gyftopoulos, "General Reactor Dynamics," p. 175 in *The Technology of Nuclear Reactor Safety,* Vol. 1, *Reactor Physics and Control,* T. J. Thompson and J. G. Beckerley, Eds., The MIT Press, Cambridge, Massachusetts (1964).
9. M. Becker, "A Generalized Formulation of Point Nuclear Reactor Kinetics Equations," *Nucl. Sci. Eng.,* **31,** 458 (1968).
10. J. F. Walter and A. F. Henry, "The Asymmetric Source Method of Measuring Reactor Shutdown," *Nucl. Sci. Eng.,* **32,** 332 (1968).
11. H. W. Glauner and G. Heusener, "Eine statische Methode zur Unterkritikalitätsbestimmung an Schnellen Reactoren," KfK-1148, Kernforschungszentrum Karlsruhe, FRG (Jan. 1970).
12. R. C. Kryter, N. J. Ackermann, Jr., and A. R. Buhl, "Measurement of Subcriticality in Large Fast Reactors by Combining Noise and Multiplication Techniques," *Trans. Am. Nucl. Soc.,* **14,** 42 (1971).
13. B. E. Simmons and J. S. King., "A Pulsed-Neutron Technique for Reactivity Determination," *Nucl. Sci. Eng.,* **3,** 595 (1958).
14. T. Gozani, "A Modified Procedure for Evaluation of Pulse Source Experiments in Sub-Critical Reactors," *Nukleonik,* **4,** 348 (1962).

15. E. Garelis and J. L. Russell, Jr., "Theory of Pulsed Neutron Source Measurements," *Nucl. Sci. Eng.*, **16,** 263 (1963).
16. H. C. Corben, "The Computation of Excess Reactivity from Power Traces," *Nucl. Sci. Eng.*, **5,** 127 (1959).
17. K. O. Ott, "Theorie verzögert überkritischer Exkursionen zur Messung der Doppler-Koeffizienten schneller Reaktoren," *Nukleonik,* **5,** 285 (1963); or see English translation, KFK-153, Kernforschungszentrum Karlsruhe, FRG (1963).
18. S. G. Carpenter, "Reactivity Measurements in the Advanced Epithermal Thorium Reactor (AETR) Critical Experiments," *Nucl. Sci. Eng.*, **21,** 429 (1965).
19. S. G. Carpenter and R. W. Goin, "Rod Drop Measurements of Subcriticality," ANL-7710, p. 206, Argonne National Laboratory (Jan. 1971).
20. C. E. Cohn, "Experience with Subcriticality Determination by Rod Drop in the FTR-3 Critical Experiments," *Trans. Am. Nucl. Soc.*, **14,** 29 (1971).
21. C. E. Cohn, "Subcriticality Determined by Rod Drop in the FTR-3 Critical Experiments," ANL-7910, p. 203, Argonne National Laboratory (Jan. 1972).
22. R. L. Murray, C. R. Bingham, and C. F. Martin, "Reactor Kinetics Analysis by an Inverse Method," *Nucl. Sci. Eng.*, **18,** 481 (1964).
23. C. Y. Yang and R. W. Albrecht, "Subcriticality Determination by a Novel Inverse Kinetics Technique," *Trans. Am. Nucl. Soc.*, **16,** 297 (1973).
24. W. Häfele, K. Ott, L. Caldarola, W. Schikarski, K. P. Cohen, B. Wolfe, P. Greebler, and A. B. Reynolds, "Static and Dynamic Measurements on the Doppler Effect in an Experimental Fast Reactor," *Proc. Third Int. Conf. Peaceful Uses of Atomic Energy,* Vol. 6, paper 644, International Atomic Energy Agency, Vienna (1964).
25. L. B. Luck, D. J. Malloy, F. J. Martin, and K. O. Ott, "Exact Inverse Space Energy Corrected Kinetics," unpublished report, Purdue University (Aug. 1974).

Ten

DYNAMICS WITH PROMPT REACTIVITY FEEDBACK

Reactivity feedback is the phenomenon that occurs when an originally applied reactivity changes the state of the system, normally via a change in the neutron flux and subsequent effects. This change in the reactor causes a change in reactivity, $\delta\rho_{fb}$, which in turn feeds back to and modifies the original reactivity.

Reactivity feedback is called "inherent" if its occurrence is based on an unavoidable and thus totally reliable physical phenomenon. An example is the Doppler broadening of resonances that is directly associated with the fuel temperature. Doppler broadening of resonances automatically leads—through reduction in resonance self-shielding—to an increase in neutron absorption and thus to a negative and inherent reactivity feedback (note that resonance capture dominates resonance fission in all power reactors).

Such feedback is called "prompt" if it directly follows the changing fuel temperature. If additional physical phenomena such as material motion or heat transfer are required to produce a certain feedback effect, a delay exists between the energy production in the fuel and the feedback. The delay of axial fuel expansion is small enough that this expansion can also be considered a prompt feedback effect.

Power transients under the influence of temperature-stimulated counteraction (feedback) have been investigated since the realization of the first chain reaction. Feedback, especially prompt and inherent reactivity feedback, is vitally important for the safety of nuclear reactors. If a reactor should become superprompt critical, the neutron flux could—in the absence of feedback—rise so rapidly that mechanical shutdown devices could not reduce the power before it reached a destructive level. Fortunately, immediately counteracting shutdown phenomena that are *inherent* in the fuel rapidly reduces the superprompt-critical reactivity to subprompt-critical values. If the reactivity quickly returns to subprompt-critical values, a rapid decrease in the reactor power results. Consequently, the total energy release of such a transient can stay below damaging levels or cause only limited damage. Since the counteracting reactivity feedback is inherent, the resulting shutdown effect is physically inevitable and cannot fail to occur. The impossibility of failure distinguishes *inherent* phenomena from *man-made* shutdown devices; the latter have finite failure probabilities. The presence of strong inherent shutdown phenomena is extremely important for any commercial nuclear power production.

Several types of feedback reactivities occur in nuclear reactors. If the initial power is virtually zero, the feedback reactivity can be modeled in most cases as being linearly proportional to the energy produced.[1] The applicability of this "linear energy feedback model" to superprompt-critical transients in a fast critical assembly such as Godiva[a] was shown by Wimmett et al.[2] More complicated feedback models are required to analyze power reactors.[3]

The initial power in the Godiva experiments[2] was practically zero. It was therefore possible to insert a given amount of reactivity prior to the shutdown phase of the transient. Then the externally applied reactivity is practically constant during the shutdown phase; it can be approximated by a "step" insertion. For superprompt-critical transients in operating reactors, "ramp" reactivity insertions must be investigated. The basic theory for such transients was developed by Nyer.[4] The theories developed by Nyer and Forbes were tested and adjusted in the SPERT (Special Power Reactor Tests) program.[5]

Normally, safety investigations require the treatment of transients below and slightly above prompt critical. In such transients, the effect of the delayed neutrons must be considered in more detail than was required in Refs. 2 to 5. Delayed prompt and superprompt-critical transients in fast reactors have been extensively investigated in the SEFOR experimental program.[6–8] The Doppler reactivity effect in SEFOR transients was experimentally established as a reliable mechanism to reduce a superprompt reactivity inserted into an operational core to a subprompt-critical value.

10-1 The Prompt Feedback Reactivity

10-1A The Fuel Temperature Rise

The rise of the fuel temperature during a transient is due to the energy produced directly or indirectly in fission reactions, which amounts

---

[a]Godiva is a 54-kg bare uranium metal spherical assembly, ~90% enriched in ^{235}U.

to ~200 MeV per fission. The directly produced fission energy appears in the form of two fission products. Small additional contributions are due to scattering and capture of fission neutrons. The γ quanta emitted along with neutron capture amount to ~5 to 7 MeV per capture (~10 MeV per fission). Most of the original energy of the fission neutrons is lost in scattering (~5 to 6 MeV per fission). A further contribution is due to the radioactive decay of fission products. For an accurate treatment of feedback during transients, these contributions should be subdivided into the instantaneously released energy and the delayed released energy. Both components are further subdivided into a part deposited in the fuel and a part deposited outside of the fuel material, such as in the cladding, coolant, moderator, and blanket or reflector.

The major part of the energy released in the fission process (~83%) appears as the kinetic energy of the two fission fragments. The fission fragments emerge as highly ionized atoms. They lose their energy through ionization within a very small range ($\lesssim 10^{-2}$ mm in the fuel). Ion pairs produced directly by the fission products also quickly dissipate their energy. Thus, after a negligibly short time ($\delta t \ll \Lambda$) the energy of the fission fragments is converted into a temperature rise of a very small volume (fission track). A smooth spatial distribution of the heat production from fission is obtained from superposition of many fission tracks and by rapid heat transfer over the very short distances between the tracks. The range of the fission products is so small compared to typical fuel rod diameters that almost all of the energy of the fission fragments is deposited in the fuel.

A smaller part of the total energy per fission (~11%) consists of practically "instantaneous" gamma rays and of the energy directly deposited by neutrons in elastic scattering. "Instantaneous" in this context means that the energy is deposited with an insignificant delay, i.e., during about one neutron generation time after its production. The "instantaneous" gamma-ray energy results directly from the fission process, from inelastic neutron scattering or from radiative capture. As the mean free paths of neutrons are much larger than the fuel rod diameter, this part of the energy per fission is produced in all components of the core and even transported into the reflector or blanket. Still, the largest part of it is produced in the fuel—either directly from fission or from inelastic scattering and radiative capture processes as both reactions occur predominantly in the fuel. Furthermore, since the fuel has the highest reaction cross section[b] for gamma rays in the core, a large portion of this gamma-ray energy is also deposited in the fuel.

---

[b]The cross sections for Compton effect, pair production, and photo effect increase with Z, the number of protons, as Z, Z^2, and Z^4 to Z^5, respectively.

A third part of the fission energy, the decay heat, appears with a delay that is significant in dynamics. As a result of these delays and of the transport of some of the energy out of the fuel, ~10% less energy per fission is deposited in the fuel during a rapid transient than during steady-state operation. That fraction is either directly deposited in the cladding and coolant or set free with delay in the form of beta particles and gamma rays at the time of fission product beta decays.

The conversion of the number of fissions into the various components of heat must reflect the distinctions discussed above. If C_{fsh} represents the heat per fission in a stationary system where *fsh* indicates "*f*ission energy → *s*tationary *h*eat," the stationary total power P_0^{total} is written as:

$$P_0^{total} = C_{fsh} \int_V \int_E \Sigma_{f0} \phi_0 \, dE \, dV \quad , \tag{10.1}$$

with

$$C_{fsh} \simeq 0.35 \times 10^{-10} \quad , \text{W·s/fission} \quad . \tag{10.2}$$

These equations include the contribution of gammas from neutron capture and the delayed energy release.

The power increase during a rapid transient, however, must be calculated with a different conversion factor, C_{fph}, which describes the rapid or *p*rompt conversion of *f*issions into *h*eat. If C_{fdh} describes the production of *d*elayed *h*eat (through decay of fission products), the following balance exists:

$$C_{fsh} = C_{fph} + C_{fdh} \quad . \tag{10.3}$$

Typical values obtained from experimental investigations of the fission process are:

$$\frac{C_{fph}}{C_{fsh}} \simeq 0.94 \tag{10.4a}$$

and

$$\frac{C_{fdh}}{C_{fsh}} \simeq 0.06 \quad . \tag{10.4b}$$

During rapid transients, only the heat promptly produced needs to be considered. In very slow transients, a portion of the delayed heat also has to be considered, which complicates the treatment of these transients. A further complication results from the explicit treatment of the heat that is directly deposited in the cladding and coolant through gamma rays or neutrons. The gross space distribution of the heat deposition by gammas and neutrons differs from that of fission fragments and beta particles. The treatment of these complications is beyond the scope of this book. Only the prompt heat production in the fuel will be treated explicitly. The conversion factor C_{fhf} denotes the corresponding constant, which converts the number of *f*issions into promptly available *h*eat in the *f*uel. The promptly deposited heat amounts to 90% of the stationary heat. Thus, the conversion constant C_{fhf} has the value

$$C_{fhf} \simeq 0.9 \cdot C_{fsh} \quad . \tag{10.5}$$

The remaining 10% of the heat production, which is delayed and/or is deposited outside of the fuel rods, is not explicitly considered in this chapter because of the emphasis on dynamics with prompt reactivity feedback.

Safety considerations and accident analyses are concerned primarily with the behavior of the power in the reactor core. The ~3 to 10% contribution that leaves the core in the form of neutrons or gammas or that is produced by fission in a fast breeder reactor (FBR) blanket normally has no effect on an accident progression. Thus, in this chapter, $P(t)$ denotes the power of the core.

Generally, transients that start from a stationary state are considered. The power transient $P(t)$ can then be split into the stationary component P_0 and the additional power $\delta P(t)$:

$$P(t) = P_0 + \delta P(t) \quad , \tag{10.6}$$

with

$$P_0 = C_{fsh} \int_{\text{core}} \int_E \Sigma_{f0}\phi_0 \; dE \; dV \tag{10.7a}$$

and

$$\delta P(t) = C_{fhf} \int_{\text{core}} \int_E (\Sigma_f \phi - \Sigma_{f0}\phi_0) \; dE \; dV \quad . \tag{10.7b}$$

The distinction between prompt and delayed heat release is to be considered in the calculation of the feedback effects for rapid transients.

The increase in the fuel temperature results from the additionally produced fission energy. Subsequent heat transfer leads to a delayed increase of cladding and coolant temperature; this increase is in addition to the instantaneous and delayed energy deposition through gamma rays and neutrons. Since the transfer of heat out of a material with a low heat conductivity such as UO_2-PuO_2 is a fairly slow process [see Eq. (10.12)], simplified heat transfer models are used in this chapter.

During rapid transients, most of the energy that is deposited in the fuel raises the fuel temperature and is not transferred to the cladding or coolant. In survey calculations, the heat transfer out of the fuel, therefore, can be treated by simple approximations; for example, it can be partially neglected. The neglect of heat transfer of the fuel, i.e., the adiabatic boundary condition of thermodynamics at the fuel cladding boundary, can be applied either to all of the heat or only to the incrementally produced heat. This results in the following two formulas for the average fuel temperature rise $\delta T(t)$ after the onset of a transient (the effect of the space distribution of the temperature is discussed below):

$$\delta T(t) = C_{QT} \int_0^t P(t')\, dt' \quad \begin{cases} \text{completely} \\ \text{adiabatic} \\ \text{fuel} \end{cases} \tag{10.8}$$

or

$$\delta T(t) = C_{QT} \int_0^t [P(t') - P_0]\, dt' \quad \begin{cases} \text{adiabatic} \\ \text{fuel for} \\ \text{power} \\ \text{increment} \end{cases} , \tag{10.9}$$

where P denotes the total power in the core; C_{QT}, the conversion factor between energy and temperature, is calculated below (see also Table 10-I).

The approximation, Eq. (10.8), is the same as Eq. (10.9) if the initial

TABLE 10-I
Typical Thermodynamics Data for Several Reactor Fuel Materials*

	Typical Density	Specific Heat, c_p			Thermal Conductivity		Heat Capacity per mm^3, c_h
	$\rho_d\left[\frac{g}{mm^3}\right]$	$\left[\frac{Btu}{lb\cdot°F}\right]$	$\left[\frac{cal}{g\cdot K}\right]$	$\left[\frac{W\cdot s}{g\cdot K}\right]$	$\left[\frac{Btu}{ft\cdot h\cdot °F}\right]$	$\left[\frac{W\cdot s}{mm\cdot s\cdot K}\right]$	$\left[\frac{W\cdot s}{mm^3\cdot K}\right]$
UO_2-PuO_2	10.0×10^{-3}	0.85	0.076	0.32	1.7	0.003	3.2×10^{-3}
UC-PuC	12.2×10^{-3}	0.70	0.062	0.26	11.0	0.019	3.2×10^{-3}
U-Pu-Zr (metal)	16.0×10^{-3}	0.56	0.050	0.21	13.9	0.024	3.4×10^{-3}
UN-PuN	12.9×10^{-3}	0.66	0.059	0.25	13.3	0.023	3.2×10^{-3}

*The specific heat data were taken from Refs. 11 through 14; the remaining data were inferred from data presented in Ref. 15.

Note: 1 cal = 0.0039666 Btu = 4.185 W·s;
1 lb = 454 g;
$1\ \frac{Btu}{ft\cdot h\cdot °F} = 733 \times 10^{-3}\ \frac{W\cdot s}{mm\cdot s\cdot K}$.

power is negligible compared to the power during the transient. Equation (10.8) was applied in the Manhattan Project, particularly by Fuchs,[1] Nordheim,[9] and Hansen.[10] This approximate model of complete neglect of heat transfer is often referred to as the "linear energy model," or it is named after the above authors.

The adiabatic model for the incremental energy, Eq. (10.9), is based on the following argument: The heat current out of the fuel rod is proportional to the negative gradient of the temperature distribution. The distribution of the additionally produced heat is essentially flat over the rod; thus, during a rapid transient, the temperature distribution is merely lifted up without any significant change in the shape of the distribution. Thus the derivative is approximately maintained and with it the initial heat release rate out of the fuel rod.

The adiabatic model can be improved in a first approximation by treating the heat release out of the fuel with a single time constant, λ_H:

$$\delta Q_{\text{remaining in fuel}} = \int_0^t \delta P(t') \exp[-\lambda_H(t - t')]\, dt' \quad , \tag{10.10a}$$

$$\delta Q_{\text{released from fuel}} = \int_0^t \delta P(t')\{1 - \exp[-\lambda_H(t - t')]\}\, dt' \quad , \tag{10.10b}$$

and

$$\delta Q_{\text{total}} = \int_0^t \delta P(t')\, dt' \quad . \tag{10.10c}$$

The heat release constant λ_H for a cylindrical fuel rod of radius R with uniform heat production rate and temperature-independent heat conductivity is proportional to $1/R^2$:

$$\lambda_H \propto \frac{1}{R^2} \quad . \tag{10.11}$$

For a 6-mm (~0.25-in.)-diam oxide fuel rod, one has

$$\lambda_H \simeq \frac{1}{2\ s} \quad . \tag{10.12}$$

The energy-temperature conversion constant for the fuel C_{QT} [K/MW·s] in Eqs. (10.8) and (10.9) is given by

$$C_{QT}\left[\frac{\text{K}}{\text{MW·s}}\right] = \frac{1}{c_h}\left[\frac{\text{K·mm}^3}{\text{W·s}}\right]\frac{1}{V_f\,[\text{mm}^3]} \cdot \frac{\text{W}}{\text{MW}} \quad , \tag{10.13}$$

where V_f is the fuel volume. The fuel heat capacity per unit volume, c_h, is given by the product of the specific heat, c_p, and density, ρ_d:

$$c_h \left[\frac{\mathrm{W\cdot s}}{\mathrm{K\cdot mm^3}}\right] = c_p \left[\frac{\mathrm{W\cdot s}}{\mathrm{K\cdot g}}\right] \rho_d \left[\frac{\mathrm{g}}{\mathrm{mm^3}}\right] \quad . \tag{10.14}$$

Typical values for these quantities are given in Table 10-I in international and British units.

The conversion factor C_{QT} represents the rise of the average fuel temperature in the core per MW·s heat energy. Thus, if δQ_{cf} is the energy deposited in the core fuel, the corresponding temperature increase is given by:

$$\delta T(t) = C_{QT}\delta Q_{cf}(t) \quad . \tag{10.15}$$

According to Eq. (10.13), C_{QT} is inversely proportional to the fuel volume, whereas δQ_{cf} is proportional to the fuel volume for a given rise in the specific power, or in the power per unit fuel volume. Equation (10.15) can then be rewritten as:

$$\delta T(t) = C_{QT}V_f \frac{\delta Q_{cf}(t)}{V_f} \quad , \tag{10.16a}$$

$$\delta T(t) = \frac{1}{c_h}\left[\frac{\mathrm{K\cdot mm^3}}{\mathrm{W\cdot s}}\right] \delta q_{cf} \left[\frac{\mathrm{W\cdot s}}{\mathrm{mm^3}}\right] \quad , \tag{10.16b}$$

and

$$\delta T(t) = C_{qT}\delta q_{cf}(t) \quad , \tag{10.16c}$$

where q_{cf} is the energy *density* deposited in the fuel and C_{qT} the corresponding conversion factor.

For the calculation of $\delta T(t)$, it is convenient to introduce a separation of the power into the nominal power, P_n (usually called "full power"), and the *flux* amplitude function $p(t)$,

$$P(t) \simeq P_n p(t) \quad , \tag{10.17}$$

The "$\simeq$" sign needs to be used in Eq. (10.17) since $p(t)$ describes only approximately the amplitude of the power. In this approximation then, p_0 represents the ratio of initial to nominal power. Inserting this separation into Eq. (10.15) with δQ taken from Eq. (10.10c) gives

$$\delta T(t) = \frac{1}{c_h}\frac{P_n\,[\mathrm{W}]}{V_f\,[\mathrm{mm^3}]}\int_0^t \delta p(t')\,dt' \quad . \tag{10.18}$$

The integral over the amplitude difference, $\delta p(t) = p(t) - p_0$, was introduced earlier as:

$$I(t) = \int_0^t [p(t') - p_0]\, dt' \quad . \tag{10.19}$$

The integral $I(t)$ describes the additional energy in dimensionless units, which are called here "relative full-power seconds." For example, $I(t) = 1.5$ gives an energy release of 1.5 full-power seconds (fp-s). The conversion factor in Eq. (10.18) is denoted by C_{IT}, in analogy with other notations:

$$\delta T(t) = C_{IT} I(t) \quad . \tag{10.20}$$

The important conversion factor C_{IT} has the dimension of K/s; it is the rate of temperature rise per second of (uncooled) nominal power application:

$$C_{IT}\ [\text{K/fp-s}] = \left(\frac{\partial T}{\partial t}\right)_{P_n} \quad . \tag{10.21}$$

Note that "fp-s" has the dimension "time" and not the dimension of "energy" as the product P_n·second or fp·s:

$$C_{IT}\ [\text{K/s}] = C_{IT}\ [\text{K/fp-s}] \quad . \tag{10.22}$$

For example, for oxide fuel with an operational power density of 0.13 or 1.3 MW/ℓ of fuel (0.13 or 1.3 W/mm^3), typical of boiling water reactors (BWRs) or FBRs, respectively:

$$C_{IT} = \frac{1}{0.0032}\left[\frac{\text{K}\cdot\text{mm}^3}{\text{W}\cdot\text{s}}\right]\cdot\left\{\begin{array}{l}0.13\ \text{W/mm}^3\\ 1.3\ \text{W/mm}^3\end{array}\right.$$

$$C_{IT} \simeq \left\{\begin{array}{l}40\ \text{K/fp-s}\\ 400\ \text{K/fp-s}\end{array}\right. ; \tag{10.23}$$

i.e., 40 or 400 K per full-power second. For pressurized water reactors, C_{IT} is about twice as large as for BWRs.

The relations presented above hold for the average fuel temperature in the core, which was denoted by T for simplicity. The fuel temperature within a fuel rod varies strongly around this average. The fuel rod averaged temperature also depends on space. The "average" core fuel temperature, T, was used above as a basis for the calculation of prompt reactivity feedback.

For the fairly flat power distribution across the fuel rods, the temperature distribution within the rod is approximately given by a parabola in the radial coordinate r:

$$T(r) = T_0 - \frac{r^2}{R^2}(T_0 - T_R) \quad , \tag{10.24}$$

where R designates the radius of the fuel, and T_0 and T_R the temperatures at $r = 0$ and $r = R$, respectively. The rod average value, $\overline{T}_{rod}$, can be obtained from Eq. (10.24) as:

$$\overline{T}_{rod} = \frac{1}{2}(T_R + T_0) \quad . \tag{10.25}$$

Equation (10.24) is derived under the assumptions of temperature-independent thermodynamic properties and a space-independent heat production rate within a fuel rod: the latter assumption is only applied in survey investigations. The temperature variation within a fuel rod described by Eq. (10.24) is quite significant due to the high power density and due to the low heat conductivity of the oxide fuel. In fast reactors, $T_0 - T_R$ may be larger than 1000 K across an oxide fuel rod, but only about one-tenth of it for metal fuel.

The Doppler reactivity feedback effect is generally calculated for media with a uniform temperature distribution and not for the strongly space-dependent temperature profile within a fuel rod. An effective temperature therefore must be found that yields the correct Doppler reactivity when it is inserted into the formula based on a uniform temperature distribution. The effect of rod temperature profile on the Doppler effect of neutrons was first investigated for individual resonances by Reichel[16,17] and for a $1/T$ variation of a local Doppler reactivity effect by Greebler and Goldman.[18] Even though the analytical formulas that are derived in both investigations are different due to the different physical assumptions, the numerical results are essentially in agreement for temperatures that correspond to normal power operation. Both investigations yield an effective temperature that is ~3% smaller than the fuel rod averaged temperature. A correction of this type can be applied easily in practical calculations. In this text, the linear averaged temperature $T(= \overline{T})$ is used for simplicity.

The temperature distribution in the core, i.e., the r and z dependence of the rod averaged temperatures, could also be approximated by an effective "average" in analogy to that in the fuel rod. The reactivity effect for "isothermal" cores, $\delta\rho(T_{iso})$, would have to be determined such that the effective "average" temperature change gives the same $\delta\rho$ that would be obtained from the gross temperature distribution in the core. This approach was generally applied in earlier investigations.[18,19]

In more recent investigations, the Doppler feedback reactivity for the gross temperature distribution, $T(r,z)$, is calculated and a detailed feedback reactivity, $\delta\rho[T(r,z)]$ (normally in the form of a regional dependency), is applied directly instead of an isothermal quantity. Consequently, an effective temperature is unnecessary. In space-energy dependent dynamics, the reactivity effect of the local temperature is described by temperature-dependent group constants (see Chapter 11).

10-1B Prompt Feedback

There is only one really prompt feedback effect in thermal as well as in fast reactors, the Doppler reactivity effect. The corresponding feedback loop is as short as possible since the Doppler broadening of resonances occurs simultaneously with the dissipation of the kinetic energy of the fission products into temperature.

A reactivity insertion, $\delta\rho_{in}$, causes a change in power, $\dot{P}(t)$. The time integration of the kinetics equation yields $\delta P(t)$. The subsequent accumulation of energy, $\delta Q(t)$, changes the temperature, $\delta T(t)$, which in turn leads to a change in reactivity, $\delta\rho_{Dop}(t)$, that feeds back to $\delta\rho_{in}$. This is illustrated in Fig. 10-1.

The Doppler reactivity effect is negative in thermal and in large fast reactors.[20] It represents the most important inherent shutdown mechanism and thus helps to assure the safety of nuclear reactors.

The axial expansion of the fuel is the other feedback effect that may be considered as "prompt" in most transients. Axial expansion causes a reduction in the average fuel density in the core due to an elongation of the fuel rods. This then leads to a decrease in the reactivity. The reactivity feedback through axial expansion is not as prompt as the

DOPPLER FEEDBACK LOOP

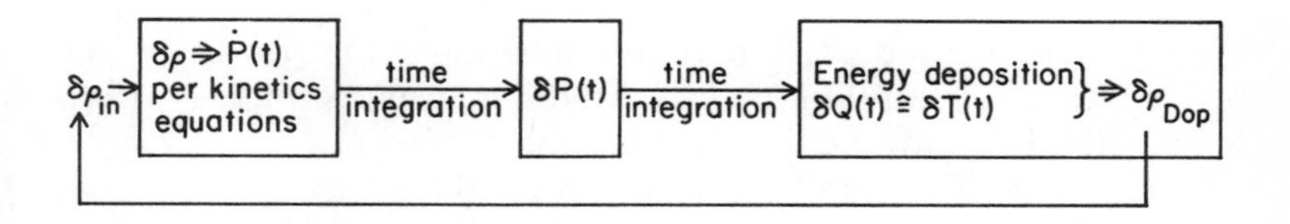

DISPLACEMENT FEEDBACK LOOP

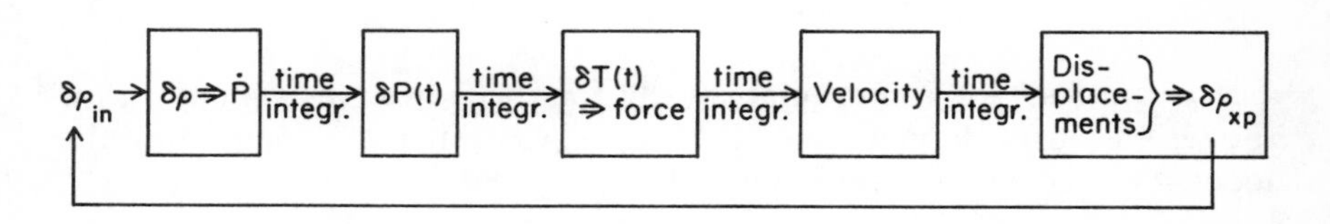

Fig. 10-1. A schematic of Doppler and displacement feedback loops.

Doppler feedback since a temperature rise does not directly cause a reactivity feedback; δT is only the basis of a force. Two further time integrations are required to obtain the axial displacement that causes the reactivity feedback, as illustrated in Fig. 10-1.

The displacement is established almost at the speed of sound in the fuel material (typically ~1000 m/s); the corresponding delay is then ~1 ms. This delay can be neglected in all thermal and fast reactor transients. The only exceptions are in disassembly investigations, such as cases where transients are eventually shut down by the displacement of fuel. The delay of the displacements and thus the difference of the time behavior of feedback through the Doppler effect versus expansion plays a characteristic role in such transients. For all transients discussed here, the axial expansion is considered a prompt feedback effect and appears added to the Doppler feedback.

The reactivity effect of the Doppler broadening of resonances is calculated in the form of a temperature coefficient (compare Sec. 11-5). For the calculation of feedback reactivity acting in rapid transients, it is convenient to convert the temperature coefficient into an energy coefficient by using one of the temperature-energy conversion formulas, Eq. (10.15) or (10.20):

$$C_{QT}\left(\frac{\partial\rho}{\partial T}\right)_{\text{Dop}} = \gamma_e^D\left[\frac{\delta\rho}{\text{MW}\cdot\text{s}}\right] = \text{energy coefficient} \quad , \qquad (10.26\text{a})$$

$$C_{IT}\left(\frac{\partial\rho}{\partial T}\right)_{\text{Dop}} = \gamma^D\left[\frac{\delta\rho}{\text{fp-s}}\right] = \text{full-power second coefficient} \quad . \qquad (10.26\text{b})$$

The latter formula is used predominantly here since the transients are described by the flux amplitude $p(t)$. If the power as such is used, Eq. (10.26a) should be applied to describe the feedback.

Analogous formulas hold for the energy coefficient due to axial fuel expansion, γ^{exp}, in either of the units of Eqs. (10.26). Normally, the different time characteristics of the Doppler and expansion feedback loops need not be distinguished. Then, both energy coefficients appear as a sum:

$$\gamma = \gamma^D + \gamma^{exp} \quad . \qquad (10.27)$$

If the fuel is treated adiabatically for the power increment and if the temperature dependence of the energy coefficient is neglected, the feedback reactivity is simply given by

$$\delta\rho(t) = \gamma\int_0^t [p(t') - p_0]\, dt' \quad . \qquad (10.28)$$

Inclusion of first-order heat transfer yields:

$$\delta\rho(t) = \gamma \int_0^t [p(t') - p_0] \exp[-\lambda_H(t - t')]\, dt' \quad . \tag{10.29}$$

The temperature dependence of the Doppler coefficient, $(\partial\rho/\partial T)_{\mathrm{D}}$, shows approximately a $1/T^x$ behavior:

$$\left(\frac{\partial\rho}{\partial T}\right)_{\mathrm{D}} = \frac{A_D}{T^x} \quad . \tag{10.30}$$

For thermal reactors, x is close to 0.5 and for fast reactors, close to 1.0.

The expansion coefficient is essentially independent of temperature. The temperature dependence of the sum of the two coefficients results primarily from the Doppler effect. If the temperature dependence of the temperature coefficient is taken into account, feedback reactivity in the simple case of the adiabatic approximation for the incremental heat is calculated from

$$\delta\rho(t) = \int_{T_0}^{T(t)} \frac{\partial\rho}{\partial T}\, dT = \int_0^t \frac{\partial\rho}{\partial T}\frac{\partial T}{\partial t'}\, dt' \quad , \tag{10.31}$$

with

$$\frac{\partial T}{\partial t'} = C_{IT}[p(t') - p_0] \quad . \tag{10.32}$$

If the Doppler coefficient is proportional to $1/T$, as in large oxide-fueled fast reactors, a "Doppler constant," $A_D = T(\partial\rho/\partial T)_D$, can be introduced. Its values are in the range of

$$0.003 \lesssim |A_D| \lesssim 0.009 \quad . \tag{10.33a}$$

Division by an approximate operating reactor average fuel temperature of $T \simeq 1500$ K gives

$$2 \times 10^{-6}/\mathrm{K} \lesssim \left|\left(\frac{\partial\rho}{\partial T}\right)_D\right| \lesssim 6 \times 10^{-6}/\mathrm{K} \quad . \tag{10.33b}$$

The values of A_D and $(\partial\rho/\partial T)_D$ for a sodium-cooled fast reactor are in the upper part of the ranges given in Eqs. (10.33); the corresponding values for gas-cooled fast reactors are a factor of ~2 smaller than for sodium-cooled reactors.

In the numerical fast reactor examples presented below, the following coefficient is used without specifying precisely the contributions of Doppler effect and expansion or the value of β:

$$\gamma_T = \frac{\partial\rho}{\partial T} = -0.002\$/\mathrm{K} \quad \text{for FBRs} \quad , \tag{10.34a}$$

Doppler coefficients in thermal reactors with low enrichment fuel are generally much larger than in fast reactors by about one order of magnitude. This is due to the fact that the cross-section resonances that strongly contribute to Doppler feedback (the ones around and below neutron energies of several kiloelectron volts) are exposed to a higher portion of the neutron spectrum in thermal than in fast reactors. Thus,

$$\gamma_T = \frac{\partial \rho}{\partial T} = -0.02\$/\mathrm{K} \quad \text{for LWRs} \quad . \tag{10.34b}$$

Using C_{IT} = 400 K/fp-s or 40 K/fp-s, Eq. (10.23), with Eqs. (10.34), one obtains

$$\gamma = C_{IT}\left(\frac{\partial \rho}{\partial T}\right) = -0.8\$/\mathrm{fp\text{-}s} \quad . \tag{10.35}$$

Since the value of the feedback coefficient γ given in Eq. (10.35) is fairly typical for FBR and LWR oxide-fueled power reactors, it is used throughout the examples and semi-quantitative evaluations presented here. The two values for C_{IT} of Eq. (10.23) are applied for simplicity to distinguish LWR from FBR transient results.

10-2 Transients in the Subprompt-Critical or Subcritical Reactivity Domain

For simplicity, the initially inserted reactivity is approximated in this section by a step, ρ_1. The treatment of transients following a subprompt reactivity step is greatly simplified by employing some of the approximations discussed in previous sections. Even with such simplifications, however, a complete analytical solution has not yet been found. To obtain a semi-quantitative understanding of this important class of transients, the flux at small times, the asymptotic flux behavior, and the time integral, i.e., the total energy release, are investigated. The approximate results obtained from these investigations are compared with complete calculations (normally using six groups of delayed neutrons) presented in the graphs of this chapter.

In many of the approximate evaluations in this chapter, the prompt jump approximation (PJA) is applied. Though most comparisons with complete calculations are for the small Λ values of fast reactors ($\Lambda \simeq 4 \times 10^{-7}$ s), the typical Λ values for LWRs ($\Lambda \simeq 10^{-5}$ s) are small enough to allow the neglect of $\Lambda\dot{p}(t)$. Thus, the PJA is generally a good approximation in reactivity domain $\rho < \beta$ (better, $\rho \lesssim 0.9\beta$).

Some of the approximate evaluations employ only one instead of six delayed neutron groups. The applicability of this approximation or the resulting inaccuracies are largely independent of the neutron generation time as it can be deduced from the inhour equation evaluation of Chapter 6.

Since the feedback coefficients γ measured in dollars per full-power second are similar in magnitude for all oxide-fueled power reactors (see end of Sec. 10-1), the results that depend primarily on γ are applicable to LWRs and FBRs as well.

10-2A The Transient at Small Times

The one-delay-group kinetics provides a good approximation for small times ($t \lesssim 0.2/\bar{\lambda}$) as was shown in Sec. 6-2 (see, for example, Fig. 6-12). The point kinetics equations with feedback reactivity are given by (ρ_1 denotes the initial step reactivity):

$$\Lambda \dot{p} = \left\{ \rho_1 + \gamma \int_0^t [p(t') - p_0]\, dt' - \beta \right\} p + \sum_k \lambda_k \zeta_k \qquad (10.36a)$$

and

$$\dot{\zeta}_k = -\lambda_k \zeta_k + \beta_k p \quad . \qquad (10.36b)$$

The PJA (neglect of $\Lambda\dot{p}$) is employed to further simplify the problem. This leads to the approximate kinetics equation for the flux,

$$0 = \left\{ \rho_1 - \beta + \gamma \int_0^t [p(t') - p_0]\, dt' \right\} p(t) + \sum_k \lambda_k \zeta_k \quad , \qquad (10.37a)$$

which is solved with the initial condition

$$p(0) = p^0 = \frac{\beta}{\beta - \rho_1} p_0 \quad . \qquad (10.37b)$$

Transients analyzed with the PJA may "start" with a pseudo-initial flux p^0, which eliminates the rapid flux change during the prompt jump. The flux variations after the prompt jump are comparatively slow; they can be described by a Taylor series for small t. Note that without the PJA a Taylor expansion at $t = 0$ would result in a tedious description of the prompt jump (see Fig. 6-3) and would not allow a description of $p(t)$ after the prompt jump with only a few terms.

The competitive action of the increasing delayed neutron source and the negative prompt feedback is already exhibited in the second term of the Taylor expansion.[21] Thus, only the first two terms of the Taylor expansion of p, ρ, and ζ are investigated in the following:

$$p(t) = p^0 + p't + \cdots$$
$$\rho(t) = \rho_1 + \rho' t + \cdots$$
$$\zeta_k(t) = \zeta_{k0} + \zeta_k' t + \cdots \quad . \tag{10.38}$$

With $\delta\rho(t)$ from Eq. (10.28), one obtains for $\dot{\rho}$:

$$\dot{\rho} = \gamma[p(t) - p_0]$$
$$= \gamma[p^0 - p_0] + \cdots \quad . \tag{10.39}$$

Inserting the appropriate expansions into Eq. (10.36b) gives the rate of increase of the precursor density immediately after the prompt jump as

$$\zeta_k' = -\lambda\zeta_{k0} + \beta_k p^0$$
$$= \beta_k(p^0 - p_0) \quad . \tag{10.40a}$$

From this rate of precursor increase, the following rate of increase of the delayed neutron source results:

$$s_d' = (p^0 - p_0)\sum_k \beta_k\lambda_k = (p^0 - p_0)\beta\overline{\lambda} \quad . \tag{10.40b}$$

The precursor density and thus the delayed neutron source increase linearly right after the prompt jump. In the PJA, the increase in the delayed neutron source causes an instantaneous linear increase in the flux level. On the other hand, the instantaneous counteraction of the prompt feedback tends to decrease the flux. Let p_{del}' and p_{fb}' denote the two parts of the total slope that correspond to delayed neutrons and prompt feedback, respectively:

$$p' = p_{del}' + p_{fb}' \quad . \tag{10.41}$$

Both parts can be obtained by inserting the Taylor expansion into Eq. (10.37a); using s_d' from Eq. (10.40b) gives:

$$0 = [\rho_1 - \beta + \gamma(p^0 - p_0)t]\,(p^0 + p't)$$
$$+ \beta p_0 + (p^0 - p_0)\beta\overline{\lambda}t \quad . \tag{10.42}$$

At $t = 0$, this equation reduces to

$$(\rho_1 - \beta)p^0 + \beta p_0 = 0 \quad , \tag{10.43}$$

which is satisfied by the definition of p^0, Eq. (10.37b). The slope is obtained from the terms proportional to t, which are given by:

$$[\rho_1 - \beta]p' + \gamma(p^0 - p_0)p^0 + (p^0 - p_0)\beta\overline{\lambda} = 0 \quad . \tag{10.44}$$

Omitting the feedback yields the contribution of the increasing precursor population to the slope; Eq. (10.44) gives for $\gamma = 0$,

$$p'_{del} = \bar{\lambda} \frac{\beta}{\beta - \rho_1} (p^0 - p_0) \quad . \tag{10.45a}$$

Inserting $p^0 = \beta p_0/(\beta - \rho_1)$ from Eq. (10.37b) brings p'_{del} into the form

$$p'_{del} = \bar{\lambda} \frac{p^0}{p_0} (p^0 - p_0) \quad . \tag{10.45b}$$

Thus, the linear *increase* of the flux due to the increase in the delayed neutron source is proportional to the product of flux and flux jump.

Omitting the increase of precursors in Eq. (10.44) yields the influence of feedback on the variation of the flux after the prompt jump:

$$\begin{aligned} p'_{fb} &= \gamma \frac{(p^0 - p_0)}{\beta - \rho_1} p^0 \\ &= \frac{\gamma}{\beta} \frac{p^0}{p_0} (p^0 - p_0) p^0 \quad . \end{aligned} \tag{10.46}$$

Thus, the counteraction of prompt reactivity feedback leads to a linear *decrease* of the flux, which contains one further flux factor, p^0, compared to p'_{del}, since $\dot{p}$ is influenced by the product of $\delta\rho \cdot p$.

Adding p'_{del} and p'_{fb} and bracketing out the common factor gives

$$p' = \left[\bar{\lambda} + \frac{\gamma}{\beta} p^0 \right] \frac{p^0}{p_0} (p^0 - p_0) \quad , \tag{10.47}$$

which shows that for small p^0 the slope p' can be (and practically is) positive, but for large prompt jumps, the feedback starts to turn the flux down immediately after the prompt jump.

For one special prompt jump, p^{00}, the linear flux change after the prompt jump vanishes. From Eq. (10.47) follows

$$-\frac{\gamma}{\beta} p^{00} = \bar{\lambda} \quad . \tag{10.48}$$

If p is considered as the power, then Eq. (10.48) can be interpreted as the influence of feedback and delayed neutrons on the flux change after the prompt jump is cancelled, i.e., when the energy coefficient times the prompt jump power is equal to the average precursor decay constant.

Subprompt-critical transients of this type were proposed by one of the authors for the measurement of the Doppler coefficient in the SEFOR experimental fast reactor.[6,21] Figure 10-2 shows a computer

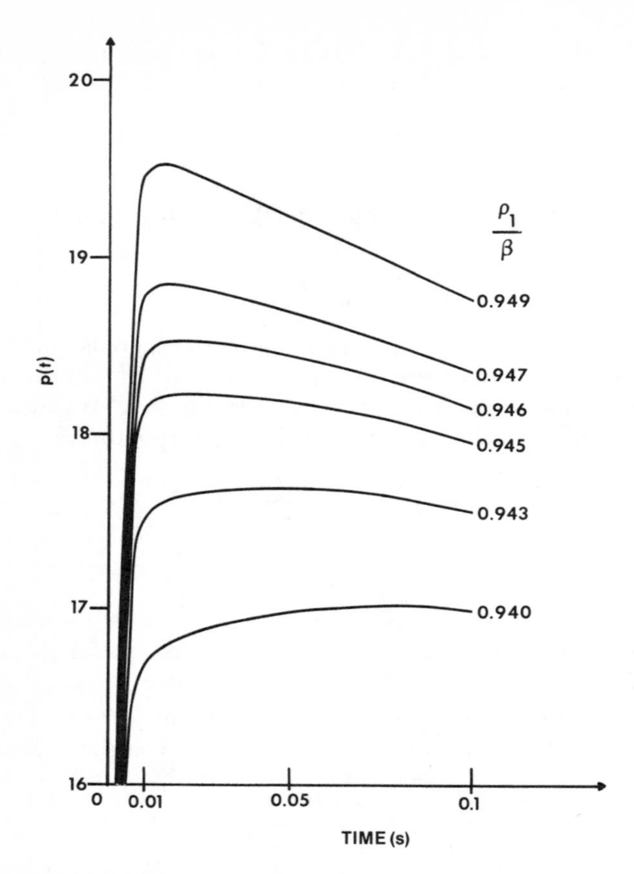

Fig. 10-2. Delayed supercritical ρ-step-induced transients.[21]

simulation[21] of transients introduced by a delayed supercritical reactivity step around the special transient for which the slope after the prompt jump is practically zero. The lower curve ($\rho_1/\beta = 0.940$) is still rising after the prompt jump; the upper curves in Fig. 10-2 are clearly decreasing immediately after the prompt jump. Around $\rho_1/\beta \simeq 0.945$, the slope is practically zero after the prompt jump. The corresponding value of p^{00} is ~18 times the initial flux. The energy coefficient obtained from Eq. (10.48) is:

$$\frac{\gamma}{\beta} = -\frac{\bar{\lambda}}{p^{00}} \quad . \tag{10.49}$$

With $\bar{\lambda} \simeq 0.40$ (^{235}U was used as fuel and $P_0 = 10$ MW in the computer simulation of Ref. 21) and $p^{00} \simeq 18\, p_0$, Eq. (10.49) yields

$$\gamma \simeq -0.022\$/\text{fp-s} \tag{10.50a}$$

or

$$\gamma \simeq -2.2 \times 10^{-3} \frac{\$}{\text{MW·s}} \quad . \tag{10.50b}$$

The value 2.2×10^{-3}, which is obtained from the simulated analysis, is 2% different from the value of 2.15×10^{-3} \$/MW·s actually used to calculate the flux transients of Fig. 10-2 from the kinetics equations with six delayed groups. This comparison also shows the accuracy of the PJA with a single delayed group in a transient with feedback. The value of γ, Eq. (10.50a), is ~35 times lower than that given in Eq. (10.35) since the power density in this computer simulation is ~16 times smaller and the β of the ^{235}U fuel is about two times larger than in a fast power reactor.

Note that the knowledge of ρ_1 is not required in such a measurement of the Doppler coefficient.[c] The special transient—zero slope after the prompt jump—may even be used to determine ρ_1 (in addition to the Doppler coefficient). SEFOR was equipped with a fast rod extraction device (FRED), which allowed the rapid insertion of well-defined reactivities.[22] In practice, however, it is too tedious to find the specific transient (with p^{00}). If inverse kinetics is applied in analyzing such transients (as was also proposed in Ref. 21 and further investigated in Ref. 23), the special transient p^{00} is not required. The actual analysis of the delayed supercritical transients in SEFOR was performed with inverse kinetics.

The results for the two SEFOR cores were

$$T \frac{d\rho}{dT} \simeq -0.008 \text{ Core I} \quad \text{(Refs. 24 and 25)}$$

$$T \frac{d\rho}{dT} \simeq -0.006 \text{ Core II} \quad \text{(Refs. 25 and 26)} \quad .$$

Both results are in close agreement with theoretical predictions as discussed in the references given above [compare these values also with Eq. (10.35)]. The neutron spectrum in the small SEFOR was softened by BeO rod additions to simulate closely the spectrum in a large reactor. The fuel rod diameter in SEFOR was 24.5 mm (1 in.), which allows the high fuel temperatures of an operating power reactor with the low power

[c]The SEFOR fuel elements were specifically designed such that the expansion effect is suppressed. Thus, virtually the entire energy coefficient results from the Doppler effect.

density of SEFOR. Since the neutron spectrum in SEFOR as well as the fuel temperature distribution closely simulate the corresponding properties in a large power reactor, the measured Doppler coefficients are also representative for large reactors.

10-2B The Asymptotic Transients

In addition to the expansion at $t = 0$, Eqs. (10.38), an asymptotic expansion of Eq. (10.37a) can be derived.[27] It follows from such a derivation that Eq. (10.37a) has a constant asymptotic ("*as*") solution. With $\dot{p} = 0$, the kinetics equation in the PJA, Eq. (8.10), assumes the form

$$0 = \frac{\lambda\rho_{as} + \dot{\rho}_{as}}{\beta - \rho_{as}} \quad . \tag{10.51a}$$

This requires that the numerator vanish, i.e.,

$$\lambda\rho_{as} + \dot{\rho}_{as} = 0 \quad . \tag{10.51b}$$

The asymptotic time-independent solution of Eq. (10.51b) is

$$\rho_{as} = 0 \text{ since } \dot{\rho}_{as} = 0 \quad . \tag{10.52}$$

The approach to the asymptotic solution above is described by the time-dependent solution of Eq. (10.51b), i.e.,

$$\rho_{as}(t) \propto \exp(-\lambda t) \quad . \tag{10.53}$$

Since $\rho_{as} = 0$, the inserted reactivity is asymptotically compensated for by the feedback:

$$-\gamma \int_0^{\infty} [p(t') - p_0]\, dt' = \rho_1 \quad , \tag{10.54}$$

in which the integral has to converge, which requires $p(t)$ to converge sufficiently fast toward p_0. Thus, a feedback reactivity given by the left side of Eq. (10.54) requires as the asymptotic flux:

$$p_{as} = p_0 \quad . \tag{10.55}$$

Under the assumptions on which Eqs. (10.37) are based, the prompt feedback forces the power back to its initial value.

The fact that the asymptotic flux, p_{as}, with feedback described by the left side of Eq. (10.54), cannot be smaller than p_0 can be shown by the following physical argument. If $p(t) < p_0$ in a certain time interval, the contribution of this interval to the integral is negative. This physically means that the fuel temperature is decreasing. Decreasing fuel temperature together with a negative temperature coefficient leads to a *reactivity increase,* which is followed by an increase in the flux.

Consider as an example the highest of the curves in Fig. 10-2 (P_0 = 10 MW). According to Eqs. (10.55) and (10.54), the transient is supposed to converge to p_0 and its integral is approximately given by

$$\int_0^\infty [p(t') - p_0]\, dt'$$

$$= -\frac{\rho_1}{\gamma}\left(= \frac{0.949}{2.15 \times 10^{-3}}\ \text{MW·s} \simeq 430\ \text{MW·s}\right) \quad . \tag{10.56}$$

Figures 10-3 and 10-4 show two additional examples calculated with a typical power reactor energy coefficient, $\gamma = -0.8\$/\text{fp-s}$ [see Eq. (10.35)]. The dashed lines show two transients for $\rho_1 = 0.95\$$ and $\rho_1 = 0.5\$$ with reactivity feedback given by Eq. (10.37a). The energy releases according to Eq. (10.56) are 1.19 and 0.625 fp-s, respectively. The dashed lines of Figs. 10-3 and 10-4 cross the initial flux levels at ~3 s and the flux approaches the asymptotic value, p_0, from below. This overshooting of the final flux level is required as can be seen from the following consid-

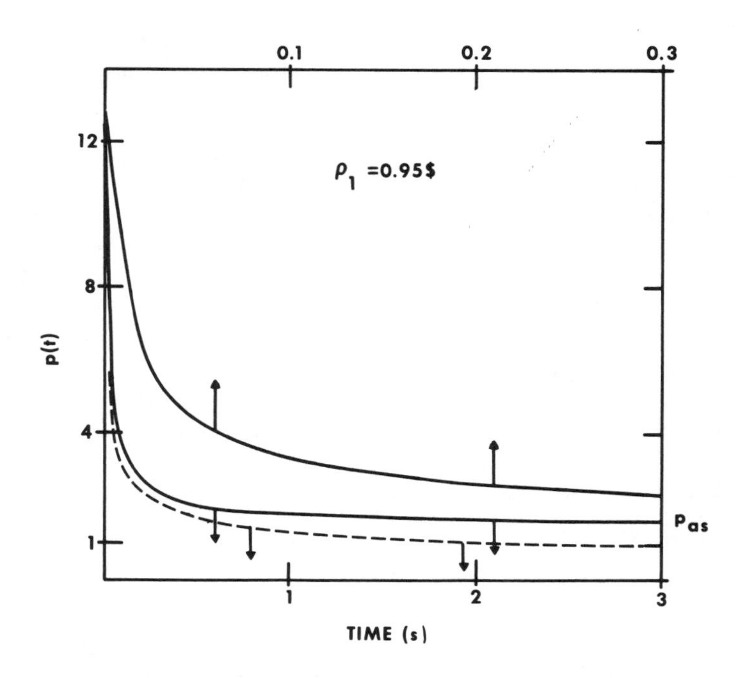

Fig. 10-3. Subprompt-critical transient (ρ_1 = 0.95\$, p_0 = 1) with reactivity feedback including first-order heat transfer (solid line) and no heat transfer (dashed line).

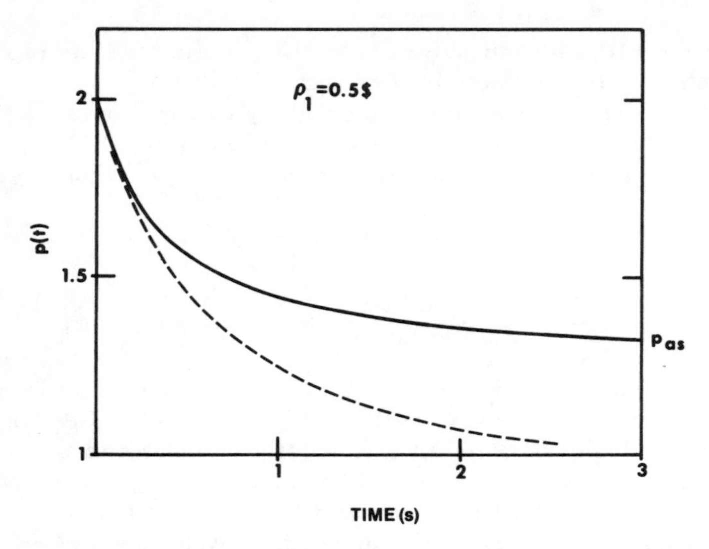

Fig. 10-4. Subprompt-critical transient (ρ_1 = 0.5$, p_0 = 1) with reactivity feedback including first-order heat transfer (solid line) and no heat transfer (dashed line).

eration. According to Sec. 6-1A, the stationary flux level is related to the delayed neutron source and β by the source multiplication formula, Eq. (6.11),

$$p_0 = \frac{s_{d0}}{\beta} \quad .$$

This relation is certainly fulfilled initially, and it must also hold asymptotically since $p_{as} = p_0$. If $p(t) \geqslant p_0$ and $\rho(t) \geqslant 0$ during the entire transient, the reactor would be permanently "loaded" with additional precursors (see Sec. 7-4). The above source multiplication equation would then require the asymptotic flux to be larger than p_0 in contradiction to Eq. (10.55). To resolve this contradiction, the flux must be $<p_0$ for some period of time to eliminate the additional precursors that are produced in the early part of the transient in order to achieve asymptotically $s_d = s_{d0}$.

Equations (10.51) are based on three assumptions:

1. the PJA
2. one-delay-group kinetics
3. neglect of transient heat transfer.

The first assumption should be accurate asymptotically since it is violated only in the immediate vicinity of fast changes.

The use of one-delay-group kinetics might be questionable since it has been shown in Sec. 6-1B that the kinetics parameters cannot be lumped uniquely into a one-delay-group λ. The one-delay-group λ, however, does not even appear in the asymptotic relations (10.55) and (10.56). Therefore, replacing a λ value by a different one that might asymptotically be more appropriate cannot improve the results in Eqs. (10.55) and (10.56). The reason is that the asymptotic state is a stationary state and therefore the flux level is independent of the precursor decay constants altogether.

The third assumption, the neglect of heat transfer, is violated during the transient and even more so asymptotically. The influence of heat transfer on the asymptotic flux can be semi-quantitatively investigated by using Eq. (10.29). With $\rho_{as} = 0$, the following equation is obtained instead of Eq. (10.54):

$$\lim_{t\to\infty} - \gamma \int_0^t [p(t') - p_0] \exp[-\lambda_H(t - t')]\, dt' = \rho_1 \quad . \qquad (10.57)$$

Asymptotically, p_{as} can be inserted as the value of the flux amplitude in performing the limit to large t. The exponential function practically eliminates the contribution of the integrand, which is far away from the upper limit. Asymptotically, the lower limit "0" is infinitely far away from "t"; it can therefore be replaced by $-\infty$. This gives:

$$-\gamma \int_{-\infty}^t (p_{as} - p_0) \exp[-\lambda_H(t - t')]\, dt' = \rho_1 \quad , \qquad (10.58)$$

or

$$-\frac{\gamma}{\lambda_H}(p_{as} - p_0) = \rho_1 \quad . \qquad (10.59)$$

Solving Eq. (10.59) for the asymptotic flux amplitude yields:

$$p_{as} = p_0 - \rho_1 \frac{\lambda_H}{\gamma} > p_0 \quad . \qquad (10.60)$$

Thus, the prompt negative feedback does *not* shut the reactor *down* in the sense of making it subcritical with an asymptotically vanishing flux. Therefore, the control or shutdown system must be employed to restore the initial power or shut the reactor down. Only if p_0 is practically zero and λ_H is very small, i.e., if there is practically no heat release, can p_{as} be very small. An example of an uncooled core with transients starting at $p_0 = 0$ is provided by the Godiva experiments.[2]

Figures 10-3 and 10-4 show as solid lines two transients ($\rho_1 = 0.95\$$ and $0.5\$$) as they develop with feedback and including first-order heat transfer ($\lambda_H = 0.5\ \mathrm{s}^{-1}$).

The basic effect of heat transfer is that the flux is not forced back to its initial value. It assumes asymptotically a value above the initial flux. The numerical calculation of the transient yields the same asymptotic flux value as Eq. (10.60). The formula based on a single delay constant λ_H provides the correct qualitative understanding of why an asymptotically constant flux above the initial flux is obtained.

Independent of the simplifications employed in describing the heat transfer, the negative reactivity feedback restores criticality. The inserted reactivity ρ_1 is compensated by a temperature rise:

$$\Delta T = -\rho_1 \left(\frac{\partial \rho}{\partial T}\right)^{-1} . \qquad (10.61)$$

With $\partial\rho/\partial T \simeq -0.002\$/\mathrm{K}$ [Eqs. (10.34)], a 0.5$ (or 0.95$) reactivity insertion would lead to a temperature rise of 250 K (or 475 K) if the control system did not start its counteraction during the transient, where T is the average fuel temperature. The temperature rise in the center of the hottest pellet may be substantially larger than that given by Eq. (10.61).

The following is a summary of basic conclusions for model transients in power reactors that can be drawn from the results presented in this section:

1. Prompt negative reactivity feedback reduces a subprompt reactivity insertion to zero; i.e., it restores criticality and does not *shut* the reactor *down*.
2. A permanently inserted reactivity is compensated by a *permanent temperature rise,* given by Eq. (10.61).
3. The flux approaches asymptotically a constant value. An asymptotic power level higher than the original power is required to sustain a permanent temperature rise.
4. If a core is originally at near zero power and "uncooled," the Doppler feedback resulting from a "permanent" temperature increase will reduce the neutron flux to near zero and lead to a "shutdown" in this limited sense. However, the asymptotic core would still be critical.

10-3 Superprompt-Critical Excursion Following a Step Reactivity Insertion

10-3A Investigation of the Differential Equation

Superprompt-critical transients induced by a reactivity step with feedback described by the linear energy model can be treated analytically if the prompt kinetics approximation of Sec. 8-2A is applied, thus providing an important semi-quantitative understanding of the power burst.

The differential equation and the initial condition in the prompt kinetics approximation are given by

$$\dot{p} = \frac{\rho - \beta}{\Lambda} p \quad , \tag{10.62}$$

with

$$p(0) = p^0 = \frac{\rho_1}{\rho_1 - \beta} p_0 \quad . \tag{10.63}$$

The reactivity contains the Doppler feedback as described by an energy coefficient γ, Eqs. (10.26). Again, γ is assumed to be independent of temperature so that simple theoretical relations between the kinetics and feedback parameters can be derived. The temperature rise is calculated on the basis of the assumption of adiabatic boundary conditions (linear energy model):

$$\rho(t) = \rho_1 + \gamma \int_0^t p(t')\, dt' \quad . \tag{10.64}$$

The stationary cooling is neglected, i.e., the term "$-p_0$" is omitted in the integrand of Eq. (10.64). This neglect is physically not well justified in most transients, but it greatly simplifies the mathematical treatment. The stationary cooling during a transient is equivalent to a positive ramp reactivity insertion, i.e.,

$$\gamma \int_0^t (-p_0)\, dt' = -\gamma p_0 t \quad . \tag{10.65}$$

Thus, simple formulas describing the *step-induced* excursion cannot be obtained without neglecting the stationary cooling. The stationary cooling is included in the investigation of ramp-induced transients, which are discussed in the next section.

For convenience, the prompt reactivity $\rho - \beta$ is designated by a single quantity ρ_p, standing for the reactivity above β:

$$\rho - \beta = \rho_p \quad . \tag{10.66}$$

The prompt kinetics equation then assumes the form

$$\dot{p} = \frac{\rho_p}{\Lambda} p \quad , \tag{10.67}$$

with

$$p(0) = p^0 = \frac{\rho_1}{\rho_{p1}} p_0 \quad . \tag{10.68}$$

After the onset of the transient, the flux rises rapidly or promptly. The corresponding flux integral causes, through the negative feedback, the reactivity to decrease. At $t = t_m$, the decreasing reactivity becomes equal to β, i.e., $\rho_p = 0$. Then the derivative in Eq. (10.67) becomes equal to zero. It follows that the prompt reactivity vanishes at the same time when the flux has a maximum:

$$\rho_p(t_m) = 0 \qquad (10.69a)$$

or

$$\rho(t_m) = \beta \quad . \qquad (10.69b)$$

Substituting into Eq. (10.64) gives

$$-\gamma \int_0^{t_m} p(t')\, dt' = \rho_1 - \beta \quad . \qquad (10.70)$$

Consequently, *the prompt part of the inserted reactivity is just compensated when the flux has reached its maximum.*

Equation (10.70) provides directly the energy release during the first part of the transient, i.e., from the beginning to the maximum of the excursion (with γ and p replaced by energy coefficient and power, γ_e and P):

$$\Delta Q_m = \int_0^{t_m} P(t')\, dt' = -\frac{\rho_{p1}}{\gamma_e} \quad . \qquad (10.71)$$

Thus, the energy release during the rising part of the transient is given by the ratio of the prompt reactivity inserted and the energy coefficient. Note the important fact that the *energy release ΔQ_m is independent of the neutron generation time.*

Equation (10.71) as well as the corresponding equation for the temperature rise may be rewritten as difference quotients. They then become special cases of the corresponding feedback coefficients:

$$\frac{\Delta\rho}{\Delta Q} = \frac{\partial\rho}{\partial Q} = \gamma_e = \text{energy coefficient}$$

and

$$\frac{\Delta\rho}{\Delta T} = \frac{\partial\rho}{\partial T} = \gamma_T = \text{temperature coefficient} \quad . \qquad (10.72a)$$

Equations (10.72a) hold at all times. At the maximum of the transient, one has $\Delta\rho = -\rho_{p1}$. Then Eqs. (10.72a) can be solved for ΔQ_m and ΔT_m:

$$\Delta Q_m = -\frac{\rho_{p1}}{\gamma_e} \quad ,$$

$$\Delta T_m = -\rho_{p1}\left(\frac{\partial\rho}{\partial T}\right)^{-1} \quad . \tag{10.72b}$$

Inserting the typical feedback coefficient of Eqs. (10.35) and (10.34), the following values are obtained for $\rho_1 = 1.1\$$ as an example:

$$\Delta Q_m \simeq \frac{0.1}{0.8} P_n \text{ seconds} \simeq 300 \text{ MW·s (for } P_n = 2500 \text{ MW)}$$

and

$$\Delta T_m \simeq 5 \text{ K (for LWRs) or } \Delta T_m \simeq 50 \text{ K (for FBRs)} \quad . \tag{10.73a}$$

The *rate* of the reactivity reduction by feedback is largest at the power maximum:

$$\dot{\rho}_m = \gamma_e P_m \quad . \tag{10.74}$$

Thus the reactivity quickly passes through the value of β and into the range below prompt critical.

The applicability of the prompt kinetics approximation (PKA) below $\rho = \beta$ is questionable (see Sec. 8-2). During a prompt jump or in cases in which the delayed neutron source practically determines the flux level through source multiplication, the application of the PKA gives unrealistic results since it cannot account for the behavior of the delayed neutron source, except in the initial condition. Below $\rho = \beta$, the prompt kinetics model only describes the prompt die-away of an existing neutron population. Since it is just the die-away of the large neutron population built up before the maximum of such transients, the PKA yields quite accurate results during this phase of the transient. During this phase, the flux level is reduced from its maximum value to a level determined by source multiplication of the delayed neutrons. After this phase, however, the PKA is not applicable (see Sec. 10-3C).

In terms of the microkinetics of Chapter 7, the superprompt transient consists of buildup of diverging fission chains. With ρ turned back below β through feedback, these chains cease to diverge and begin to die away. The die-away is described by the prompt period, which is the quantity appearing as the decay constant in the prompt kinetics equation, Eq. (10.62).

In conclusion, it is emphasized that the applicability of the PKA model is normally restricted to the reactivity range above β. Below β, it always yields a die-away of the neutron population that is normally incorrect. But this allows the exceptional applicability of the PKA to part of the range $\rho < \beta$ if $\rho(t)$ enters this range quickly from the superprompt domain because then the ensuing transient consists merely of the die-away describable by the PKA until finally the delayed neutrons take over. The quantitative comparisons below illustrate these aspects in detail.

10-3B Investigation of the First Integral

The consideration of the differential equation for superprompt-critical transients with linear energy feedback, Eq. (10.62), has already provided important information about the reactivity, energy release, and temperature rise between the onset and the maximum of the excursion. Further information is obtained by considering the "first integral" of the original equation, which can be brought in the form of a differential equation of second order.

Equations (10.62) and (10.64) are rewritten into a second-order differential equation. By denoting the integral in Eq. (10.64) with q, which gives $\dot{q}$ for p and $\ddot{q}$ for $\dot{p}$, one obtains instead of Eq. (10.62):

$$\Lambda\ddot{q} = (\rho_{p1} + \gamma q)\dot{q}$$

and

$$q(t) = \int_0^t p(t')\,dt' \quad . \tag{10.75}$$

An even simpler equation is obtained by introducing the prompt reactivity as the unknown function. The reactivity ρ_p and p are related as [see Eqs. (10.64) and (10.66)]:

$$\dot{\rho}_p = \gamma p = \gamma\dot{q} \tag{10.76a}$$

and

$$\ddot{\rho}_p = \gamma\dot{p} = \gamma\ddot{q} \quad . \tag{10.76b}$$

Multiplying Eq. (10.67) with $\Lambda\gamma$ and inserting Eqs. (10.76) yields:

$$\Lambda\ddot{\rho}_p = \rho_p\dot{\rho}_p \quad , \tag{10.77}$$

i.e., a nonlinear second-order differential equation.

The first integral of Eq. (10.77) is readily found:

$$\Lambda\int_0^t \frac{d\dot{\rho}_p}{dt'}\,dt' = \int_0^t \rho_p\frac{d\rho_p}{dt'}\,dt' \quad , \tag{10.78}$$

which gives a first integral in the form

$$\Lambda[\dot{\rho}_p(t) - \dot{\rho}_p(0)] = \frac{1}{2}\,[\rho_p^2(t) - \rho_p^2(0)] \quad . \tag{10.79}$$

The initial values are given by

$$\rho_p(0) = \rho_{p1} = \rho_1 - \beta$$

and

$$\dot{\rho}_p(0) = \gamma p^0 \quad . \tag{10.80}$$

Inserting the initial values of Eqs. (10.80) and also inserting Eq. (10.76a) results in the following forms for the first integral:

$$\Lambda[\dot{\rho}_p(t) - \gamma p^0] = \frac{1}{2}[\rho_p^2(t) - \rho_{p1}^2] \tag{10.81a}$$

and

$$\Lambda\gamma[p(t) - p^0] = \frac{1}{2}[\rho_p^2(t) - \rho_{p1}^2] \quad . \tag{10.81b}$$

As in the case of the differential equation, the first integral is investigated at the time of the maximum of the transient. Application of Eq. (10.81b) to the maximum, where $\rho_p = 0$, yields:

$$\Lambda\gamma(p_m - p^0) = -\frac{\rho_{p1}^2}{2} \quad . \tag{10.82}$$

This gives a formula for the maximum flux amplitude or power of the excursion:

$$p_m = p^0 - \frac{\rho_{p1}^2}{2\Lambda\gamma} \quad . \tag{10.83}$$

The maximum power rise in megawatts with an energy coefficient γ_e [\$/MW·s] is

$$P_m - P^0 = -\frac{\rho_{p1}^2}{2\Lambda\gamma_e} \quad . \tag{10.84}$$

Note that the power rise $P_m - P_0$ is inversely proportional to the generation time and the energy coefficient, and proportional to ρ_{p1}^2. But it is independent of the initial power if the temperature dependence of the energy coefficient is neglected.

As an example, again inserting the power reactor feedback coefficient of Eq. (10.35) yields:

$$\rho_1 = 1.1\$; \; \beta/\Lambda = 0.62 \times 10^4 \text{ s}^{-1}; \; \gamma_e = -0.8\$/P_n \text{ s}$$

$$P_m - P^0 = \frac{1}{2}(0.1)^2 0.62 \times 10^4 \text{ s}^{-1} \cdot \frac{1}{0.8} P_n \text{ s} = 38.8 \, P_n \quad . \tag{10.85a}$$

Using $P^0 = 11 \, P_0$ from Eq. (10.63) gives:

$$P_m = 49.8 \, P_n$$

$$(\simeq 1.25 \times 10^5 \text{ MW for } P_n = P_0 = 2500 \text{ MW}) \tag{10.85b}$$

for a transient starting at the nominal power P_n. If the transient starts at a low power, e.g., $P_0 = 0.1\, P_n$, then $P_m - P^0 = 388\, P_0$ and P_m in the example becomes equal to $399\, P_0$.

In the example, the rate of reactivity change at the maximum power, given by Eq. (10.74), is

$$\dot{\rho}_m = -\frac{0.8}{P_n}\$/\text{s} \cdot 49.8\, P_n = -40\$/\text{s} \quad . \tag{10.85c}$$

The first integral, Eq. (10.79), also provides information on the *total* reactivity feedback during the transient. Consider, for example, the transient terminated when $p(t)$ has returned to the pseudo-initial flux, p^0, say at $t = t_2$. Then the left side of Eq. (10.81b) vanishes initially as well as when $p(t)$ is passing through p^0 at t_2, i.e., at the "end" of the transient. Thus,

$$\rho_p^2(t) = \rho_{p1}^2 \text{ for } t = 0 \text{ and } t_2 \quad ,$$

with the two solutions

$$\rho_p(0) = +\, \rho_{p1} \text{ at } t = 0$$

and

$$\rho_p(t_2) = -\, \rho_{p1} \text{ at } t = t_2 \quad . \tag{10.86}$$

Thus, *the prompt feedback compensates the prompt reactivity twice during the transient* (see Fig. 10-5). In other words, the reactivity swing is just two times the initial prompt reactivity.

Equations (10.86), (10.64), and (10.71) show that the total energy release during a superprompt-critical transient (i.e., up to t_2) is just twice the energy release up to the maximum of the transient:

$$\Delta Q(t_2) = 2\Delta Q_m = -2\,\frac{\rho_{p1}}{\gamma_e} \quad . \tag{10.87}$$

Thus, *the total energy release during a superprompt-critical transient is independent of the generation time* Λ, as it is for delayed supercritical transients.

In the $\rho_1 = 1.1\$$ example, the average fuel temperature during the superprompt-critical transient is increased by 10°C or 100°C, respectively, assuming the temperature coefficients of Eqs. (10.34).

10-3C The Flux Transient During a Superprompt-Critical Excursion

The first integral in the form of Eq. (10.81a) is a first-order differential equation for $\rho_p(t)$ that may be solved readily. Then $p(t)$ can be

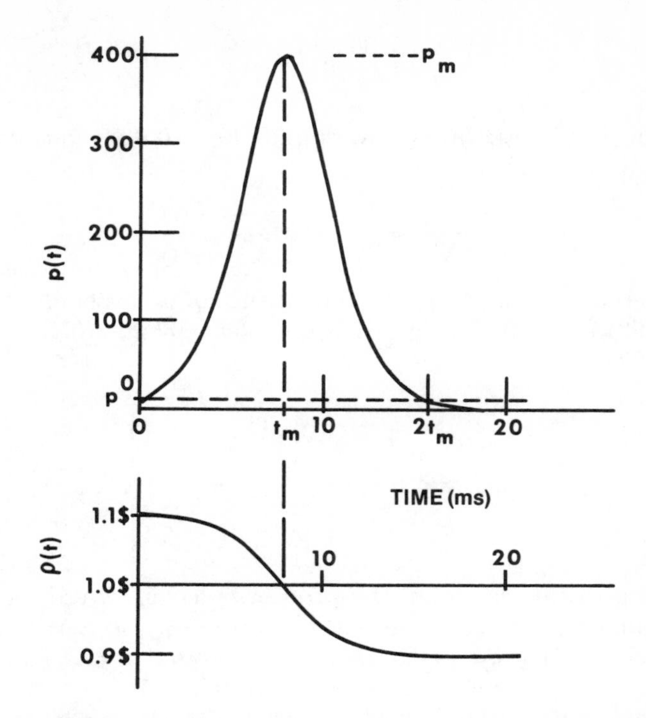

Fig. 10-5. Superprompt-critical transient (ρ_1 = 1.1\$, p_0 = 0.1 p_n) from the PKA with linear energy feedback model.

derived from $\rho_p(t)$. Rearranging the first integral, Eq. (10.81a) yields

$$2\Lambda\dot{\rho}_p(t) = \rho_p^2 - \rho_b^2 \quad , \tag{10.88a}$$

with

$$\rho_b^2 = \rho_{p1}^2 - 2\Lambda\gamma p^0 > \rho_{p1}^2 \quad . \tag{10.88b}$$

Dividing Eq. (10.88a) by its right side and integrating with respect to time from 0 to t yields

$$t = 2\Lambda \int_{\rho_{p1}}^{\rho_p(t)} \frac{d\rho_p}{\rho_p^2 - \rho_b^2} = 2\Lambda \int_{\rho_p(t)}^{\rho_{p1}} \frac{d\rho_p}{\rho_b^2 - \rho_p^2} \tag{10.89}$$

or

$$\frac{t}{\Lambda} = \frac{1}{\rho_b}\left[\ln\frac{\rho_b + \rho_{p1}}{\rho_b - \rho_{p1}} - \ln\frac{\rho_b + \rho_p(t)}{\rho_b - \rho_p(t)}\right] \quad . \tag{10.90}$$

By setting $\rho_p(t)$ in Eq. (10.90) equal to zero, one obtains the time, t_m, at which the flux maximum of the transient occurs:

$$\frac{t_m}{\Lambda} = \frac{1}{\rho_b} \ln \frac{\rho_b + \rho_{p1}}{\rho_b - \rho_{p1}} \quad . \tag{10.91}$$

Equation (10.91) may be used to simplify Eq. (10.90), which can then be written in the form

$$\frac{t - t_m}{\Lambda} = -\frac{1}{\rho_b} \ln \frac{\rho_b + \rho_p(t)}{\rho_b - \rho_p(t)} \quad . \tag{10.92}$$

The desired solution, i.e., the t dependence of ρ_p, is obtained by solving Eq. (10.92) for $\rho_p(t)$. This gives, as the *solution of the differential equation,* Eq. (10.77):

$$\rho_p(t) = \rho_b \frac{1 - \exp\left[\frac{\rho_b}{\Lambda}(t - t_m)\right]}{1 + \exp\left[\frac{\rho_b}{\Lambda}(t - t_m)\right]} \quad . \tag{10.93}$$

According to this equation, the prompt reactivity is reduced from its initial value $\rho_p(0) = \rho_{p1}$ to zero at t_m and to $-\rho_{p1}$ at $t_2 = -2t_m$. It can also be shown that $\rho_p(t)$ is antisymmetric around $t = t_m$ [compare Fig. 10-5 for $\rho(t)$ in a transient induced by $\rho_1 = 1.1\$$].

Inserting Eq. (10.93) into Eq. (10.81b), using the definition of ρ_b of Eq. (10.88b) and the value of the maximum flux p_m of Eq. (10.83), i.e.,

$$p_m = -\frac{\rho_b^2}{2\Lambda\gamma} \quad , \tag{10.94}$$

yields for the flux transient:

$$p(t) = \frac{p_m}{\left\{\cosh\left[\frac{\rho_b}{2\Lambda}(t - t_m)\right]\right\}^2} \quad . \tag{10.95}$$

Thus, the PKA together with the linear energy feedback model gives a transient that is symmetrical around $t - t_m$. Figure 10-5 shows the flux transient described by Eq. (10.95) for a transient induced by a step $\rho_1 = 1.1\$$, $p_0 = 0.1\ p_n$, and $\beta/\Lambda = 0.62 \times 10^4\ s^{-1}$.

The simplest formula for the "width" of the flux pulse (or burst) is obtained when the width $\overline{\Delta t}$ is defined by the flux integral divided by the maximum value:

$$\overline{\Delta t} = \frac{1}{p_m} \int_0^{2t_m} p(t)\, dt = -\frac{2}{p_m} \frac{\rho_{p1}}{\gamma} \quad . \tag{10.96}$$

The flux integral in Eq. (10.96),

$$\int_0^{2t_m} p(t)\, dt = -2\,\frac{\rho_{p1}}{\gamma} \quad , \tag{10.97}$$

follows directly from Eq. (10.87). The maximum flux, p_m, is given by Eq. (10.82). By using the identity,

$$\frac{1}{p_m} = \frac{1}{p_m - p^0}\left(1 - \frac{p^0}{p_m}\right) \quad ,$$

and Eq. (10.82), $\overline{\Delta t}$ is obtained as

$$\overline{\Delta t} = \frac{4\Lambda}{\rho_{p1}}\left(1 - \frac{p^0}{p_m}\right) \quad . \tag{10.98}$$

If p_m is much greater than the pseudo-initial flux p^0, the width becomes equal to four initial prompt periods:

$$\overline{\Delta t} = \frac{4\Lambda}{\rho_{p1}} \text{ for } p_m \gg p^0 \quad . \tag{10.99}$$

The proportionality of $\overline{\Delta t}$ to Λ compensates the $1/\Lambda$ dependence of p_m and thus yields a Λ-independent energy production. Figure 10-6 illustrates the definition, Eq. (10.96), of the width of the power burst ($\rho_{p1} = 0.1\$$ and $\beta/\Lambda = 0.62 \times 10^4\ \text{s}^{-1}$, $\overline{\Delta t} \simeq 6.5$ ms).

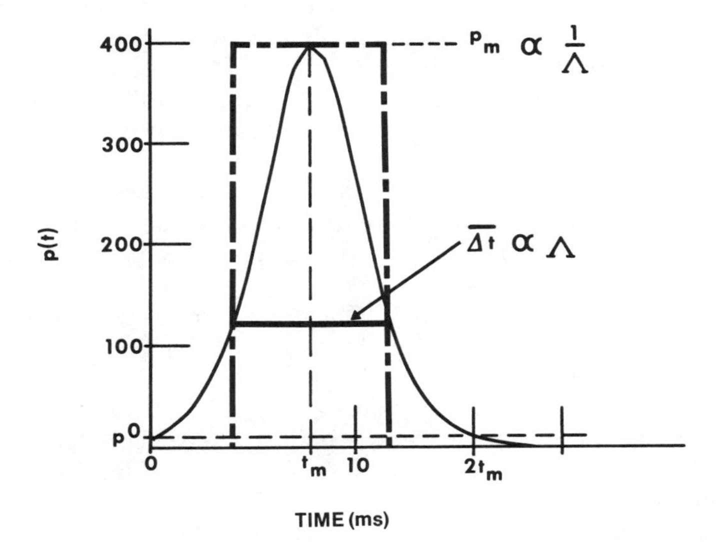

Fig. 10-6. Illustration of the width, Eq. (10.96), of the superprompt-critical transient presented in Fig. 10-5.

Figure 10-6 also shows that t_m, the time to reach the maximum, is not much larger than the burst width if the transient starts at the nominal power. The maximum is quickly reached after the onset of the burst. The value of t_m can be derived for a small initial power by simplifying Eq. (10.91):

$$t_m \simeq \frac{\Lambda}{\rho_{p1}} \left(\ln 4 \frac{p_m}{p^0} \right) \quad \text{for } p^0 \ll p_m \quad . \tag{10.100}$$

If P_0 is 10^{-6} times the nominal power, then the example of Eqs. (10.85) gives

$$t_m \simeq 16.5 \frac{\Lambda}{\rho_{p1}} \simeq 4\overline{\Delta t} \quad . \tag{10.101}$$

That is, even for very low initial power, the time to reach the maximum in a step-induced transient is only several times the width of the power burst.

10-3D The Post-Burst Flux Transient

The neutron flux as described by Eq. (10.95) decreases rapidly after it passes through its maximum value (see also Fig. 10-5). Formally, this rapid die-away continues beyond $t_2 = 2t_m$, the time that was used above as the "end" of the flux pulse. This further decrease of the flux and its asymptotic approach to zero is physically unrealistic; the reactor is still supercritical (if ρ_1 is smaller than 2β, which should always apply in reactor problems). The reason for this unrealistic behavior of the solution of Eq. (10.95) is the neglect of the delayed neutron source except as a modification of the initial condition.

At $t = t_m$, the neutron flux level is almost completely determined by the large number of neutrons that were produced during the time the reactor was superprompt critical. After the reactivity is reduced below β, the neutron flux of the superprompt reactor dies away rapidly, and the flux is eventually determined by the source multiplication of the delayed neutrons. The continuing die-away as it follows from the PKA then becomes totally unrealistic.

After the rapid flux changes during the burst are terminated, the further changes of the flux are slow so that $\Lambda\dot{p}$ can be neglected (prompt jump approximation). An estimate of the flux level can then be obtained by estimating the delayed neutron source and applying the time-dependent source multiplication formula, Eq. (8.7). A first approximation is obtained by using the initial value of the delayed neutron source and the reactivity after the burst, i.e.,

$$\rho(t_2) = \beta - \rho_{p1} \quad . \tag{10.102}$$

From the source multiplication consideration of the delayed neutrons, it follows that the flux is "held up" at least at a level p_{ab} (*a*fter the *b*urst) given by:

$$p_{ab} \simeq \frac{s_{d0}}{\beta - \rho} = \frac{s_{d0}}{\rho_{p1}} = \frac{\beta}{\rho_{p1}} p_0 \quad . \tag{10.103}$$

Thus, in the example above, the flux is held up at a level of ten times the initial power.

An improved estimate of the delayed neutron source multiplication requires accounting for the increase of the precursor population during the flux burst. During the short time of the burst, the $\bar{\lambda}$ kinetics approximation is applicable. Since the time interval considered is very small, the precursor decay can be neglected in the treatment of the precursor balance ($\bar{\lambda}t \ll 1$; see the precursor accumulation model in Sec. 6-1B). Thus, $\zeta(t_2)$ can be approximated by:

$$\zeta(t_2) \simeq \zeta_0 + \beta \int_0^{t_2} p(t')\, dt' \quad . \tag{10.104}$$

With $\bar{\lambda}\zeta_0 = \beta p_0$, and the flux integral given by Eq. (10.97), $\bar{\lambda}\zeta$ is obtained as

$$\bar{\lambda}\zeta(t_2) = \beta p_0 \left(1 - 2\,\frac{\rho_{p1}\bar{\lambda}}{p_0\gamma}\right) \quad . \tag{10.105}$$

In the example treated above, i.e., $p_0 = 0.1\ p_n$, the delayed neutron source is increased by ~140%:

$$-2\,\frac{\rho_{p1}\bar{\lambda}}{p_0\gamma} = \frac{-2(0.1\$)(0.565/\text{s})}{(0.1)(-0.8\$/\text{s})} \simeq 1.41 \quad . \tag{10.106}$$

If the initial power was very low, the fraction of the precursors produced during the flux burst would be even larger.

Instead of Eq. (10.103), the flux after the burst with the precursor increase included is given by:

$$p_{ab} \simeq \frac{\beta p_0}{\rho_{p1}} \left(1 - 2\,\frac{\rho_{p1}\bar{\lambda}}{p_0\gamma}\right) \quad . \tag{10.107}$$

Figure 10-7 shows the flux according to Eq. (10.95) (dashed line) and the result of the $\bar{\lambda}$ kinetics model (solid line). The six-delay-group results are indistinguishable from the one-delay-group results at the small times of Fig. 10-7. The delayed neutron source holds the flux at $p_{ab} \approx$

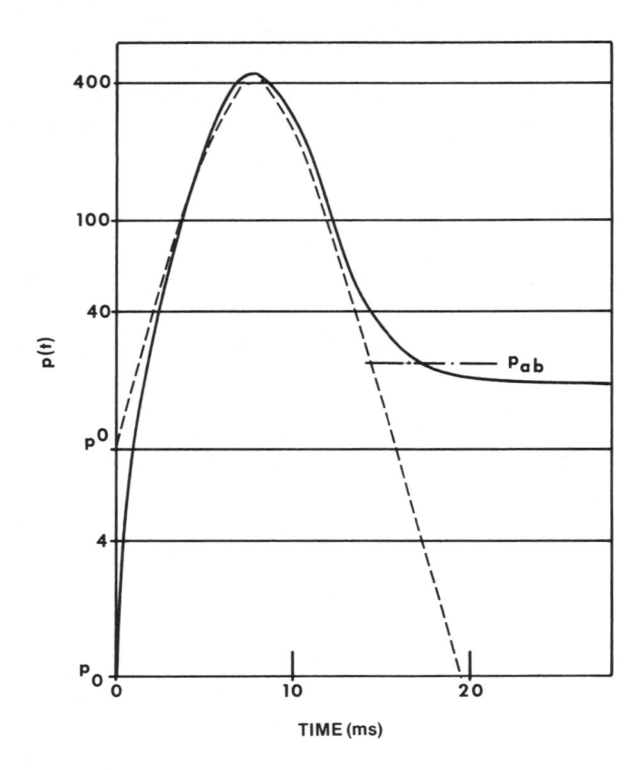

Fig. 10-7. Superprompt-critical transient (ρ_1 = 1.1\$, p_0 = 0.1 p_n) with reactivity feedback: λ kinetics model (solid line) and prompt kinetics of Eq. (10.95) (dashed line).

2.41/0.1 in accordance with Eq. (10.107). The deviations between the prompt kinetics approximation and the actual calculational results during and after the burst, shown in Fig. 10-7, give a good indication of the applicability of the prompt jump approximation.

The further progression of the excursion is essentially the same as in a transient following a subprompt-critical step reactivity insertion (compare Sec. 10-2B). The reactor power after the prompt burst is much larger than the nominal power; this causes a further increase in the fuel temperature and a corresponding decrease in the reactivity due to the negative energy coefficient. Eventually, the reactivity is reduced to zero and the flux approaches an asymptotic level above the initial flux.

This further decrease of the flux is indicated in Fig. 10-7. Compared to Fig. 10-4, the flux decrease in Fig. 10-7 appears very slow due to the strongly expanded scale.

The temperature increase during the entire transient is composed of the part produced during the prompt burst and the part produced during the subsequent relatively gradual power decrease:

$$\Delta T_{\text{total}} = \Delta T_{\text{burst}} + \Delta T_{\text{post-burst}}$$

and

$$\Delta T_{\text{total}} = -\frac{1}{\gamma_T}[2(\rho_1 - \beta)] - \frac{1}{\gamma_T}[\rho_1 - 2(\rho_1 - \beta)] = -\frac{\rho_1}{\gamma_T} \quad . \tag{10.108}$$

In reactor accident analysis, ρ_1 is almost never much larger than β. Then, only the small part of the temperature rise occurs during the prompt burst; the larger part of the energy is released during the sub-prompt-critical phase of the excursion, before or after a prompt burst. For example, for $\rho_1 = 1.1\$$, ~18% of the energy is released during the prompt burst and ~72% during the slower phase of the transient.

10-4 Superprompt-Critical Transients Induced by Reactivity Ramps

10-4A Investigation of the Differential Equation

Reactivity steps in the superprompt reactivity domain of reactors are only an idealization. Realistic simulation of superprompt-critical reactivity insertions in a reactor must account for the reactivity insertion rate, and thus for the rapid flux response that occurs *during* the reactivity insertion (and not *after* the reactivity insertion as in the case of step approximations). Therefore, this section investigates more realistic, gradual reactivity insertions, though in the idealized form of a continuing unterminated ramp. In fully realistic situations, ramps are always terminated at some maximum available reactivity. Much of the material presented in this section was originally derived in Refs. 4 and 28 through 30.

The PKA for reactivity ramps is applied (as for the step-induced transients) only after the time, t_p, when the reactor became superprompt critical:

$$\dot{p}(\tau) = \frac{\rho_p(\tau)}{\Lambda} p(\tau) \text{ with } \tau = t - t_p \quad , \tag{10.109}$$

which is solved with $p(0) = p^0$, and $\rho_p(0) = 0$ for $\tau = 0$. For reactivity ramps, the pseudo-initial flux, p^0, is approximately two times the actual flux at prompt critical [see Eq. (8.51)]:

$$p^0 \simeq 2p_{pc} \simeq p_0\beta\left(\frac{2\pi}{\Lambda a}\right)^{1/2} \quad . \tag{10.110}$$

The prompt reactivity, with the adiabatic approximation applied to the energy above p_0, is given by

$$\rho_p(\tau) = a\tau + \gamma \int_0^{\tau} [p(\tau') - p_0]\, d\tau' \qquad (10.111a)$$

or

$$\rho_p(\tau) = a'\tau + \gamma \int_0^{\tau} p(\tau')\, d\tau' \quad , \qquad (10.111b)$$

with

$$a' = a - \gamma p_0 \quad . \qquad (10.111c)$$

The stationary cooling described by the term "$-p_0$" in the integrand of Eq. (10.111a) is now included in the feedback reactivity since this term only modifies the ramp rate, Eq. (10.111c). In the case of step-induced transients, stationary cooling was neglected for mathematical simplification.

Initial results about essential characteristics of the transients are deduced directly from the differential equation as it was done for the step-induced transient. It follows from Eq. (10.109) that $\dot{p}(\tau)$ and $\rho_p(\tau)$ pass through zero simultaneously, i.e.,

$$\dot{p}(\tau) = 0 \text{ when } \rho_p(\tau) = 0 \quad .$$

Since $\rho_p(\tau)$ is zero initially, $p(\tau)$ in the PKA starts with a zero slope and has a *minimum* at $\tau = 0$. Subsequently, $\rho_p(\tau)$ and $p(\tau)$ both increase. With increasing accumulation of energy, the reactivity feedback overcomes the ramp reactivity and $\rho_p(\tau)$ is again reduced to zero; say at $\tau = \tau_{m1}$. The flux has its first maximum at this time. After passing through zero, ρ_p approaches a negative value, corresponding to $-\rho_{p1}$ in the case of a step-induced transient. By this time, the high flux during the prompt burst is reduced so that further reactivity reduction through feedback occurs relatively slowly (compare Fig. 10-5). A similar reduction of the reactivity is to be expected in the case of a ramp reactivity insertion. However, as the ramp continues to increase the reactivity, it again becomes equal to β. At this time, a second flux minimum occurs and the cycle repeats; see Fig. 10-8 where $p_0 = p_n$, $a = 50\$/\text{s}$, $\gamma = -0.5\$/\text{fp-s}$, and $p^0 = 28.0\, p_0$ according to Eq. (10.110).

The discussion of maxima and minima showed that the ramp-induced transient treated in the PKA consists of a repetition of superprompt flux bursts. The maxima and minima of the flux occur when $\rho(t)$ passes through β:

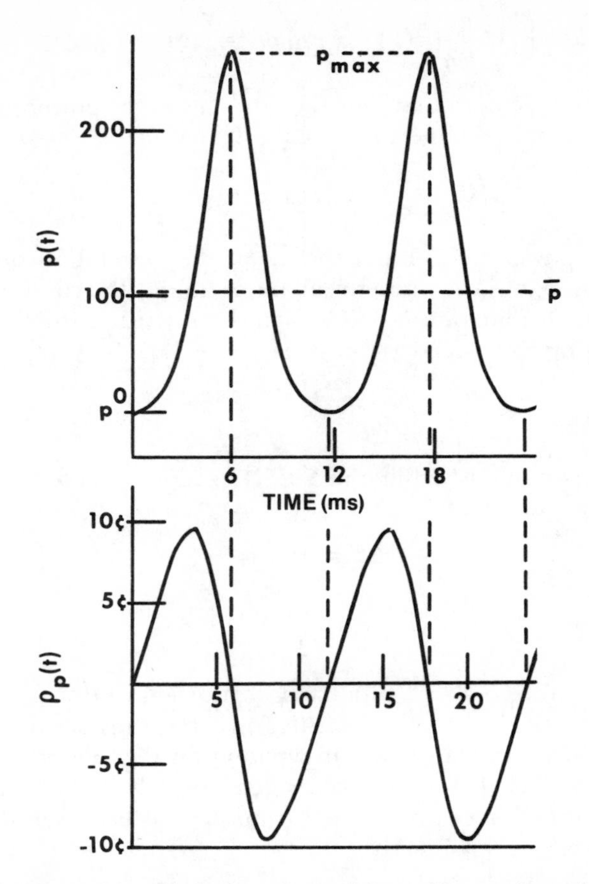

Fig. 10-8. The flux, $p(t)$, and the prompt reactivity, $\rho_p(t)$, are shown for a reactivity-ramp-induced transient with temperature-independent energy coefficient where $p_0 = p_n$, $a = 50\$/\mathrm{s}$, $\gamma = -0.5\$/\mathrm{fp\text{-}s}$, and $p^0 = 28.0\, p_0$.

$$p(t) = p_{max} \text{ or } p_{min} \text{ for } \rho_p(t_{m\ell}) = 0 \quad . \tag{10.112}$$

The succession of bursts can be characterized by an averaged power that can be determined as follows: Let Δt_ℓ be the duration of the ℓ'th burst and ΔQ_ℓ the corresponding energy release:

$$\Delta Q_\ell = \int_{\Delta t_\ell} [P(t') - P_0]\, dt' \quad . \tag{10.113}$$

Since $\rho_p(t)$ is zero at the beginning and at the end of each burst, a relation between the reactivity insertion and the feedback reactivity during a burst can be obtained:

$$0 = a\Delta t_\ell + \gamma_e \int_{\Delta t_\ell} [P(t') - P_0]\, dt' = a\Delta t_\ell + \gamma_e \Delta Q_\ell \quad . \qquad (10.114)$$

The energy release per burst time, i.e., the "average" power increase $\Delta\overline{P}$ during the burst, is obtained from Eq. (10.114) as:

$$\frac{\Delta Q_\ell}{\Delta t_\ell} = \Delta\overline{P} = \overline{P} - P_0 = -\frac{a}{\gamma_e} \quad . \qquad (10.115a)$$

The average power $\Delta\overline{P}$ is independent of the generation time and of the initial power; it is the same for all bursts under the conditions of the feedback model. Equation (10.115a) can be rewritten for the flux amplitude by using the effective ramp rate a' of Eq. (10.111c):

$$\overline{p} = -\frac{a'}{\gamma} \quad . \qquad (10.115b)$$

The horizontal dashed line in Fig. 10-8 represents $\overline{p}$. For this transient, Eqs. (10.115) yield:

$$\Delta\overline{p} = 100\ p_n$$

and

$$\overline{p} = 101\ p_n \quad . \qquad (10.116)$$

Note, for a *step-induced* supercritical transient, information on the *energy* release was obtained by investigating the transition of ρ through β; for a *ramp-induced* transient, information on the *average power* is obtained instead. This difference occurs because only the *rate* of reactivity insertion is given for ramp-induced transients, which then determines the *rate* of energy release:

$$-\frac{\rho_{p1}}{\gamma_e} \Rightarrow \text{energy, for reactivity steps} \qquad (10.117a)$$

and

$$-\frac{a}{\gamma_e} \Rightarrow \text{average power, for reactivity ramps} \quad . \qquad (10.117b)$$

Additional information is obtained by investigating the extrema of $\rho_p(\tau)$. Setting the time derivative of Eqs. (10.111) equal to zero gives

$$\dot{\rho}_p(\tau) = a + \gamma(p - p_0) = a' + \gamma p = 0 \quad . \qquad (10.118)$$

From Eq. (10.118), it follows that the flux at the extrema of the reactivity is equal to the average flux, $\overline{p}$:

$$p(\rho_p^{max,min}) = p_0 - \frac{a}{\gamma} = -\frac{a'}{\gamma} = \overline{p} \quad , \qquad (10.119)$$

i.e., the extrema of the reactivity coincide with $p(t)$ passing through $\bar{p}$ (see Fig. 10-8).

The extrema of the reactivity determine the inflection points of the flux transient in a semi-logarithmic presentation:

$$\frac{1}{\Lambda}\frac{d}{d\tau}\rho_p(\tau) = \frac{d}{d\tau}\frac{\dot{p}}{p} = \frac{d^2}{d\tau^2}\ln p(\tau) \quad . \tag{10.120}$$

Thus, the second derivative of the logarithm of the flux is zero at the extrema of the reactivity.

10-4B Investigation of the First Integral[d]

The integration of the kinetics equation, Eq. (10.109), is facilitated by multiplying it with $\dot{\rho}_p$; on the left side, $\dot{\rho}_p$ is expressed in terms of p as given by Eq. (10.118):

$$(a' + \gamma p)\dot{p} + = \frac{1}{\Lambda}\rho_p\dot{\rho}_p p \quad , \tag{10.121}$$

with a' given by Eq. (10.111c). After dividing Eq. (10.121) by p, the resulting equation may be readily integrated:

$$\int_0^\tau \left(\frac{a'}{p} + \gamma\right)\dot{p}\, d\tau' = \frac{1}{\Lambda}\int_0^\tau \rho_p\, \dot{\rho}_p\, d\tau' \quad .$$

The derivatives of p and ρ_p may be conveniently combined with $d\tau'$. Then, the two integrals are carried out over p and ρ_p, respectively:

$$\int_{p(0)}^{p(\tau)} \left(\frac{a'}{p} + \gamma\right) dp = \frac{1}{\Lambda}\int_{\rho_p(0)}^{\rho_p(\tau)} \rho_p\, d\rho_p \quad .$$

The integrations may be readily carried out. Inserting $p(0) = p^0$ and $\rho_p(0) = 0$ as the lower limits gives the "first integral," which represents a relation between the flux and the reactivity:

$$a' \ln\frac{p}{p^0} + \gamma(p - p^0) = \frac{1}{2\Lambda}\rho_p^2 \quad . \tag{10.122}$$

The first integral is used to find the maximum and minimum values of both the reactivity and the flux transients (compare Fig. 10-8). Inserting $\bar{p}$ from Eq. (10.119) into Eq. (10.122) and using Eq. (10.119) to simplify the result yields the maximum and minimum values of ρ_p:

[d]This section may be deleted in an introductory course.

$$(\rho_p^{max,min})^2 = 2\Lambda a' \left(\ln \frac{\bar{p}}{p^0} - 1 + \frac{p^0}{\bar{p}} \right) \quad , \tag{10.123}$$

which gives the two values

$$\frac{\rho_p^{max,min}}{\beta} = \pm \frac{\sqrt{2\Lambda a'}}{\beta} \sqrt{\ln \frac{\bar{p}}{p^0} - 1 + \frac{p^0}{\bar{p}}} \quad . \tag{10.124}$$

The positive and negative roots represent the reactivity maxima and minima, respectively. Apparently, the maxima and minima of ρ are located symmetrically about β.

Maximum reactivity values, calculated from Eq. (10.124) for a = 50\$/s ($\Lambda a' = 2.49 \times 10^{-5}$) are plotted in Fig. 10-9 as a function of $\bar{p}/p^0$. The abscissa in Fig. 10-9 covers all transients with realistic ramp rates and feedback coefficients starting at or around nominal power. The lower end of the scale is $\bar{p}/p^0 = 1$ since Eq. (10.124) has real roots only if $\bar{p}/p^0 \geq 1$. If Eq. (10.124) has no real roots, i.e., if

$$\bar{p} < p^0 \quad , \tag{10.125}$$

with $\bar{p}$ from Eq. (10.115b), then the PKA is fundamentally inapplicable. The inequality (10.125) may be physically realized for small ramp rates, large feedback coefficients (small $\bar{p}$), or a large pseudo-initial flux (large p^0). In such cases, the feedback effect during the subprompt-critical part

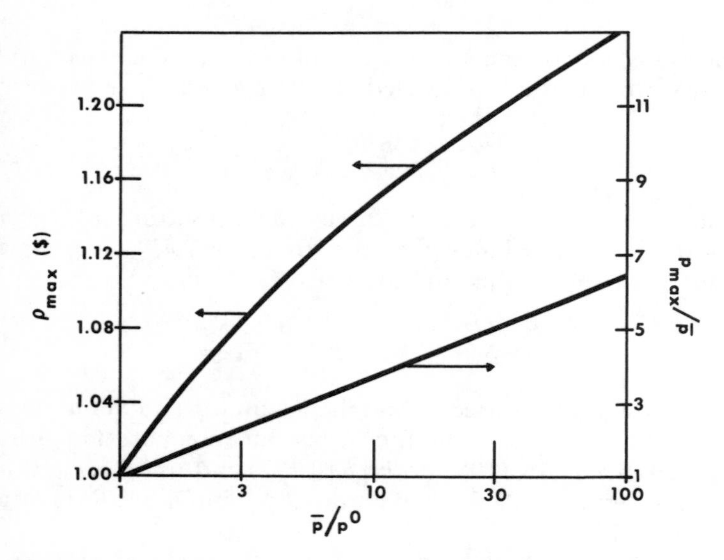

Fig. 10-9. Maximum reactivity values, Eq. (10.124), and $p_{max}/\bar{p}$, Eq. (10.126), versus $\bar{p}/p^0$ for a reactivity-ramp-induced transient where a = 50\$/s and $\gamma = -0.5$\$/fp-s.

of the transient is strong enough that the reactivity does not reach the superprompt-critical domain. The upper end of the scale in Fig. 10-9 may be exceeded only at very low starting power (e.g., accidents during reloading of the core).

Figure 10-9 shows that a ramp rate of 50$/s can only push a fast reactor beyond prompt critical by a small fraction of a dollar. Results for different ramp rates may be obtained readily from Fig. 10-9 by changing the factor in front of the square root of Eq. (10.124). For example, an increase of the ramp rate from 50 to 100$/s increases the maximum of ρ_p by a factor of ~1.4. Since positive ramp rates beyond 100$/s are very unlikely, it follows that realistic ramp rates can drive the reactivity in a fast reactor by only a fraction of a dollar beyond prompt critical. Much higher reactivity values are attainable, in principle, in thermal reactors due to the larger generation time. However, thermal reactors do not have the positive reactivity insertion mechanisms of fast reactors, as they may result from coolant boiling or fuel motion.

When the prompt reactivity passes through zero, the flux has a maximum or a minimum, $p = p_m$. Setting $\rho_p = 0$ and $p = p_m$ in the first integral, Eq. (10.122), yields a transcendental equation for the flux maxima and minima (if properly expanded, $\bar{p}/p^0$ appears as the only parameter):

$$\frac{\bar{p}}{p^0}\ln\left(\frac{p_m}{\bar{p}}\frac{\bar{p}}{p^0}\right) - \frac{p_m}{\bar{p}}\frac{\bar{p}}{p^0} + 1 = 0 \quad , \tag{10.126}$$

which is satisfied for $p_m = p^0$ and gives the flux minima as p^0:

$$p_{min} = p^0 \quad . \tag{10.127}$$

The flux maxima are given by the second positive solution of Eq. (10.126).

The ratio of the flux maxima to the average flux, i.e., $p_{max}/\bar{p}$, is also plotted in Fig. 10-9 as a function of $\bar{p}/p^0$. The figure shows that p_{max} does not become very much larger than $\bar{p}$. Although the ramp rate does not explicitly appear in Eq. (10.126), $p_{max}/\bar{p}$ increases with increasing ramp rate since $\bar{p}/p^0$, is approximately proportional to $(a)^{3/2}$.

The application of Fig. 10-9 is illustrated by deriving results for the transient presented in Fig. 10-8. With $p_0 = p_n$, $p^0 = 28\,p_0$, and $\bar{p} = 101\,p_n$, then $\bar{p}/p^0 = 3.6$. At $\bar{p}/p^0 = 3.6$, Fig. 10-9 yields

$$\frac{p_{max}}{\bar{p}} \simeq 2.4 \text{ and } \rho_p^{max} = 9.5¢ \quad .$$

Both values are in complete agreement with the results of Fig. 10-8.

The investgation of the first integral for ramp-induced transients provided important information such as $\bar{p}$, ρ_{max}, and p_{max}. The latter two quantities are depicted in Fig. 10-9 as functions of $\bar{p}/p^0$ while $\bar{p}$ is given by Eq. (10.119) and p^0 may be obtained, approximately, from Eq. (10.110). The two quantities, $\bar{p}$ and p^0, are shown in Fig. 10-10 as functions of the ramp rate. The pseudo-initial flux p^0 decreases and the average flux increases with increasing ramp rate. The PKA can only be applied for ramp rates larger than the crossover point (a^*) of the $p^0(a)$ and $\bar{p}(a)$ curves. If the reactivity insertion rate is smaller than a^*, the transient is comparatively slow. Thus, enough time is allowed to accumulate the energy, and thus the feedback reactivity, that is needed to prevent the reactor from becoming superprompt critical.

Figure 10-10 shows that in the domain of applicability of the PKA, the simple formulas for p^0 and $\bar{p}$,

$$p^0 = p_0\beta \sqrt{\frac{2\pi}{\Lambda a}} \tag{10.128a}$$

and

$$\bar{p} = -\frac{a'}{\gamma} \; , \tag{10.128b}$$

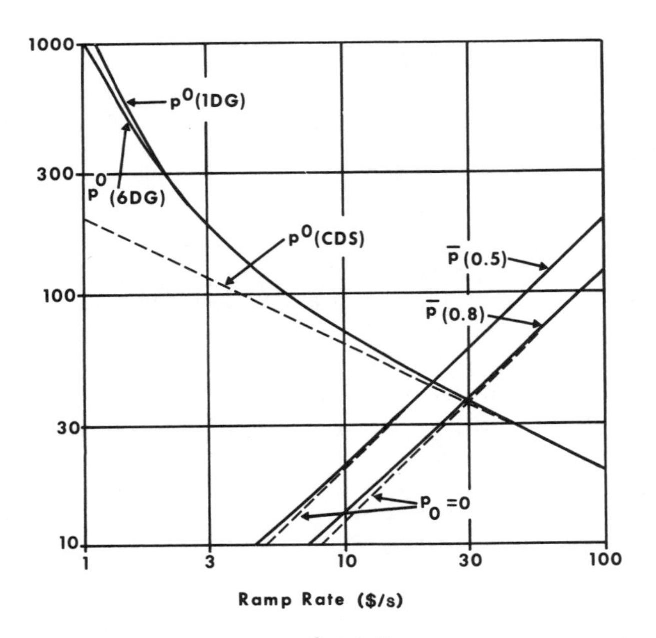

Fig. 10-10. For various approximations, p^0 and $\bar{p}$ ($\gamma = -0.5$ and -0.8\$/fp-s) versus reactivity ramp rate; the dashed lines are from Eqs. (10.128).

which are represented by the straight dashed lines, are sufficiently accurate for semi-quantitative estimates.

10-4C Discussion of the Flux Transient

The calculation of the flux as a function of time requires the complete solution of the differential equation (10.109). An analytical solution of this equation with the reactivity from Eqs. (10.111) has not been found yet. It is, however, possible to calculate the time between the bursts[28,30] and then deduce the energy per burst from the ratio of the average power, $\overline{P}$, Eq. (10.115a), and the duration of a burst. The calculation of the burst duration, however, for realistically mild transients is very complicated.[30] Therefore, the simple solution of the step-induced transients with a proper ρ_{p1} value is frequently related to ramp-induced transients.

The similarity of ramp- and step-induced superprompt-critical bursts was observed experimentally by Nyer et al.[31] Theoretical relations for ramp- and step-induced bursts were first introduced by Forbes.[29] Equivalence relations derived by Forbes are based on matching the values of the burst maxima; i.e., p_m of Eq. (10.126) is set equal to p_m of Eq. (10.83):

$$[p_m(a) \text{ from Eq. (10.126)}] = [p_m(\rho_{p1}) \text{ from Eq. (10.83)}] \quad . \qquad (10.129)$$

The resulting equivalence relation is relatively simple but only because the ramp rate was assumed to be very large and thus ρ_1 is much larger than β. The investigations in the previous sections showed that this assumption is not realistic for large fast power reactors; it may, in principle, be realized in thermal reactors.

An equivalence relation with a wider range of applicability can be based on the temperature rise during the burst. Such a relation was introduced by Canosa.[30] It applies the formula for the energy production per burst as derived in the same reference.

The equivalence relation presented here is based on the maximum reactivity that can be obtained from the first integral:

$$[\rho_p^{max}(a) \text{ from Eq. (10.124)}] = [\rho_{p1} \text{ for steps}] \quad . \qquad (10.130)$$

The ρ_p^{max}-based equivalence relation gives approximately the same results as Canosa's relation,[30] which requires the energy release per burst.

The equivalence relation, Eq. (10.130), makes it possible to apply the simple formulas for step-induced transients to the much more complicated ramp-induced bursts. Figure 10-11 shows the maximum reactivity calculated from Eqs. (10.124) and (10.128) as a function of the ramp rate. Since $\rho_{p,max}$ is set equal to ρ_{p1}, Fig. 10-11 yields directly the ramp rate equivalent to the step reactivity ρ_{p1}. Other quantities such as

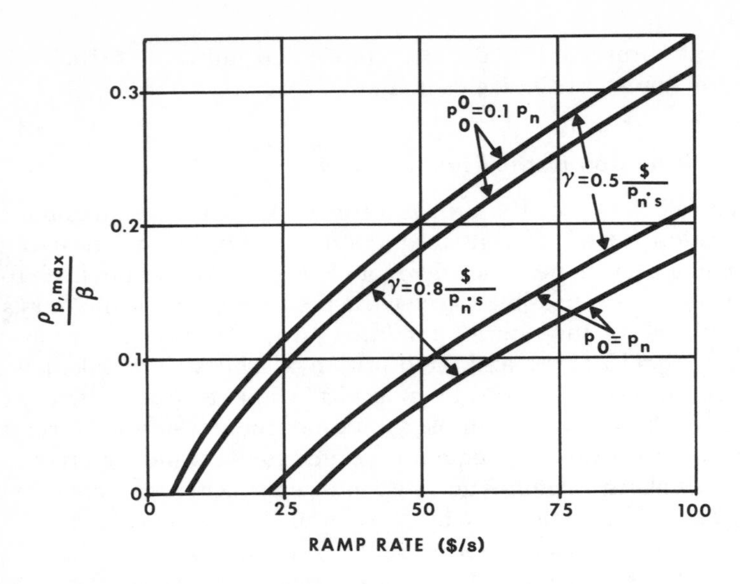

Fig. 10-11. The maximum reactivity versus reactivity ramp rate for two different energy coefficients and initial conditions.

the energy per burst can then be found from the corresponding equations in Sec. 10-3.

The previous investigations in this section as well as the numerical solution presented in Fig. 10-8 are based on the PKA; i.e., the delayed neutron source is omitted and the resulting error is approximately corrected by starting the transient from a pseudo-initial flux, p^0. The value of p^0 is determined by matching the PKA solution with a more correct result on the rising wing of the first burst. This adjustment yields a fairly accurate PKA solution during the major part of the burst. The comparison of the PKA for step-induced bursts and the corresponding numerical solution showed that the PKA becomes invalid "after" the burst, i.e., after the flux is reduced to a value comparable to the flux sustained by the delayed neutron source [see Fig. 10-7 and Eq. (10.106)]. The same phenomenon is to be expected for ramp-induced bursts, i.e., the flux is *not* expected to decrease to the p^0 value as in Fig. 10-8.

Figure 10-12 shows a comparison between the flux transient in the PKA (dashed line) with the actual solution. The striking difference is the occurrence of significant damping, which results from the proper treatment of the delayed neutrons. The important fact is that even though the oscillations are damped, the flux still oscillates around the *same* average value $\bar{p}$ as in the PKA. Because of the strong damping, the PKA for the burst shape is only applicable during the first burst.

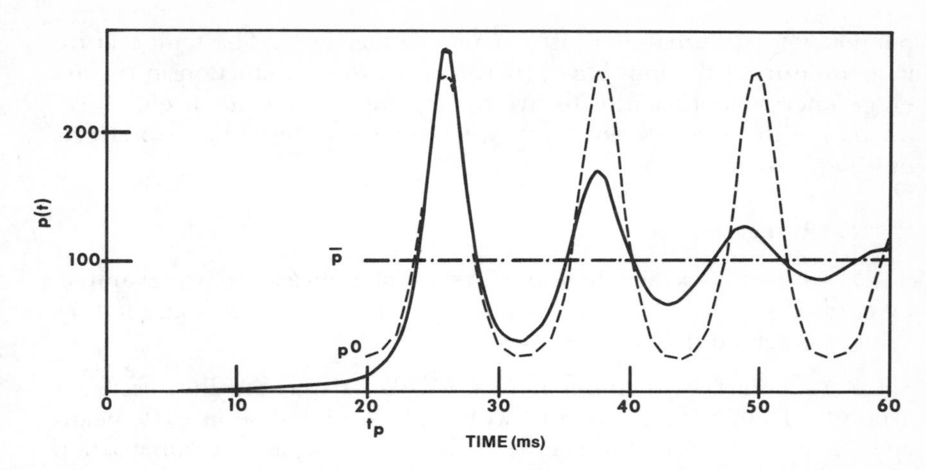

Fig. 10-12. The flux transient induced by a reactivity ramp in the PKA (dashed line) and the actual solution (solid line) with a temperature-independent energy coefficient.

During the first burst, the PKA yields fairly accurate results. The small deviation shown in Fig. 10-12 is similar to that exhibited in Fig. 10-7 for the step-induced transient. In both cases, the PKA flux in the initial phase of the transient is high due to the artificially high starting value, p^0. The additional power produced during this phase results in an earlier reduction of the reactivity through feedback. The flux maxima in the PKA are consequently somewhat lower than the correct solution.

This situation is reversed for subsequent bursts in ramp-induced transients. The actual flux is higher than the PKA flux due to source multiplication of the delayed neutron source with a reactivity close to β. This results in a substantial reactivity feedback even during the phase in which $p(t) < \bar{p}$. Therefore, subsequent bursts are driven by a smaller effective ramp rate. A further contribution to the damping effect of the delayed neutrons stems from the maximum flux being smaller if p^0 is larger, even for the same ramp rate; it follows from Fig. 10-9 that p_{max} approaches $\bar{p}$ from above if p^0 approaches $\bar{p}$ from below.

The energy coefficient was assumed independent of the temperature in all of the solutions presented in this section. The results for single bursts are fairly insensitive to this assumption. A temperature dependence can be taken into account approximately by using a proper average value. In a succession of bursts, as it occurs with high ramp rate reactivity insertions, the temperature dependence of γ has to be taken into account more explicitly.

According to Eq. (10.30), the Doppler effect contribution to the energy coefficient has approximately a $1/\sqrt{T}$ or a $1/T$ temperature dependence for thermal and fast reactors, respectively. The temperature increase during the first burst, therefore, causes a reduction in the average energy coefficient effective during the second burst, etc. A reduction of the energy coefficient, γ, from burst to burst has two effects on the transient:

1. It increases $\bar{p} \simeq -\frac{a'}{\gamma}$.

2. It increases the maximum flux, which is measured, for example, by $p_m/\bar{p}$. Figure 10-9 shows, for any given p^0 value, that a larger $p_m/\bar{p}$ value is obtained if $\bar{p}$ is larger.

Since $\bar{p}$ increases with temperature, it follows that the flux oscillates around a $\bar{p}$ curve that increases with time. An increase in $p_m/\bar{p}$ means an increase in the amplitude of the oscillation, i.e., a partial compensation of the damping effect that appears in Fig. 10-12.

Figure 10-13 shows the same transient as in Fig. 10-12 (50$/s) with an energy coefficient proportional to $1/T$. The temperature increase in the constant power removal model is obtained from Eqs. (10.19) and (10.20):

$$\delta T(t) = T(t) - T_0 = C_{IT} \int_0^t [p(t') - p_0] \, dt' = C_{IT} I(t) \quad . \qquad (10.131)$$

The reactivity feedback is found from Eq. (10.31) as:

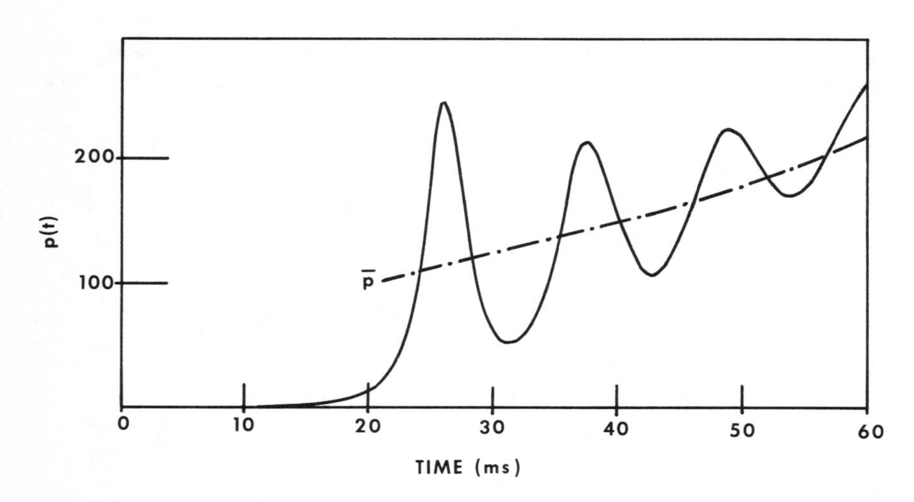

Fig. 10-13. The transient in Fig. 10-12 with an energy coefficient proportional to $1/T$.

$$\delta\rho(t) = \int_{T_0}^{T(t)} \frac{A_D}{T}\, dT = A_D \ln \frac{T(t)}{T_0} \quad , \tag{10.132}$$

or by inserting $T(t)$ from Eq. (10.131) as:

$$\delta\rho(t) = A_D \ln\left\{\frac{T_0 + C_{IT} \int_0^t [p(t') - p_0]\, dt'}{T_0}\right\} \quad . \tag{10.133}$$

Eq. (10.133) replaces the simplified feedback reactivity calculated with a temperature-independent Doppler coefficient:

$$\delta\rho(t) = \gamma \int_0^t [p(t') - p_0]\, dt' = \gamma\, I(t) \quad . \tag{10.134}$$

For a comparison of the two feedback reactivities, Eqs. (10.133) and (10.134), the contribution of expansion to the prompt feedback is neglected. (This neglect is frequently made in conservative safety calculations.) Then the values of γ and A_D in Eqs. (10.134) and (10.133), respectively, are directly related.

Dividing Eqs. (10.134) and (10.131) yields as the temperature coefficient (due to Doppler effect):

$$\frac{\delta\rho}{\delta T} = \left(\frac{d\rho}{dT}\right)_{\mathrm{D}} = \frac{\gamma}{C_{IT}} \quad . \tag{10.135}$$

For comparison with the T-dependent Doppler coefficient, the right side of Eq. (10.135) is set equal to the average coefficient during the first burst:

$$\left(\frac{d\rho}{dT}\right)_{\mathrm{D}} = \frac{\gamma}{C_{IT}} = \frac{A_D}{\overline{T}^{(1)}} \quad , \tag{10.136}$$

using the $1/T$ dependence of the fast reactor example presented here. The γ value from Fig. 10-12 was used together with the C_{IT} value estimated in Eq. (10.23):

$$C_{IT} = 400 \text{ K/fp-s}$$

and

$$\gamma = -0.5\$/\text{fp-s} \quad . \tag{10.137}$$

In addition, a value for $\overline{T}^{(1)}$ is needed to estimate A_D from Eq. (10.136). The following value is used as a close approximation:

$$\overline{T}^{(1)} \simeq T_0 + \frac{1}{2}\,\overline{\Delta T}^{(1)} \quad . \tag{10.138}$$

The temperature increase during the first burst, $\overline{\Delta T}^{(1)}$, in the example of Fig. 10-12 ($T_0 = 1500$ K), was estimated above as ~200 K, which gives $\overline{T}^{(1)} \simeq 1600$ K. With these values, the following magnitude of A_D is obtained:

$$A_D = \frac{\gamma}{C_{IT}}\,\overline{T}^{(1)} = \frac{-0.5\$/\text{fp-s}}{400\ \text{K/fp-s}}\,1600\ \text{K} = -2\$ \quad . \tag{10.139}$$

Using Eq. (10.139) yields, for the temperature coefficient for both approximations,

$$\gamma_T = \frac{A_D}{\overline{T}^{(1)}} = -1.25 \times 10^{-3}\,\frac{\$}{\text{K}} = -0.125\,\frac{¢}{\text{K}} \quad . \tag{10.140}$$

If the temperature coefficient is constant, the same $\delta\rho$ is obtained for the same value of δT, independent of the temperature at which the increment δT is applied. For example, $\delta T = 160$ K yields $\delta\rho = 20¢$ independent of T.

For a $1/T$-dependent coefficient, the same $\delta\rho$ is obtained for the same *relative* change of the temperature. Let T_2/T_1 be the ratio of the temperatures before and after the change ($T_2 = T_1 + \delta T$). The corresponding reactivity is given by

$$\Delta\rho = A_D \ln\frac{T_2}{T_0} = A_D \ln\frac{T_1}{T_0} + A_D \ln\frac{T_2}{T_1} \quad . \tag{10.141}$$

If, for example, $T_2/T_1 = 1.1$ ($\ln T_2/T_1 \simeq 0.1$), then

$$\delta\rho = A_D \ln\frac{T_2}{T_1} \simeq A_D\,0.1 = 20¢ \quad .$$

The same value of $20¢$ is obtained when T changes by 10% about any base temperature.

The transient depicted in Fig. 10-13 clearly shows the expected features, i.e., oscillation of the flux around an increasing $\overline{p}$ (dashed-dotted line) and a lessened damping effect as compared to Fig. 10-12.

The presentation of transients with feedback follows an inductive path: It starts with the most idealized reactivity insertion, the step, and the most idealized feedback, the adiabatic linear energy model, and proceeds by including—in well-defined increments—more and more reality in the transient treatment. Quantitative comparisons show the effect of the various changes in the physical model. Practical applications require detailed numerical solutions. However, the semi-quantitative feature developed here can serve as a useful guide toward setting up the problems and evaluating the numerical solutions.

Homework Problems

1. Take for the temperature coefficient of the Doppler effect the T dependence given in Sec. 10-1B, i.e.,

$$\frac{\partial \rho}{\partial T} = \frac{A_D}{T^x} \quad ,$$

with x between 0.5 and ~1.2. Find the total reactivity change $\Delta\rho$ between $T_0 = 1000$ K and $T_1 = 1900$ K and $T_2 = 2800$ K, the latter being about the melting temperature of oxide fuel. Compare $\Delta\rho_1$ (1000, ..., 1900) and $\Delta\rho_2$ (1900, ..., 2800) for the three x values of 0.5, 1.0, and 1.2. Discuss your results.

2. Find the numerical value of p^{00}, the flux after a prompt jump for which the increase due to delayed neutrons is just compensated by Doppler feedback, for an LWR from the typical λ and γ/β values given in the text. Discuss why p^{00} is different from the value for the SEFOR reactor.

3. Consider the reactivity-step-induced transient with an adiabatic boundary condition for the incremental heat, as considered in Sec. 10-2B. List the asymptotic results derived in the text and apply them to estimate

$$\int_0^\infty [s_d(t') - s_{d0}]\, dt = ?$$

Present the physical arguments as to why the integral of the additional delayed neutron source must be finite although $s_{das} = s_{d0}$.

4. Calculate the burst width in a superprompt-critical transient with the linear energy feedback model from the formula derived in the text. Use LWR data: $\beta = 0.0075$, $\Lambda = 10^{-5}$ s; let $\rho_1 = 1.1\$$. Also find the energy release

$$Q(t_2) = \int_0^{t_2} P(t)\, dt \quad .$$

Estimate the magnitude of the effects of the following two approximations:

a. The neglect of the heat release (i.e., the assumption of $\lambda_H = 0$). Consider an oxide fuel rod of 0.6 cm in diameter.
b. Neglect of P_0 under the feedback integral, i.e., the integral of P_0 over the burst width with $Q(t_2)$.

5. Explain and estimate quantitatively the damping effect for the reactivity-ramp-induced transient depicted in Fig. 10-12 by considering the source multiplication of the increasing delayed neutron source (see the p_{ab} evaluation for the reactivity-step-induced transient; the transient data are given in the text and in the caption of Fig. 10-8).

6. Explain and estimate quantitatively the two effects of a temperature coefficient decreasing as $1/T$ on the maxima of the reactivity-ramp-induced transient depicted in Fig. 10-13.

Review Questions

1a. What are energy, temperature, and power coefficients?
 b. Give examples of physical phenomena involved.
 c. Give an example of a transient for the application of energy, temperature, and power coefficients, respectively.
2a. Give the formula for the adiabatic feedback model for the entire and the incremental energy respectively.
 b. Give a typical application for each of the two models.
 3. Describe the general concepts of the use of reactivity coefficients and their application (for constant and variable coefficients).
 4. Describe qualitatively the transient at small times (including the two competing effects). Which approximation allows an accurate quantitative evaluation of those effects?
5a. Derive (as briefly as possible) the "slope" of the flux as it results from the increased delayed neutron population.
 b. Derive (as briefly as possible) the "slope" of the flux as it results from feedback.
 c. Compare both slopes and find the reactivity insertion at which they compensate each other.
 6. Find the value of the asymptotic flux (derived from the assumption that it is constant): (a) without heat transfer and (b) with heat transfer described by a single exponential function.
 7. Find the energy deposited in the fuel without and with heat transfer.
 8. Why is the feedback reactivity approximated by the Nordheim-Hansen model (i.e., neglecting the stationary heat removal) in the analytical treatment of step-induced superprompt-critical transients? Questions 9 through 12 pertain to these transients.
 9. Give the differential equation in the prompt kinetics approximation, including the initial condition.

10. Find the reactivity at the maximum of the transient.
11. Find the energy release up to the maximum of the transient.
12. Derive the formula for ρ at the maximum of the flux.
13. Transform the original equation for superprompt-critical transients with linear energy feedback into a second-order differential equation for the prompt reactivity ρ_p.
14. Find the first integral.

15a. Find ρ_m, the flux amplitude at the maximum.
  b. Find the reactivity at $2t_m$.

16. Find the energy release at $2t_m$. Interpret the energy release and the corresponding temperature rise in terms of the respective reactivity coefficients.
17. Indicate the way the solution for $\rho_p(t)$ is derived from the first integral in a flux transient during a superprompt-critical excursion.
18. How do you find $p(t)$ from $\rho_p(t)$?

19a. Define the width of the flux pulse.
  b. Give the main factors of the formula for the width of the flux pulse and show why the energy release is independent of Λ.

20a. Discuss the principal effect of the delayed neutron on the transient as treated in the prompt kinetics approximation.
  b. Discuss the applicability of the prompt kinetics approximation (for $\rho > \beta$ and for $\rho < \beta$).

21a. Estimate the power after the burst, say, at $t = 2t_m$:
  b. Give the estimate without additional precursors.
  c. How would the additional precursors that are produced during the burst affect this flux estimate?
  d. Describe qualitatively the continuation of the transient after $t = 2t_m$.

22a. Discuss the "shutdown capability" of the prompt negative reactivity feedback coefficients.
  b. How does this "shutdown capability" depend on the magnitude of the energy coefficient?

23. Describe the "initial" condition as it is used in the prompt kinetics approximation for superprompt critical transients induced by reactivity ramps.
24. Give the differential equation, the initial condition, and the reactivity including prompt feedback.
25. Sketch qualitatively the time dependence of the reactivity and the flux in the prompt kinetics approximation and relate key points.
26. Find and discuss the information on energy release that can be extracted from the differential equation.

27a. Describe qualitatively the impact of the increase of the delayed neutron source on the first post-burst minimum.

b. Explain the damping effect of the delayed neutrons.

c. How do the delayed neutrons affect $\overline{P}$?

28. Suppose $\lambda(T)$ decreased with T; describe the resulting two changes in the transient.

REFERENCES

1. K. Fuchs, "Efficiency for Very Slow Assembly," LA-596, Los Alamos Scientific Laboratory (1946).
2. T. F. Wimmett, L. B. Engle, G. A. Graves, G. R. Keepin, Jr., and J. D. Orndoff, "Time Behavior of Godiva through Prompt Critical," LA-2029, Los Alamos Scientific Laboratory (1956).
3. S. G. Forbes, "Simple Model for Reactor Shutdown," IDO-16452, p. 38, Reactor Physics Branch, Idaho Operations Office, U.S. Atomic Energy Commission (1958).
4. W. E. Nyer, "Ramp-Induced Bursts," IDO-16452, p. 65, Reactor Physics Branch, Idaho Operations Office, U.S. Atomic Energy Commission (1958).
5. W. E. Nyer and S. G. Forbes, "SPERT Program Review," IDO-16634, Idaho Operations Office, U.S. Atomic Energy Commission (1960).
6. W. Häfele, K. Ott, L. Caldarola, W. Schikarski, K. P. Cohen, B. Wolfe, P. Greebler, and A. B. Reynolds, "Static and Dynamic Measurements on the Doppler Effect in an Experimental Fast Reactor," *Proc. Third Int. Conf. Peaceful Uses of Atomic Energy,* Vol. 6, paper 644, International Atomic Energy Agency, Vienna (1964).
7. L. D. Noble, G. Kussmaul, and S. L. Derby, "SEFOR Core 1 Transients," GEAP-13837, General Electric Co. (Apr. 1972).
8. D. D. Freeman, "SEFOR Experimental Results and Applications to LMFBR's," GEAP-13929, General Electric Co. (Jan. 1973).
9. L. W. Nordheim, "Pile Kinetics," MDDC-35, U.S. Atomic Energy Commission (1946).
10. G. E. Hansen, "Burst Characteristic Associated with the Slow Assembly of Fissionable Materials, LA-1441, Los Alamos Scientific Laboratory (1952).
11. "1000-MW(e) Fast Breeder Reactor Follow-On Study," Task 1 Final Report, WARD-2000-33, pp. 3–14, Westinghouse Electric Corp. (June 1968); revision (Jan. 1969).
12. J. Leitnaker and T. G. Godfrey, "Thermodynamics Properties of Uranium Carbides," *J. Nucl. Mater.,* **21,** 175 (1967).
13. F. L. Oetting and J. M. Leitnaker, "The Chemical Thermodynamics Properties of Nuclear Materials," *J. Chem. Thermodynamics,* **4,** 199 (1972).
14. H. Savage, "The Heat Content and Specific Heat of Some Metallic Fast Reactor Fuels Containing Plutonium," *J. Nucl. Mater.,* **25,** 249 (1968).

15. *Fuels and Materials,* Vol. 7 of *Liquid Metal Fast Breeder Reactor Program Plan,* WASH-1107, U.S. Atomic Energy Commission (Aug. 1968).
16. A. Reichel, "The Effect of Non-Uniform Fuel Rod Temperatures on Effective Resonance Integrals," AEEW-R-76, U.K. Atomic Energy Authority (June 1961).
17. A. Reichel, "Trends in the Temperature Dependence of Effective Resonance Integrals," AEEW-R-102, U.K. Atomic Energy Authority (Nov. 1961).
18. P. Greebler and E. Goldman, "Doppler Calculations for Large Fast Ceramic Reactors—Effects of Improved Methods and Recent Cross Section Information," GEAP-4092, General Electric Co. (1962).
19. R. Froelich, K. Ott, and J. J. Schmidt, "Dependence of Fast Reactor Doppler Coefficients on Nuclear Data Uncertainties," *Proc. Conf. Breeding, Economics and Safety in Large Fast Power Reactors,* ANL-6792, Argonne National Laboratory (Oct. 1963).
20. Karl O. Ott and W. A. Bezella, *Introductory Nuclear Reactor Statics,* American Nuclear Society, La Grange Park, Illinois (1983).
21. K. O. Ott, "Theorie verzögert überkritischer Exkursionen zur Messung der Doppler-Koeffizienten schneller Reaktoren," *Nukleonik,* **5,** 285 (1963); or see English translation, KFK-153, Kernforschungszentrum Karlsruhe, FRG (1963).
22. R. A. Meyer, A. B. Reynolds, S. L. Stewart, M. L. Johnson, and E. R. Craig, "Design and Analysis of SEFOR Core 1," GEAP-13598, General Electric Co. (June 1970).
23. B. A. Hutchins and K. Ott, "An Analysis of Errors Involved in the Subprompt Critical Transient Experiments in SEFOR," KFK-578, Kernforschungszentrum Karlsruhe, FRG (1965).
24. L. D. Noble, G. Kussmaul, and G. R. Pflasterer, "Sub-Prompt Critical Transients in SEFOR," *Trans. Am. Nucl. Soc.,* **14,** 741 (1971).
25. L. D. Noble, B. U. B. Sarma, S. Derby, M. Plummer, G. R. Pflasterer, M. Nielson, and M. L. Johnson, "SEFOR Core II Zero-Power Experiments," *Trans. Am. Nucl. Soc.,* **15,** 500 (1972).
26. D. D. Freeman, "SEFOR Experimental Results and Applications to LMFBR's," GEAP-13929, General Electric Co. (Jan. 1973).
27. L. Caldarola, "Analysis of Power Excursions by Means of Solutions with Asymptotic Expansions," *Nukleonik,* **9,** *3,* 129 (Feb. 1967).
28. W. K. Ergen and A. M. Weinberg, "Some Aspects of Non-Linear Dynamics," *Physica,* **20,** 413 (1954).
29. S. G. Forbes, "The Dependence of Reactor Behavior on the Self-Shutdown Mode," IDO-16635, p. 19, Idaho Operations Office, U.S. Atomic Energy Commission (1961).
30. J. Canosa, "A New Method for Nonlinear Reactor Dynamics Problems," *Nukleonik,* **9,** 289 (1967).
31. W. E. Nyer, S. G. Forbes, F. L. Bentzen, G. O. Bright, F. Schroeder, and T. R. Wilson, "Experimental Investigations of Reactor Transients," IDO-16285, Idaho Operations Office, U.S. Atomic Energy Commission (1956).

Eleven

SPACE-ENERGY DEPENDENT DYNAMICS

11-1 Introduction

Solutions of reactor kinetics problems and some dynamics problems with prompt reactivity feedback were discussed in Chapters 6 and 10, respectively. The solutions were derived from the point kinetics equations without explicit consideration of the space and energy dependence of the neutron flux. The principal effect of the variation of the space-energy dependence of the flux was discussed only in the chapter on analysis of reactivity measurements, Chapter 9. In dynamics problems, the effect of the flux shape variation on the course of the transient may be even larger than in the kinetics problems of Chapter 9. The analyses of many safety problems actually require a solution utilizing space-energy dependent dynamics, or at least improved versions of the point reactor dynamics.

The importance of the space-energy dependent treatment of safety analysis problems is highlighted by the fact that the point reactor results are not only inaccurate in important cases, but also generally nonconservative. The reasons are that positive reactivity insertions, which may cause or intensify an accident, are normally underpredicted by the neglect of the flux shape variation (i.e., by the point reactor model, PRM), and that the value of a negative counteracting reactivity on the other hand is frequently overpredicted in the PRM.

The following two examples illustrate this characteristic behavior. If an accident is initiated by dropping a fuel subassembly into a near-critical reactor, the neutron flux increases within and around this fuel subassembly due to the local increase in the fission neutron source. The neglect of this local flux variation leads to an underprediction of the positive reactivity increase. The effect of a negative counteracting reactivity change can be demonstrated by considering the insertion of a scram rod. The corresponding local increase of the absorption leads to a flux depression. The neglect of this local flux depression results in an over-prediction of the negative counteracting reactivity.

The two simple examples of positive and negative reactivity insertions clearly show that the PRM results tend to be nonconservative. This nonconservative nature of the point reactor model prevails in most of the practical cases. However, special complicated sequences or superpositions of events may occur in some safety problems in which the situation is reversed. The originally nonconservative PRM results could predict a more violent reactor response in a subsequent accident phase than a correct treatment would. Then, the PRM would overestimate the eventual energy release; but this is the exception.

Methods are presented in this chapter to account for the space-energy variation of the neutron flux during transients. These variations are very important for dynamics problems. Since the solution of dynamics problems represents the neutronics part of safety analysis, the solution methods presented for space-energy dependent dynamics are discussed in conjunction with special safety application considerations. Consequently, Sec. 11-2 is devoted to a general discussion of the dynamics problem to show the interplay of neutronics and non-neutronics phenomena in safety analysis. For specific information on the non-neutronics part of thermal and fast reactor safety analysis, the reader is referred to Refs. 1 through 5.

It should be emphasized that the scram system shuts the reactor down during the "normal" course of a reactor accident if such action is required by any system malfunction. The remaining energy production at a rate of several percent of the original power is due to afterheat, which results from the radioactive decay of fission products. If required, natural convection or a special emergency cooling provision removes the afterheat. Structural damage does not occur in such accidents, and the reactor can be restarted. The core damage in the accident at Three Mile Island Unit 2 resulted from an inadvertent shutdown of the emergency core cooling system. A complete failure of the scram systems is an extremely improbable event. Still, it needs to be considered in accident progressions.

In the presentation and discussion of the methods for the treatment of space-energy dependent dynamics, emphasis is placed on flux factorizing methods that have been applied very successfully to both thermal and fast reactor problems. The emphasis is thus different from that found in other books or survey articles, which predominantly consider earlier methods; for example, Refs. 6 through 8.

The time dependence of the neutron flux is characteristically different in fast versus thermal reactors. The flux distribution in a fast reactor is much less susceptible to spatial distortions than the flux distribution in a thermal reactor of the same power. This is due to the combined effects of the smaller core size and the considerably larger mean free path of fast neutrons compared to thermal neutrons (the core volume of fast reactors is much smaller than that of thermal reactors of the same power). In other words, fast reactors are more tightly coupled than comparable thermal reactors. Fast reactors in their stationary state also have a lower built-in reactivity for compensation for fuel burnup and buildup of fission products than thermal reactors. However, the possibility of an accidental release of most of this surplus reactivity is eliminated by the use of burnable poisons. Spatial flux distortions in fast reactor transients are generally smaller by about one order of magnitude. Thus, the inclusion of the space dependence is more important for thermal reactor transient analysis than for fast reactor analysis.

The energy dependence of the neutron flux in thermal reactor transient analysis is approximated by a few-group model. In a fast reactor, however, the energy dependence of the flux during severe accidents can vary considerably. Since a variation of the spectrum changes the Doppler coefficient, which in turn affects the transient, the time dependence of the fast reactor neutron spectrum must be treated in greater detail than in the few-group model employed in thermal reactor dynamics.

11-2 General Discussion of the Dynamics Problem

The basic neutronics problem of any transient analysis is the solution of the balance equation of the neutron flux and the delayed neutron precursor concentrations (see Sec. 3-2A). The balance equations in operator notation are given by

$$\frac{1}{v}\frac{\partial \Phi}{\partial t} = (\mathbf{F}_p - \mathbf{M})\Phi + S_d + S \quad , \tag{11.1a}$$

$$\frac{\partial C_k}{\partial t} = -\lambda_k C_k + \mathbf{F}_k^d \Phi + \text{transport and loss terms} \quad , \tag{11.1b}$$

and

$$S_d = \sum_k \lambda_k C_k \chi_k^d(E) \quad . \tag{11.1c}$$

The microscopic cross sections and atom densities in the operators of Eqs. (11.1) depend on time, particularly through feedback effects. Since feedback is caused by changes in the neutron flux and the power, the operators in Eqs. (11.1) are implicitly functions of the flux Φ. The calculation of the dependence of the macroscopic cross sections on flux changes normally requires the solution of further differential equations that describe phenomena such as heat transfer, flow of coolant, cladding and fuel motion, etc. Therefore, the neutronics equations, Eqs. (11.1), must be completed by the equations that describe the variation of the macroscopic cross sections resulting from changes of the power distribution through the various feedback phenomena of power reactors. The major feedback phenomena are briefly described below. For further discussion of the mathematical equations and their solutions, see, for example, Refs. 1 through 5.

The time dependence of the *microscopic* cross sections results principally from the Doppler broadening of resonances with increasing temperature. The effect of this Doppler broadening can be expressed as a temperature dependence of "effective" cross sections or group constants[3]:

$$\delta\tilde{\sigma}_g = \delta\tilde{\sigma}_g\,[T(\mathbf{r},t)] \text{ through Doppler effect} \quad . \tag{11.2}$$

In more approximate treatments, the Doppler feedback is accounted for by "reactivity coefficients," either for the entire reactor (see Sec. 10-1) or for individual regions (see Sec. 11-5). The calculation of the Doppler feedback effect for a given temperature field requires only the consideration of neutronics; the Doppler effect leads to a "neutronic" feedback.

The temperature field, $T(\mathbf{r},t)$, varies as a result of the power transient, and is determined by the space distribution of the energy deposition and by heat transfer. The transfer of heat is strongly influenced by changes in the heat transfer media, e.g., by a reduction of coolant density or velocity:

$$\delta T(\mathbf{r},t) \text{ from } \begin{cases} \text{power distribution} \\ \text{heat transfer} \\ \text{compositional changes} \quad . \end{cases} \tag{11.3}$$

Thus, compositional changes must be treated in addition to or simultaneously with the variation of the temperature field.

The variation of the composition or the atom densities at a point $\mathbf{r}$, $N_i(\mathbf{r},t)$, occurs either because of the variation of the local temperature, $T(\mathbf{r},t)$, which leads to expansion and possibly melting or boiling, or because of forced displacement. The forced displacement is due to changes in the atom densities in spatial areas different from $\mathbf{r}$, say $\mathbf{r}'$, which then cause changes in the atom density at $\mathbf{r}$. The major modes of compositional changes are summarized below. All effects are caused by or affect temperature and/or number density changes (δT or δN_i):

$$\delta\Sigma[\delta N,\delta T]\ \begin{cases}\delta T(\mathbf{r},t)\\ \delta N_i(\mathbf{r}',t)\end{cases}\ \left\{\begin{array}{l}\text{loss of coolant}\\ \text{expansion}\\ \text{melting}\\ \text{boiling}\\ \text{cladding motion}\\ \text{fission gas ejection}\\ \text{fuel/coolant interaction}\\ \text{fuel slumping}\\ \text{fuel sweep out}\end{array}\right. . \tag{11.4}$$

In many dynamics problems, the transport of precursors is not negligible as in the kinetics problems and in the simple dynamics problems treated above. An approximate description of the transport and loss of precursors was introduced by Meneley et al.[9]

The precursor density can change through motion and/or density change of fuel. If fuel elements are damaged or melt, precursor loss can occur through the loss of volatile components of fission products, particularly bromine, krypton, iodine, and cesium. Since different precursor groups contain different fractions of volatile precursors (e.g., precursor group 1 consists almost entirely of ^{87}Br; see Chapter 2), the precursor loss can vary strongly with delay groups. The loss of precursors is not explicitly treated here. Only the major qualitative effect is briefly discussed.

First, there is no effect of the loss of precursors on β, and thus on the subprompt-critical reactivity range. The reason is that β is determined by the production rate of precursors, and not by their residing density (see Chapters 3 and 5).

Second, a reduction of the residing density of precursors generally leads to a reduction of the neutron flux. This can be shown readily in the subprompt-critical domain, where the flux amplitude $p(t)$ is approximately given by Eq. (8.7):

$$p(t) \simeq \frac{s_d(t)}{\beta - \rho(t)} \quad . \tag{11.5}$$

Clearly, if the precursor density $[\zeta_k(t)]$ is reduced, the delayed neutron source,

$$s_d(t) = \sum_k \lambda_k \zeta_k(t) \quad , \tag{11.6}$$

is decreased; thus the flux is decreased. If the transient reactivity enters the superprompt-critical domain, it will do so from lower flux levels when precursors have been lost during the preceding delayed supercritical phase.

The equations that describe the neutronic and non-neutronic feedback effects represent the description of the complete nonlinear dynamics problem. Thus, the calculation of the neutron flux during a transient is only a part, primarily the smaller part, of the treatment of the complete dynamics problem.

Due to the strong coupling of the equations of the dynamics problem, the complete set of equations must, in principle, be solved simultaneously (see Sec. 11-4c). However, not all of the phenomena listed above occur during the same phase of an accident. Several phases of an accident progression can therefore be distinguished.

An accident may have been caused by a reactivity insertion (e.g., control rod ejection) or a coolant flow reduction (resulting from a pipe rupture or loss of pump power) together with an assumed failure of the entire scram system. During the first phase of such an accident, the time dependence of the neutron flux is influenced essentially by the two feedback phenomena of Doppler effect and coolant temperature rise. The resulting density reduction of the coolant in a light water reactor (LWR) has a negative reactivity effect; in a large liquid-metal fast breeder reactor, it has mainly a positive reactivity effect, whereas the Doppler effect always provides a negative feedback.

In the next phase of an accident without scram, the coolant may begin to boil with resultant voiding of fuel subassemblies. After the onset of coolant boiling, cladding melting and cladding motion can occur. Fission gas may then be injected into the coolant and cause an increase in the voiding rate. Fuel melting could also occur during this phase. Molten fuel could be swept out of the core by the coolant. Fuel may also be driven out of the core by the pressure of fission gases enhanced by increased pressures due to fuel/coolant interaction. In some cases, the reactivity could be increased by fuel slumping (under gravity). All of these phenomena have a strong effect on the neutron flux so that the compositional, temperature, and neutron flux variations have to be treated simultaneously during this phase.

In addition to the energy released during the accident, further energy production results from the decay heat of the fission products. The removal of the decay heat requires a special cooling capability, which must operate even after core damage (postaccident heat removal). If an accident progresses through such a disruptive phase, a possibility exists for the release of radioactivity. All of these events are carefully evaluated in a complete safety analysis even though the probability of such accidents is very small.

Although the thermodynamics, i.e., the coolant and fuel dynamics phenomena, are not treated in this text, a survey of the complete problem was presented to provide background information that is helpful for the discussion of space-energy dependent dynamics. In particular, the survey showed the strong interplay between the neutron dynamics and fuel and coolant dynamics; this interplay is much stronger in fast than in thermal reactors.

11-3 Survey of Space-Energy Dependent Dynamics Approaches

11-3A Finite Difference Methods

The finite difference method is the most straightforward of the space-energy dependent dynamics approaches. It consists basically of replacing the differential operators in Eqs. (11.1), i.e., $\partial/\partial t$ and the spatial differential operators in **M**, by the corresponding finite difference quotients.[7] A significant refinement of the treatment of the time dependence in space-energy dependent kinetics was introduced by Hansen and Johnson[10] and Andrews and Hansen.[11] An exponential time variation of the flux within time intervals was assumed for each point of the space-energy domain resulting in accurate solutions with largely expanded time steps.[a]

The great advantage of the finite difference method is that for sufficiently small time steps it yields essentially the correct solution of the time-dependent multigroup equations. The disadvantage, however, is that the computation time is too large for normal practical applications. This is particularly true for fast reactors where the variation of the energy spectrum during a severe accident may require many groups ($\approx$20).[b] A space-energy dependent treatment is particularly important during phases of an accident for which the core is significantly distorted. Such distorted cores often do not have the symmetry that can be used to reduce the number of dimensions (e.g., from three to two). Thus, a large number of space points is also required to describe the spatial detail of the reactor configuration during a core-distorting accident.

Figure 11-1 illustrates the finite difference procedure. Each cross on the bottom line represents a flux value in the space-energy domain. A typical value of $\sim$20 000 space-energy points is needed for a descrip-

---

[a]The assumption of an exponential time variation of the flux within a time interval was initially employed to numerically solve the point kinetics equation (see Ref. 12).

[b]For many applications not involving significant core distortions, a lesser number of groups ($\approx$8) generally suffices.

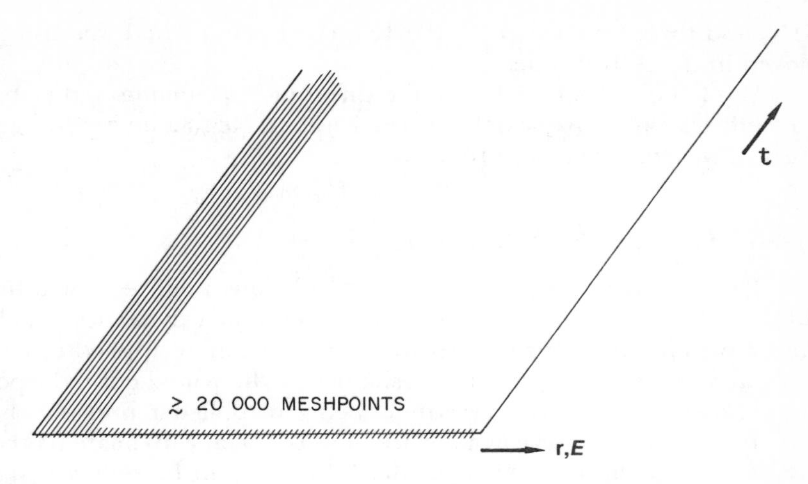

Fig. 11-1. Illustration of mesh representation in the finite difference method.

tion in two dimensions of the multigroup neutron flux (e.g., 20 groups × 40 × 25 meshpoints). In the finite difference method, each of the flux values is followed along the time axis, as indicated by the lines, in the *t* direction. The time intervals may have to be very small to avoid numerical instability and obtain sufficient accuracy. Thus, the number of flux values in the combined space-energy-time domain becomes very large, which leads to very long computation times. Computer programs based on this approach are, therefore, normally only applied for checking approximate methods.

The capabilities of the better known computer codes based on the finite difference method are summarized briefly below.

One of the first programs introduced was the WIGLE code.[13] The code was restricted to one-dimensional slab geometry and two energy groups. Subsequent versions of this code, WIGL2 (Ref. 14) and WIGL3 (Ref. 15), extended the geometrical capability to include cylindrical and spherical geometries and provided temperature, xenon, and control feedback. The WIGLE method was extended to two dimensions in the TWIGL computer code.[16] The TWIGL code solves a two-group problem in either (x,y) or (r,z) geometry. The WIGLE and TWIGL codes are widely used in the thermal reactor area where a two-group description is adequate for most applications.

An expansion of the time interval over at least parts of the transient was demonstrated by the methods of Hansen et al.[10,11] On the basis of this method of exponential flux variation, multigroup computer codes were written in one dimension,[10] GAKIN; in two dimensions,[17,18] MIT-

KIN; and three dimensions,[19] 3DKIN. The same method was also employed in the ADEP code.[20]

All of the codes based on finite difference techniques either have no feedback capability or only feedback applicable to a limited range of thermal reactor dynamics problems.

11-3B The Point Reactor Model

The extremely large number of calculations required for a finite difference solution of a two- or three-dimensional dynamics problem emphasizes the need for simpler methods that can be applied economically in routine investigations. The simplest of these methods is the point reactor model (PRM), which was described and applied in previous chapters. It is mentioned here again since most of the approximate methods are, in a sense, improvements of the PRM and can be reduced to the PRM with certain simplifying assumptions.

Figure 11-2 illustrates the essential feature of the PRM; namely, the condensation of detailed initial information into a single flux quantity $p(t)$. Only the time variation of the flux amplitude $p(t)$ is investigated under the influence of a time-dependent reactivity in the PRM. The associated reduction in computation time compared to the space-energy finite difference method is obvious; the time dependence of only one instead of, for example, 20 000 flux quantities is calculated. The limitations of the PRM were discussed earlier (see Sec. 11-1). Since the PRM generally yields nonconservative results, improvements or replacements are required for safety problem investigations.

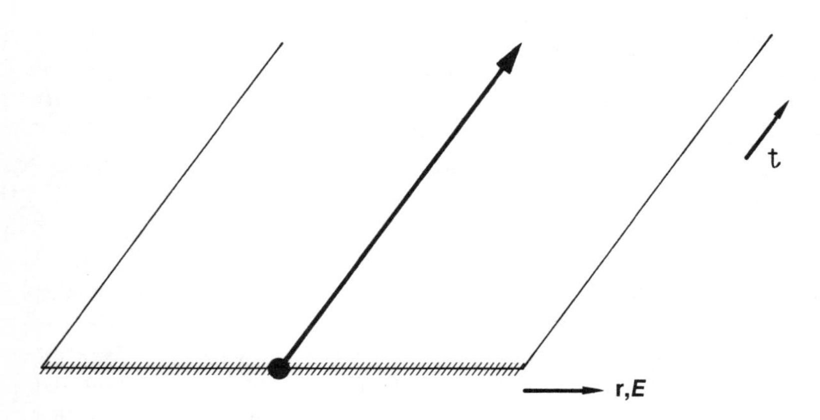

Fig. 11-2. Illustration of mesh representation in the PRM.

11-3C The Modal Approach

The conceptual idea of the modal approach is to construct or synthesize the space-, energy (group)-, and time-dependent neutron flux as a superposition of two or more components (modes). The desired savings in computation time is achieved by employing only a very few modes as well as by calculating the flux modes with simpler methods, for example, as solutions of static problems of normally lower dimensionality.

The simplest form of a time-dependent flux synthesis is given by the expression[21]:

$$\phi_g(\mathbf{r},t) = \sum_m h^{(m)}(t) f_g^{(m)}(\mathbf{r}) \quad , \tag{11.7}$$

where the flux modes, $f_g^{(m)}(\mathbf{r})$, contain the space and group dependence whereas the superposition coefficients, $h^{(m)}(t)$, are functions of time only. The use of only one term in the sum of Eq. (11.7) corresponds to the PRM if $f_g(\mathbf{r})$ is chosen as the initial multigroup flux shape.

More general and sophisticated expressions than Eq. (11.7) could be used in practical applications. For example, different coefficient functions have been used for different groups to allow for a different time dependence of the thermal and fast flux in thermal reactors; formally, $h^{(m)}(t)$ is then replaced by $h_g^{(m)}(t)$ as in Ref. 21. If a three-dimensional description is required, the coefficient often includes one of the space dimensions[22]:

$$\phi_g(\mathbf{r},t) = \sum_m h_g^{(m)}(z,t) f_g^{(m)}(x,y) \quad . \tag{11.8}$$

A further and important generalization is the subdivision of the time domain into several large intervals with different sets of flux modes in the different intervals.[23]

The use of flux expressions such as Eqs. (11.7) and (11.8) requires the derivation of a set of equations for the calculation of the time-dependent coefficient functions (the flux modes are calculated from appropriate statics computer programs). The derivation of the equations for the $h^{(m)}(t)$ is, in principle, similar to the derivation of the point kinetics equations: the flux expression of Eq. (11.7) is inserted into the time-dependent diffusion equation, Eqs. (11.1), multiplied with weighting functions and integrated with respect to space and energy. The number of weighting functions required equals the number of unknowns, $h^{(m)}(t)$, in order to generate the appropriate set of independent equations. For a more detailed discussion of the derivation of equations in the synthesis approach using linear flux superposition, the reader is referred to the literature.[6–8,21–24]

Figure 11-3 shows an illustration of the simplest form of the modal

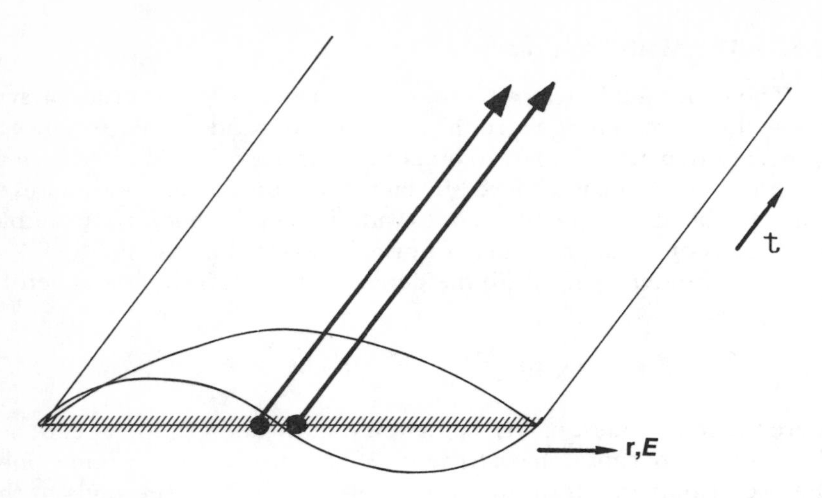

Fig. 11-3. Illustration of mesh representation in a simple form of the modal approach.

approach, two modes in a one-group approximation. The all-positive curve describes the unperturbed flux. The second mode is depicted by a sine function. The superposition of the sine function with the unperturbed flux allows the perturbed flux to shift to the left or right, depending on the sign of the coefficient. The two coefficients, $h^{(1)}(t)$ and $h^{(2)}(t)$, are to be determined as functions of time.

The modal approach applied to purely space-dependent dynamics problems is often called the "space-time" synthesis method. Sometimes modal methods employing two or more energy groups are also called "space-time" synthesis methods, a usage which indicates the dominance of spatial over spectral distortion of the flux in thermal reactor transients.

Synthesis methods have been developed in the United States for thermal reactor dynamics problems that may involve distorted spatial flux distributions. The major problem is the description of the spatial distortions of the flux and not the spectral distortion. The description of the spectrum by two energy groups suffices in most cases. The synthesis approach has also been applied for fast reactor dynamics problems employing a large number of energy groups.[24]

11-3D The Nodal Approach

The conceptual idea of the nodal approach is to approximate the time-dependent flux by non-overlapping flux branches in a set of spatial subregions or "nodes":

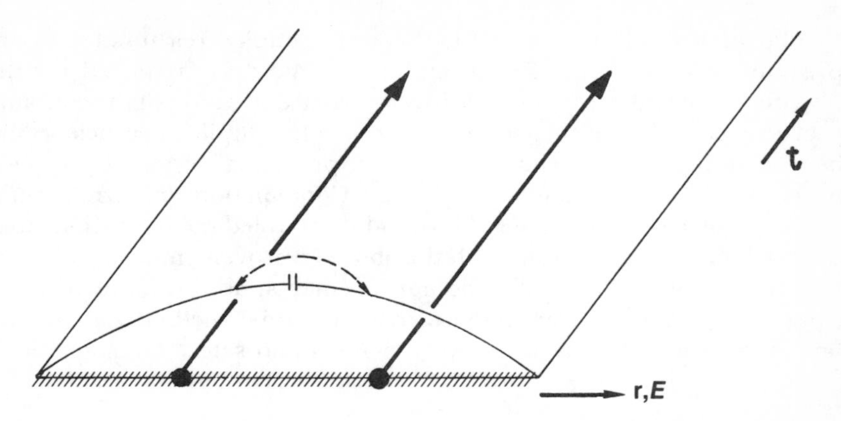

Fig. 11-4. Illustration of mesh representation in the nodal approach with two flux branches; the coupling is indicated by the dashed line.

$$\phi_g(\mathbf{r},t) = \begin{cases} h^{(1)}(t) f_g^{(1)}(\mathbf{r}) \text{ in node 1} \\ h^{(2)}(t) f_g^{(2)}(\mathbf{r}) \text{ in node 2} \\ \cdot \\ \cdot \\ \cdot \end{cases} \quad . \tag{11.9}$$

If only one node is employed and if $f_g^{(1)}(\mathbf{r})$ is equal to the initial flux distribution, the nodal method reduces to the PRM.

The nodal approach is illustrated in Fig. 11-4 for a single energy group. The spatial domain is subdivided into two nodes. Neutron transfer between neighboring nodes leads to a coupling of the nodal flux branches, which is indicated by the dashed line in Fig. 11-4. If a flux tilt develops during the transient, i.e., if the flux branch in one node is shifted compared to the other branch, the nodal method then yields a somewhat discontinuous flux distribution.

The development of the nodal approach is historically based on the kinetics of coupled reactors.[25] Avery derived a set of equations for the time-dependent coefficient functions, $h^{(n)}(t)$, with time-independent coupling coefficients describing the neutron transfer between neighboring regions/nodes. If a system consists of two or more loosely coupled reactors, the equations of Ref. 25 can describe the transient behavior fairly accurately. Two reactors or regions can be considered to be loosely coupled if the neutron transfer rate between them is very small compared to the neutron production rate in either region.

Further developments of the theory of coupled reactors led to improvements concerning, for example, the time delay required for the transfer of neutrons,[26,27] and improved formal derivation of the coupling coefficients.[28,29] A recalculation method for the coupling coefficients has been introduced for thermal reactor static problems[30,31] and also applied to thermal reactor dynamics problems.[32] Combination with other methods is summarized in Ref. 33. More tightly coupled configurations may be treated more successfully with the above improvements than with the first version of the theory.[25] The improvements, however, added more complexity, with the consequence that the nodal method has not yet been practically applied in reactor dynamics and safety investigations.

11-3E Flux Factorization Approaches

The approaches that are formally closest to the familiar PRM are those based on flux factorization (see Sec. 5-1A):

$$\phi(\mathbf{r},E,t) = p(t)\psi(\mathbf{r},E,t) \quad , \tag{11.10}$$

where $p(t)$ is the amplitude and $\psi(\mathbf{r},E,t)$ is the time-dependent shape function. Flux *factorization* is the formal generalization of flux *separation,* an approximation in which the shape function is assumed to be independent of time. Since the factorization allows for an arbitrary variation of the shape function with time, it is formally rigorous, i.e., no approximation is introduced by representing the flux in the factorized form of Eq. (11.10). Therefore, the same solution can be obtained from this approach as from the finite difference method (Sec. 11-4C).

The shape function normally varies much more slowly with time and to a lesser degree than the amplitude function. Flux factorization thus provides the opportunity to employ different intervals for the amplitude and shape function calculation. The computation time can be drastically reduced by choosing much larger time intervals between shape function calculations than between calculations of the amplitude function. Performing a shape function calculation only when needed does not practically reduce the accuracy of the solution.

There is only one type of time interval in the finite difference method, i.e., the one for the flux. This conceptually does not allow a benefit from the slow variation of the flux shape.

The accuracy of the factorization approach is in part determined by the frequency of the shape function recalculation. If the shape function is calculated often in time, the factorization and the finite difference methods become equally accurate. However, the computer time requirements are lower for the flux factorization method since a much smaller number of flux shape calculations are required for the same accuracy (see Sec. 11-4C).

In Chapter 5, the flux factorization, Eq. (11.10), was inserted into the time-dependent diffusion equation, Eq. (3.28); by integrating the resulting equation with respect to space and energy and by eliminating the explicit appearance of the shape function with a proper constraint condition, a balance equation for the amplitude function alone was derived [Eq. (5.24)]:

$$\dot{p}(t) = \frac{\rho(t) - \beta(t)}{\Lambda(t)} p(t) + \frac{1}{\Lambda(t)} s_d(t) \quad . \tag{11.11}$$

The definition of the integral parameters, ρ, β, and Λ and of the reduced delayed neutron source, $s_d(t)$, is given in Sec. 5-1B.

Since *two* functions are introduced by Eq. (11.10) instead of only one on the left side, a second equation is required to determine the complete solution. The second balance equation is obtained directly by inserting the flux factorization into the time-dependent diffusion equation given in Sec. 3-2A:

$$(\mathbf{F}_p - \mathbf{M})\phi(\mathbf{r},E,t) + S_d[\phi(\mathbf{r},E,t')] = \frac{1}{v}\frac{\partial}{\partial t}\phi(\mathbf{r},E,t) \quad . \tag{11.12}$$

The delayed neutron source results from the decay of precursors that were produced at earlier times, $t' \leq t$. The space distribution of the delayed neutron precursors at time t can be expressed in terms of the flux at earlier times by formally solving Eq. (3.32) for $C_k(\mathbf{r},t)$:

$$C_k(\mathbf{r},t) = C_k(\mathbf{r},0)\exp(-\lambda_k t)$$
$$+ \int_0^t \exp[-\lambda_k(t - t')] \int_E \nu_{dk}\, \Sigma_f(\mathbf{r},E,t')p(t')\psi(\mathbf{r},E,t')\, dE\, dt' \quad . \tag{11.13}$$

The delayed neutron source is then composed of the precursor concentration as in Eq. (3.29c):

$$S_d(\mathbf{r},E,t) = \sum_k \chi_{dk}(E)\lambda_k C_k(\mathbf{r},t) \quad . \tag{11.14}$$

The dependence of the delayed neutron source on the previous flux is indicated by

$$S_d(\mathbf{r},E,t) = S_d[p(t')\psi(\mathbf{r},E,t')] \quad . \tag{11.15}$$

Insertion of Eq. (11.10) into Eq. (11.12) and division by $p(t)$ gives the balance equation for the flux shape:

$$[\mathbf{F}_p - \mathbf{M}]\psi(\mathbf{r},E,t) + \frac{1}{p(t)} S_d[p(t')\psi(\mathbf{r},E,t')]$$
$$= \frac{1}{v}\frac{\dot{p}}{p}\psi(\mathbf{r},E,t) + \frac{1}{v}\frac{\partial}{\partial t}\psi(\mathbf{r},E,t) \quad . \tag{11.16}$$

Equations (11.11) and (11.16) represent a coupled system of equations for the two unknown functions $p(t)$ and $\psi(\mathbf{r},E,t)$. The amplitude function appears explicitly in two terms of Eq. (11.16), whereas the integral parameters and $s_d(t)$ in Eq. (11.11) contain the shape function implicitly. Equation (11.11) determines $p(t)$; it depends only weakly and implicitly on ψ. Alternately, Eq. (11.16) determines ψ; it depends only weakly on $p(t)$.

The pair of coupled equations, Eqs. (11.11) and (11.16), is solved simultaneously with the solution at all times subjected to the constraint condition, Eq. (5.6):

$$\int_V \int_0^\infty \frac{\phi_0^*(\mathbf{r},E)\psi(\mathbf{r},E,t)}{v(E)}\, dV\, dE = K_0 \quad . \tag{11.17}$$

Since Eq. (11.11) and (11.16) depend nonlinearly on p and ψ (which appear as products), the basic solution method is similar to that often applied in solving nonlinear systems of differential equations of this type. One function is assumed known; this results in a linear equation for the other function, which is solved first. The latter function is then assumed known, and the resulting linear equation for the former is solved. A detailed discussion of the solution procedure is presented in Sec. 11-4B.

Equations (11.11), (11.16), and (11.17) are equivalent to the original time-dependent diffusion equation since no approximation has yet been introduced. A sequence of approximations is introduced below leading to a sequence of methods, which ranges from the complete numerical solution to the PRM.

The first simplification ("approximation") consists of introducing different time steps for the amplitude and shape functions. Since the shape function varies slowly with time, its time derivative may be calculated by a backward difference:

$$\frac{\partial\psi(\mathbf{r},E,t)}{\partial t} \simeq \frac{1}{\Delta t_s^s}[\psi(\mathbf{r},E,t_s^s) - \psi(\mathbf{r},E,t_{s-1}^s)] \quad , \tag{11.18}$$

where t_s^s and t_{s-1}^s denote the times of the present and previous shape calculations. The time interval $\Delta t_s^s = t_s^s - t_{s-1}^s$ is generally very much larger than the corresponding time step required to solve the amplitude equation, Eq. (11.11).

Inserting Eq. (11.18) into Eq. (11.16) and collecting the terms containing only previous times on the right side yields the equation for the shape function in the so-called *improved quasistatic method*[34,35]:

$$\left(\mathbf{F}_p - \mathbf{M} + \frac{1}{v\Delta t_s^s} - \frac{1}{v}\frac{\dot{p}}{p}\right)\psi(\mathbf{r},E,t^s)$$

$$= -\frac{1}{p}\,S_d[p(t')\psi(\mathbf{r},E,t')] + \frac{\psi(\mathbf{r},E,t_{s-1}^s)}{v\Delta t_{s-1}^s} \quad . \qquad (11.19)$$

The development of the improved quasistatic method was preceded by the quasistatic method in which the shape function derivative in Eq. (11.16) is neglected. This approximation leads to the following balance equation for the shape function in the *quasistatic method*[34,36]:

$$\left(\mathbf{F}_p - \mathbf{M} - \frac{1}{v}\frac{\dot{p}}{p}\right)\psi(\mathbf{r},E,t) = -\frac{1}{p}\,S_d[p(t')\psi(\mathbf{r},E,t')] \quad . \qquad (11.20)$$

After the neglect of the derivative $\partial\psi/\partial t$, the time variable t appears only as a parameter in the balance equation for the flux shape. Thus, the problem is formally a *static* problem with cross sections depending parametrically on time. The left side of Eq. (11.20) is augmented by an α/v term ($\alpha = \dot{p}/p$ = inverse period), which represents the residue of the time derivatives in Eq. (11.16). It treats the problem *as if* the flux shape were static. The delayed neutron source, however, depends on the entire flux history and thus on earlier times, $t' \leqslant t$. The implications of this distinction of the delayed neutron source from the other terms are discussed in Sec. 11-4A.

If the delayed neutron source is approximated at time t in a quasistationary fashion as a fraction of the prompt source, and if the α/v term is neglected, the specific kinetic features in the shape equation are completely eliminated. An eigenvalue, λ, must be incorporated to obtain a nontrivial solution; this results in the balance equation for the *adiabatic method*[37]:

$$(\mathbf{M} - \lambda\mathbf{F})\Psi = 0 \quad . \qquad (11.21)$$

Neglecting also the time dependence of the operators **M** and **F** in calculating the shape function results in the balance equation for the *point reactor model:*

$$(\mathbf{M}_0 - \lambda_0\mathbf{F}_0)\Psi_0 = 0 \quad . \qquad (11.22)$$

The sequence of factorization methods is listed in Table 11-I. The first column describes certain features and the body of the table indicates whether or not these features are accounted for in the various methods.

TABLE 11-I
Sequence of Approximation in Factorization Methods Showing Effects Treated by the Various Models*

	Model		
	Direct Numerical	Improved Quasistatic	Quasistatic
$\frac{\partial\psi}{\partial t}$	Yes	Yes	No
$\frac{\alpha}{v}$	Yes	Yes	Yes
$S_d[p(t')\psi(\mathbf{r},E,t')]$	Yes	Yes	Yes
Implicit coupling via feedback	Yes	Yes	Yes
Decoupled shape distortion	Yes	Yes	Yes
Initial shape	Yes	Yes	Yes
	Model		
	Adiabatic	**Adiabatic with Precalculated Shape Functions**	**Point Reactor Model**
$\frac{\partial\psi}{\partial t}$	No	No	No
$\frac{\alpha}{v}$	No	No	No
$S_d[p(t')\psi(\mathbf{r},E,t')]$	No	No	No
Implicit coupling via feedback	Yes	No	No
Decoupled shape distortion	Yes	Yes	No
Initial shape	Yes	Yes	Yes

*From Ref. 34.

The first two methods (direct numerical and improved quasistatic) are mathematically equivalent in the sense that no approximations except finite differencing are introduced. Thus, both methods yield the same result for sufficiently small time steps (see Sec. 11-4C). The accuracy of the other methods generally decreases from left to right, i.e., with decreasing sophistication of the shape function balance equation.

Both adiabatic methods listed in Table 11-I calculate the flux shape from the same type of balance equation, Eq. (11.21). The simpler of the two methods allows only changes in the reactor for which the sequence of configurations is known in advance. Examples are the motion of control rods or fuel subassemblies. For such known sequences of motion, the flux distortion and its effect on the reactivity can be precalculated. The resulting time-dependent reactivity is inserted into the point kinetics equation. This method represents an obvious improvement over the PRM. If there are no other changes in the reactor except the precalculated configurations, the results may be fairly accurate. If, however, feedback plays an important role, the reactor may exhibit a number of changes—the combination of which cannot be known in advance, that is, without actually solving the initial value problem.

11-4 Quasistatic and Related Methods

11-4A Physical Interpretation of the Quasistatic and Adiabatic Assumptions

The evaluation of the type of inaccuracy caused by the quasistatic calculation of the flux shape is based on a physical understanding of the approach. The inherent physical deviation of quasistatic and correct results can best be illustrated by considering the transient following an asymmetrically applied step change in the reactor; for example, instantaneous withdrawal of an off-center control rod. The shape function increases in the area where the neutron absorption is reduced and decreases in the remaining area in such a way that the integral constraint condition, Eq. (11.17), remains constant.

Figure 11-5 illustrates the time behavior of the shape function near the position of the instantaneously withdrawn control rod. The increase of the *shape function* exhibits two distinct phases: a very rapid (prompt) adjustment of the shape of the neutron population to the changed configuration followed by a comparatively slow (delayed) shape change due to the adjustment of the precursor population. The prompt as well as the delayed response of the *flux* to a step change in the system correspond exactly to the prompt and delayed changes of the *amplitude function* (compare Sec. 6-2, particularly Figs. 6-14 and 6-15). The only difference between the amplitude function and shape function behavior is that asymptotically the amplitude function varies exponentially (except in a subcritical reactor with independent source), whereas the shape function becomes asymptotically constant.

The approximation to the variation of the shape function in the quasistatic approach is shown as the dashed line in Fig. 11-5. It differs from the correct solution only during the prompt adjustment phase. The prompt adjustment of the shape function in the quasistatic approach is approximated by an instantaneous adjustment. It therefore represents the prompt jump approximation (PJA) for the shape function. The PJA

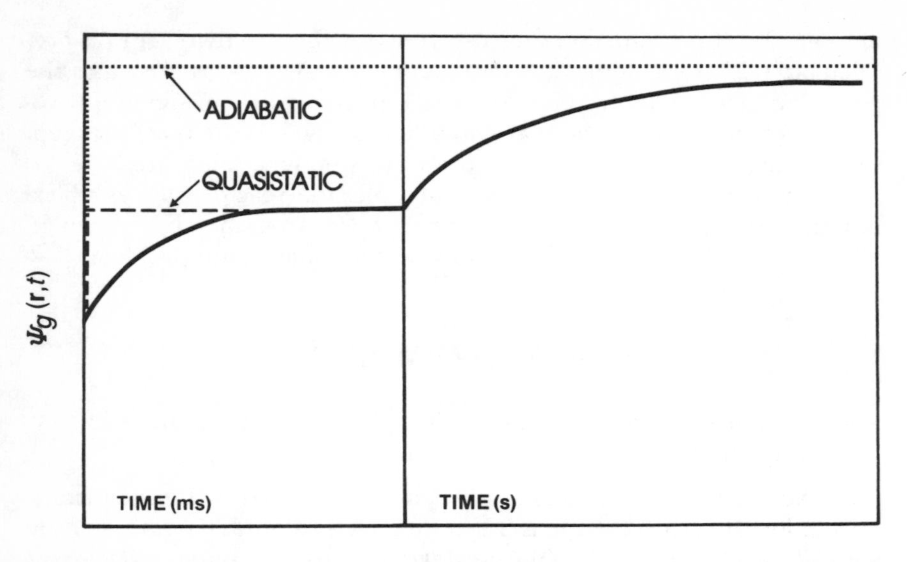

Fig. 11-5. Group flux component of the shape function following an asymmetric reactivity step change.

for the shape function can be applied in the entire reactivity domain; it is not limited to $\rho < \beta$ as in the case of the amplitude function.

The temporary plateau of the flux shape is described as accurately by the PJA as the corresponding plateau of the amplitude transient (see Fig. 6-15). The reason for this is that the ψ derivative term in Eq. (11.16) becomes practically negligible after the prompt shape jump due to the division by the large neutron velocity. Thus, after the prompt jump, the shape function can be calculated from Eq. (11.16) with the ψ derivative deleted, that is, from the balance equation of the quasistatic approach, Eq. (11.20).

Further variation of the shape function is caused by the redistribution of the precursors initiated by the prompt variation of the flux shape. The resulting time dependence of $S_d[p(t')\psi(\mathbf{r},E,t')]$ is fully incorporated in the quasistatic approach. Since the shape derivative term can be neglected during the delayed transition, the quasistatic approach describes the delayed shape transition properly.

Asymptotically, the shape derivative is exactly zero. Thus, the quasistatic approach describes the asymptotic flux shape exactly.

During the prompt jump following a step change in the reactor, the typical inaccuracy of the *adiabatic* approach is of the same type but larger than that of the *quasistatic* approach. The delayed transition is missing altogether in the adiabatic approach. Also, asymptotically, both approaches differ as indicated in Fig. 11-5.

As basis for the investigation of the typical asymptotic inaccuracy of the adiabatic approach, and also to explore the usefulness of the α mode versus λ mode shape functions, the asymptotic solution after a step change is derived and investigated in the following.

The balance equation for the asymptotic shape function, in the case of a step change in the reactor, is obtained from Eq. (11.16) by introducing the asymptotic behavior of the constituents:

$$\psi(\mathbf{r},E,t) \rightarrow \psi_{as}(\mathbf{r},E), \text{ independent of } t \quad ,$$

$$p(t) \propto e^{\alpha t} \quad ,$$

and

$$C_k(\mathbf{r},t) \propto e^{\alpha t} \quad .$$

The asymptotic precursor concentrations are obtained in the same way as the corresponding lumped quantities (compare Sec. 6-1C). The result is an eigenvalue equation with α as eigenvalue and ψ_{as} as eigenfunction:

$$\left(\mathbf{F}_p - \mathbf{M} + \sum_k \frac{\lambda_k}{\alpha + \lambda_k}\mathbf{F}_{dk} - \frac{\alpha}{v}\right)\Psi_{as} = 0 \quad , \tag{11.23}$$

with

$$\mathbf{F}_{dk}\,\Psi = \chi_{dk}(E)\int_0^\infty \nu_{dk}\Sigma_f(\mathbf{r},E')\psi(\mathbf{r},E')\,dE' \quad . \tag{11.24}$$

Equation (11.23) is very complicated since the eigenvalue α occurs in the denominators of the sum over the precursor groups. If the delayed neutrons are not distinguished from prompt neutrons, the following much simpler equation results:

$$\left(\mathbf{F} - \mathbf{M} - \frac{\alpha}{v}\right)\Psi_\alpha = 0 \quad . \tag{11.25}$$

The fundamental solution of the eigenvalue problem, Eq. (11.25), is called the "α mode." It is not actually the solution of the asymptotic flux because of the approximation of the delayed neutron source.

Since the α mode is explicitly derived from the time-dependent diffusion equation, it is sometimes proposed that the α mode is more appropriate in kinetics problems than the λ mode [the solution of Eq. (11.21), applied in the adiabatic approximation]. This proposal is discussed in the following by comparing the two balance equations, Eqs. (11.21) and (11.25), with the correct balance equation of the asymptotic shape, Eq. (11.23). If the α mode were more accurate than the λ mode, an adiabatic approach based on the α mode would be preferable to the one using the λ mode.

In the λ mode equation, Eq. (11.21), a nontrivial solution of a supercritical reactor is obtained by reducing the fission source with the factor $\lambda = 1/k$. A nontrivial solution is achieved in the α mode equation, Eq. (11.25), by subtracting the equivalent of an absorption term, α/v. In the correct equation, Eq. (11.23), both types of changes occur: the fission source is reduced (actually only the delayed part of it) and an α/v absorption term is included. The investigation of whether the λ or the α mode provides a better approximation to Ψ_{as} requires, therefore, the comparison of the fission source reduction in Eq. (11.23) with the α/v term. For the purpose of this comparison, the corresponding integrated terms as they appear in the inhour equation (see Secs. 6-1C and 6-2D) can be considered instead of the space-dependent terms in Eq. (11.23), i.e.:

$$\sum_k \frac{\lambda_k}{\alpha + \lambda_k} \beta_k - \alpha\Lambda = \beta - \rho \quad . \tag{11.26}$$

Initially, ρ and α are equal to zero and both sides of Eq. (11.26) are reduced to β. If $\rho \neq 0$, the question is whether the corresponding variation of the left side of Eq. (11.26) is achieved primarily by variation of the delayed neutron term or of the $\alpha\Lambda$ term, which corresponds to the α/v term in Eq. (11.23). For $0 < \rho < \beta$, the $\alpha\Lambda$ term is very small. Thus, the reduction of the left side of Eq. (11.26) and thus of the parentheses in Eq. (11.23) is primarily due to the reduction of the delayed neutron source. Even for $\rho = \beta$, the $\alpha\Lambda$ term itself is still small; thus its variation is small compared to the reduction of the delayed neutron term. At $\rho = \beta$, the delayed neutron source term in Eq. (11.26) is practically reduced to zero with virtually nothing compensated by the α/v absorption (represented by $\alpha\Lambda$ in this equation). At $\rho = 2\beta$, the variations of both terms are comparable: The delayed neutron term has been reduced by about β; i.e., from its original value of β to practically zero. The $\alpha\Lambda$ term has also been changed by β, from its original value of zero to about β. For $\rho > 2\beta$, the $\alpha\Lambda$ variation dominates. Since for most transients ρ is much smaller than 2β, the reduction of the delayed neutron term in Eq. (11.21) dominates the α/v term in practically all transients. This holds even more for subcritical transients.

The reduction of the delayed neutron source is far better approximated by an equivalent reduction of the prompt neutron source than by including $1/v$ absorption. Therefore, the λ mode is a better approximation to Ψ_{as} than the α mode.

The flux shape approximation resulting from the adiabatic approach is illustrated by the dotted line in Fig. 11-5. Since the delayed neutron source is not distinguished from the prompt neutron source, the delayed and the prompt transitions occur at the same time. Both are approximated by an instantaneous adjustment. In addition, the asymptotic flux shape is not correctly described in the adiabatic approach (see Fig. 11-5). The deviation in the asymptotic shape values results from the approximation of Ψ_{as} by the λ mode, Eq. (11.21).

The effect of the use of the λ mode rather than the exact flux shape can be demonstrated in an example presented by Yasinsky and Henry[38] in Fig. 11-6. In this example, an amount of fissile material ($\delta\rho < \beta$) is inserted in the left quarter of a thermal slab reactor model. The flux shape subsequently tilts toward the left and the entire flux increases. The curves in Fig. 11-6 represent the thermal flux at $t = 0.8$ s for exact, adiabatic, and point kinetics treatments. The dashed-dotted line (point

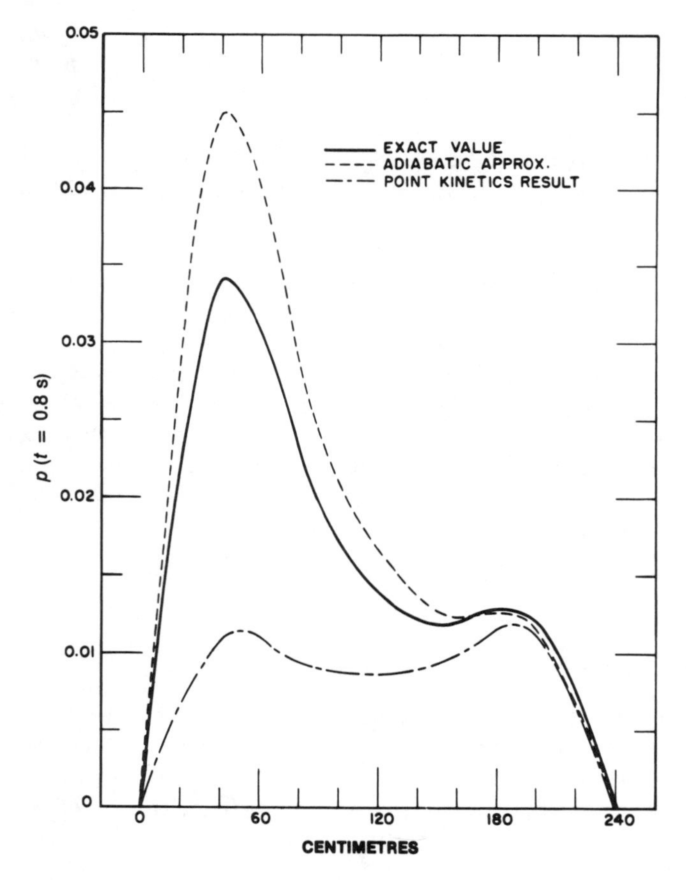

Fig. 11-6. Delayed supercritical excursion thermal flux at 0.8 s for a 240-cm core.[38]

kinetics) also shows the initial (flattened) flux shape. The solid line shows the exact flux at $t = 0.8$ s and the dashed line (adiabatic) shows the strong shape distortion resulting from the λ mode. The instantaneous redistribution of the precursors in the adiabatic approach strongly exaggerates the flux shape distortion. In discussing this result, the authors of Ref. 38 conclude that the adiabatic approach is not accurate enough for thermal reactor dynamics.

Based on source multiplication considerations, it has been suggested[36] that the entire deviation shown in Fig. 11-6 is due to the neglect of the time delay in the adjustment of the precursor shape and that the neglect of the shape derivative should have virtually no effect on the result. Madell[39] showed that the error of the neglect of the shape function derivative does not appear in the first four digits in the example of Fig. 11-6 if the correct space distribution of the precursors is used. This result proved numerically that the quasistatic approach, in which the precursor distribution is described correctly, yields an accurate treatment of spatial kinetics problems. The development of the quasistatic method itself is reported in Refs. 35 and 40 to 42.

The importance of the precursor distribution for the flux shape can be understood readily by applying the fission chain concept of Chapter 7. Although the number of delayed neutrons is much smaller than the number of prompt neutrons, the delayed neutrons are the only source of fission chains. Thus, the space distribution of the precursors determines the space distribution of the initiation of fission chains and, through it, affects the entire flux distribution.

The shape derivative in Eq. (11.16) is multiplied by $1/v$. The neutron velocities in LWRs and in fast reactors are on the average very large but larger in fast reactors than in thermal reactors. Also for this reason the neglect of the shape derivative should lead to smaller errors in fast than in thermal reactors. However, this difference in the applicability of the quasistatic approach to fast and thermal reactors is unimportant in the improved quasistatic approach where the shape function derivative is accounted for.

11-4B Factorization and Quasistatic Methods

The conceptual basis of the quasistatic method is the flux factorization, which is introduced to allow larger time intervals between shape recalculation than the small intervals required for the calculation of a rapidly changing amplitude function. Therefore, the scheme of time steps and the transfer between the various levels is an integral part of the method.

Two levels of the time steps are evident by the basic approach, i.e.,

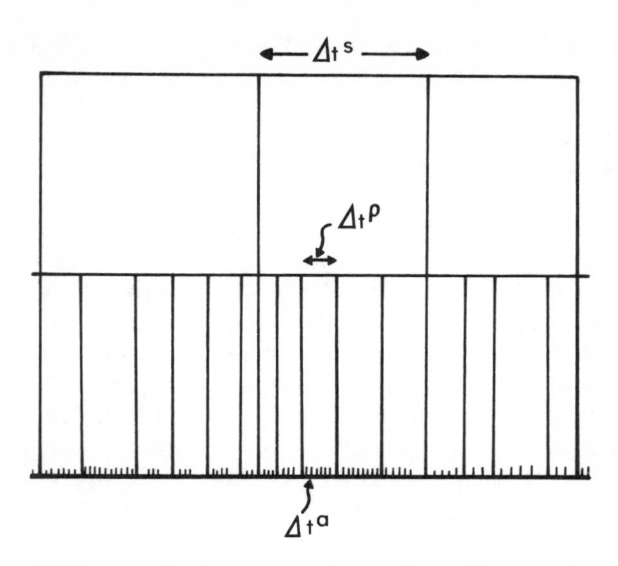

Fig. 11-7. Time-step hierarchy in the quasistatic method where Δt^a = amplitude, Δt^p = reactivity, and Δt^s = shape time steps.

amplitude function and shape function time steps. Practical application requires a third level, between the shape step and the amplitude step levels.

Variations of the spatial configuration are evaluated at the middle level as a basis for a decision about the recalculation of the flux shape. The coefficients in the exact kinetics equation, i.e., the integral parameters ρ, β, and Λ, are to be calculated as a function of time. The calculation of the integral parameters, which is required because of changes in the system, is also performed on the intermediate level.

Figure 11-7 illustrates the time step hierarchy. The amplitude, reactivity, and shape time steps are denoted by Δt^a, Δt^p, and Δt^s, respectively. The calculations to be performed on the individual levels include the solution of the exact point kinetics equation, the calculation of ρ, β, and Λ, and the computation of shape function, respectively. The computational procedures for these quantities are the same as applied in previous sections except for the shape function, for which special considerations are required.

The balance equation for the shape function is given by Eq. (11.20) for the quasistatic approach, or by Eq. (11.19) for the improved quasistatic approach. The amplitude function $p(t)$ and thus $\alpha(t) = \dot{p}/p$ is calculated from the exact point kinetics equations, based on extrapolated values for ρ, β, and Λ. At times t^p, the values of ρ, β, and Λ are recalculated, and the change in the configuration is evaluated. If a new shape calculation is required, for example, at t_s^s, the following procedure is applied.

The delayed neutron source at time t_s^s, which is $S_d(\mathbf{r},E,t_s^s)$, is obtained from $S_d(\mathbf{r},E,t_{s-1}^s)$ and the flux history $t' < t_s^s$, where $\psi(\mathbf{r},E,t')$ in Eq. (11.13) is either approximated by the flux shape at the previous shape step or an improved shape is obtained by linear extrapolation of the previous two shapes. Inserting this delayed neutron source in Eqs. (11.20) and (11.19) yields inhomogeneous problems for $\psi(\mathbf{r},E,t_s^s)$. These inhomogeneous problems have mathematically unique solutions including the magnitude of ψ. However, due to inaccuracies of $p(t)$ accrued over earlier time steps, the magnitude of ψ can become inaccurate and will not automatically satisfy the constraint condition, Eq. (11.17). This inaccuracy becomes particularly large for very high flux amplitudes encountered in superprompt-critical transients. If the amplitude function becomes very large, the delayed neutron source virtually drops out of Eqs. (11.19) and (11.20). The inhomogeneous problems then approach homogeneous ones; i.e., the operator on the left side becomes nearly singular. At this point, very small inaccuracies in α/v on the left side have a very large effect on the magnitude of ψ, but not on its space-energy variation. To overcome this problem in numerical solutions of Eqs. (11.19) and (11.20), a free parameter was introduced by Meneley[42] in front of the fission source (e.g., for the quasistatic equation):

$$\left(-\mathbf{M} + \lambda^p \mathbf{F}_p - \frac{\alpha}{v}\right)\Psi = -\frac{1}{p} S_d \quad . \tag{11.27}$$

Physically, a small change $(1 - \lambda^p)\mathbf{F}_p$ is introduced to compensate for the error in α/v. Slight variations of λ^p may change the magnitude of ψ by a large amount. Then, λ^p is determined such that the resulting ψ satisfies the constraint condition, Eq. (11.17). Equation (11.27) is solved iteratively. Ideally, the converged λ^p should be equal to 1. The deviation of the converged λ^p from 1 indicates the magnitude of the error. If $\lambda^p - 1$ exceeds a certain limit, the results in the previous interval are iteratively improved.

If $p(t)$ becomes very large and the right side in Eq. (11.27) becomes negligible, the parameter λ^p assumes the character of an eigenvalue in the "near" homogeneous problems.

Practical experience showed[43] that including the derivative into Eq. (11.27) leads to an equation that converges more rapidly than Eq. (11.27). This may be due to the fact that the second term on the right side of Eq. (11.19) does not become small for large p. The practical consequence of this finding is that the improved quasistatic method is not only more accurate than the quasistatic method but also requires less computation time. Consequently, all computer programs based on the factorized approach allow both options, Eqs. (11.19) and (11.20), for the shape calculation. The improved quasistatic approach should be preferred in practically all cases.

Several computer programs have been developed implementing the quasistatic and the improved quasistatic methods; the first program was the quasistatic one-dimensional excursion code, QX1, developed at Argonne National Laboratory.[42] It allows up to 30 energy groups and performs a detailed interpolation and extrapolation of temperature-dependent group constants to obtain proper values for all regional temperatures.

Subsequently developed programs QX2S (Refs. 44 through 47), FX2 (Refs. 9, 48, and 49), and the German program KINTIC (Ref. 50) utilize the original approach in two space dimensions. The methods applied in these programs for joining the various components of the solution procedure, including extrapolations, interpolations, decision quantities, and criteria, are very similar to those in QX1. It appears, however, that a modified procedural scheme should be developed that is more suitable for safety applications in order to reduce the lengthy computations of the complicated feedback phenomena. In the Los Alamos SIMMER program,[51] the quasistatic computational scheme has also been applied for two-dimensional neutron transport dynamics, S_n.

An important feature of the computational procedure developed for the quasistatic program QX1 is illustrated in Fig. 11-8. In the transient depicted in this figure, a typical 1000-MW(e) fast reactor is subjected to the insertion of a fuel subassembly. During the first shape interval, Δt_1^s, the initial flux shape is used for the calculation of the integral parameters ρ, β, and Λ required in the exact point kinetics equation. Thus, the first result for this interval is the same as for the point reactor model. At $t = t_1^s$, the flux shape is recalculated, and the reactivity calculated with the new flux shape is different from the previously used value at this point in time. The elimination of this reactivity error requires at least a second point kinetics run over the shape interval Δt_1^s with a properly corrected reactivity. The same procedure is applied in other shape intervals in order to reduce or eliminate the deviation of extrapolated and calculated reactivity in a Δt^ρ interval (see Fig. 11-8).

In the improved quasistatic programs, a linear extrapolation of the flux shapes is employed after the second shape interval. This reduces the correction calculated by the iterative point kinetics sweep. Nevertheless, the point kinetics iteration is needed in most cases. This general procedure is illustrated in Fig. 11-9. The dashed line indicates the reac-

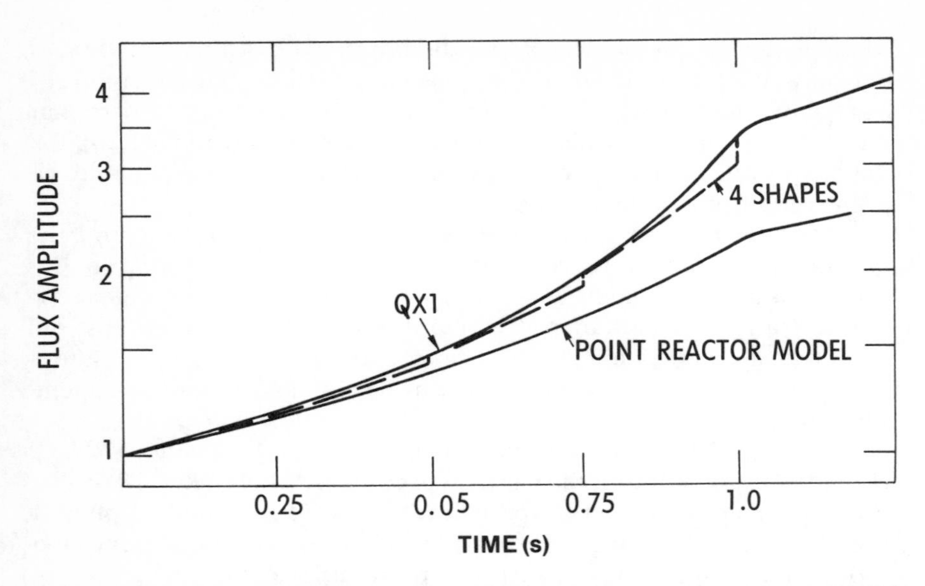

Fig. 11-8. Transient following the insertion of a fuel subassembly into a typical 1000-MW(e) fast reactor.[41]

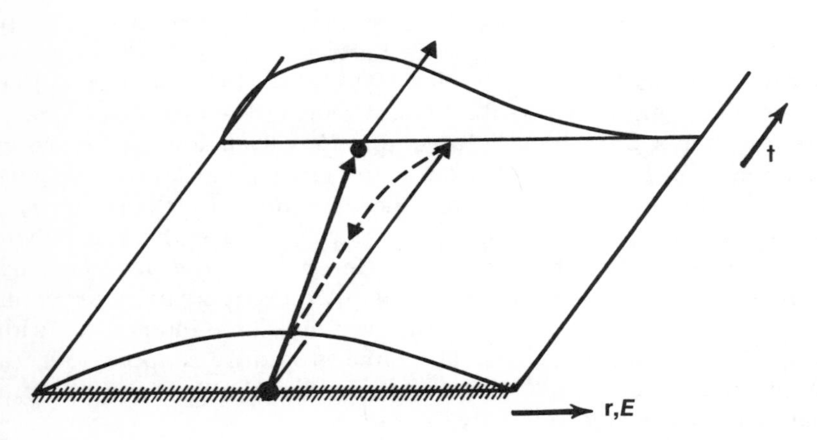

Fig. 11-9. Illustration of the quasistatic computation scheme. The arrow parallel to the t axis indicates the point kinetics run for the first interval, using the initial flux shape. The dashed arrow suggests the backward correction of the reactivity trajectory that makes use of the modified flux shape at the end of the interval. The second solid arrow indicates the modified point kinetics run over this interval, reflecting the use of a recalculated (quasistatic) flux shape.

tivity correction, which is applied for a second point kinetics run (indicated by the heavy arrow) after a recalculation of the flux shape.

The investigations presented in the following section show that the improved quasistatic approach yields results that for small shape steps are identical to those of the finite difference approach. By utilizing only the inital shape calculation, results identical to the point reactor model are obtained. Consequently, any desired accuracy between PRM results and "exact" finite difference results can be obtained by proper choice of shape steps.

11-4C Comparison of Kinetics and Dynamics Results

The following comparison of results is subdivided into a comparison of transients without feedback and into transients with feedback. Both types of transients are considered for subprompt- and superprompt-critical reactivity insertions. The comparison includes results calculated with different methods for thermal and fast reactors. This is of special methodological interest because it provides information on the effect of different neutron lifetimes on the accuracy of the quasistatic approach. The replacement of the prompt shape adjustment by an instantaneous one as it is done in the quasistatic approach may not be as good an approximation in thermal reactors as in fast reactors because of the lower thermal reactor average neutron velocity. This is, however, less important in the improved quasistatic approach.[52]

Figure 11-6 shows a comparison of the results of different methods in a subprompt-critical thermal reactor kinetics problem. The results of the point reactor model exhibit a large error due to the neglect of the considerable shape distortion. The adiabatic results show a sizable error due to the neglect of the delay in the adjustment of the precursor population. The delay of the precursor population, for $\rho < \beta$, has a strong effect on the shape function through source multiplication. Results of the quasistatic and improved quasistatic methods are not available for this specific case. Quasistatic results should be practically indistinguishable from the finite difference results since the shape variation is very slow in this case. The results of the improved quasistatic method should be identical with the finite difference results. In similar subprompt-critical transients in *fast* reactors, all results should be closer together because of smaller shape distortions.

Figure 11-10 shows various reactivity traces in a superprompt-critical ramp-induced transient for the same thermal reactor as in Fig. 11-6. The reactivity is inserted such that it appears as a linear ramp in the point reactor model. The more correct treatments of this asymmetric

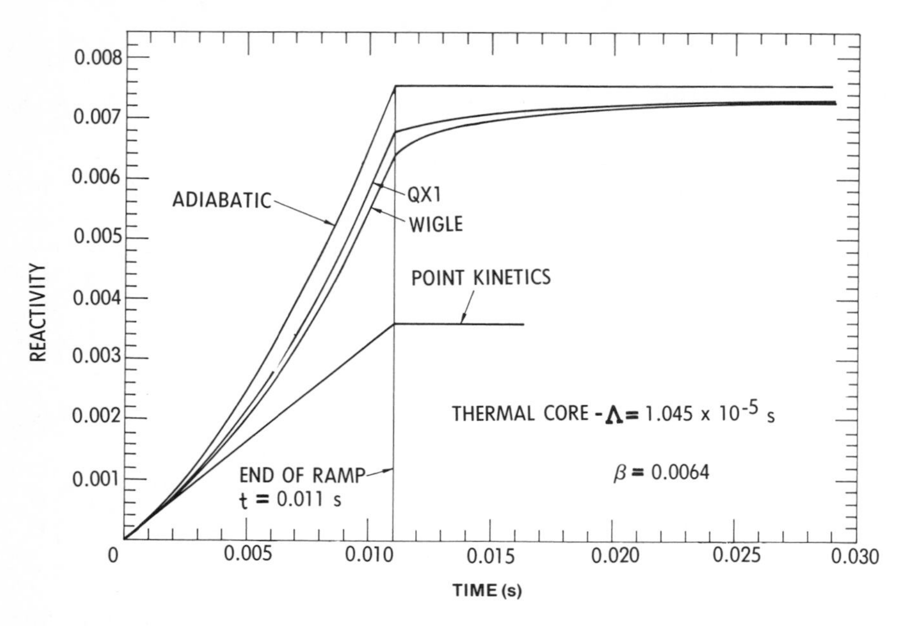

Fig. 11-10. Reactivity versus time for limited ramp insertion in a thermal reactor core.[34]

change yield a slightly curved reactivity insertion resulting from the flux tilting. The upper curve is obtained from the adiabatic treatment with an instantaneous adjustment of the *complete* flux shape. The next curve results from the quasistatic model in which only the prompt adjustment occurs instantaneously. Both the adiabatic and the quasistatic reactivities are higher than the correct reactivity on the rising part of the reactivity curve since they "anticipate" the flux deformation. This situation would be reversed for decreasing reactivities where the quasistatic and the adiabatic reactivities are below the correct reactivity; compare the similar phenomenon for the inaccuracy of the prompt jump approximation in, for example, Fig. 8-2.

For a subprompt-critical reactivity insertion, a set of curves would be obtained similar to the set shown in Fig. 11-10 for thermal as well as for fast reactors. In superprompt-critical fast reactor transients, however, adiabatic, QX1, and WIGLE results are close together because the flux shape is nearly completely determined by the prompt neutrons, since the delayed neutron source has lost its dominating effect.

The behavior of the flux amplitude, which corresponds to the reactivities of Fig. 11-10, is illustrated in Table 11-II. The amplitude values at twice the insertion time of the reactivity, $p(t = 0.022 \text{ s})$, are chosen for the comparison. The flux amplitude results are in exact agreement

TABLE 11-II

Amplitude Function at $t = 0.022$ s for Ramp-Reactivity Insertion*

(Comparison of factorization approaches)

Reactor Generation Time		Model				
		Direct Numerical	Improved Quasistatic	Quasistatic	Adiabatic	Point Reactor Model
Thermal core ($\Lambda = 1 \times 10^{-5}$ s)	$p(0.022s)$	14.92	14.92	16.18	21.27	2.253
	p/p_{WIGLE}	1.000	1.000	1.084	1.426	0.1510
Thermal core ($\Lambda = 3 \times 10^{-7}$ s)	$p(0.022s)$	2.793×10^{17}	2.793×10^{17}	2.949×10^{17}	3.947×10^{17}	2.280
	p/p_{WIGLE}	1.000	1.000	1.045	1.413	8.163×10^{-18}

*From Ref. 34; see Figs 11-8 and 11-10.

between the values of the improved quasistatic and the direct numerical (finite difference and WIGLE) methods. The quasistatic method leads to an 8.4% deviation at $t = 22$ ms. The corresponding adiabatic result is too large by ~43%. A further increase of this deviation should be expected since the adiabatic reactivity at $t = 22$ ms is still substantially larger than the correct reactivity (compare Fig. 11-10). The point reactor model result is 85% too small.

To investigate the effect of the magnitude of the generation time on this comparison, the value of Λ was reduced to 3×10^{-7} s. This reduced value is typical for a smaller fast reactor. The core composition and group constants were left unchanged. The finite difference and the improved quasistatic method results shown in Table 11-II are again the same in all digits. The deviation of the quasistatic method is reduced to ~4.5%. The deviation of the adiabatic result is about the same as in the thermal reactor case, although a lesser "asymptotic" error is to be expected than for the thermal core. The result of the point reactor model is smaller by orders of magnitude than the other results since the PRM reactivity is subprompt critical. In an actual fast reactor, the PRM results should exhibit a smaller error than in this reduced Λ model since the flux distortions are less severe.

A set of one-dimensional kinetics benchmark problems has been solved with QX1 (Ref. 42), WIGLE (Ref. 13), and RAUMZEIT (Ref. 53). The two latter programs are based on finite difference techniques whereas QX1 applies flux factorization. Some results of this comparison are presented in Fig. 11-11, which is reproduced from Ref. 54. Figure 11-11 presents the flux amplitude at 5 ms in a superprompt-critical asymmetric transient of an "increased velocity" reactor. The reactor calculated in this benchmark problem is a thermal reactor with the neutron velocity increased to correspond to that of a fast reactor. The value of the flux amplitude, $p(0.005\text{ s})$, is plotted as a function of the average length of the interval between shape calculations. Figure 11-11 shows that the accuracy of the RAUMZEIT results rapidly deteriorates if the step length is increased over $\sim 5 \times 10^{-5}$ s. In QX1, a much smaller number of shape functions is required. The point at 10^{-3} s corresponds to five shape function recalculations in the 5-ms time interval. If actual fast reactor cross sections were used, the number of shape function calculations at 10^{-3} s could have been <5 because of the smaller overall flux shape distortion in a fast reactor. It should be emphasized that the error of the QX1 results is negative and asymptotically approaches the error of the point kinetics results. Reducing the number of shape recalculations to zero yields the point kinetics results. Thus, the quasistatic results are between the exact and the PRM result.

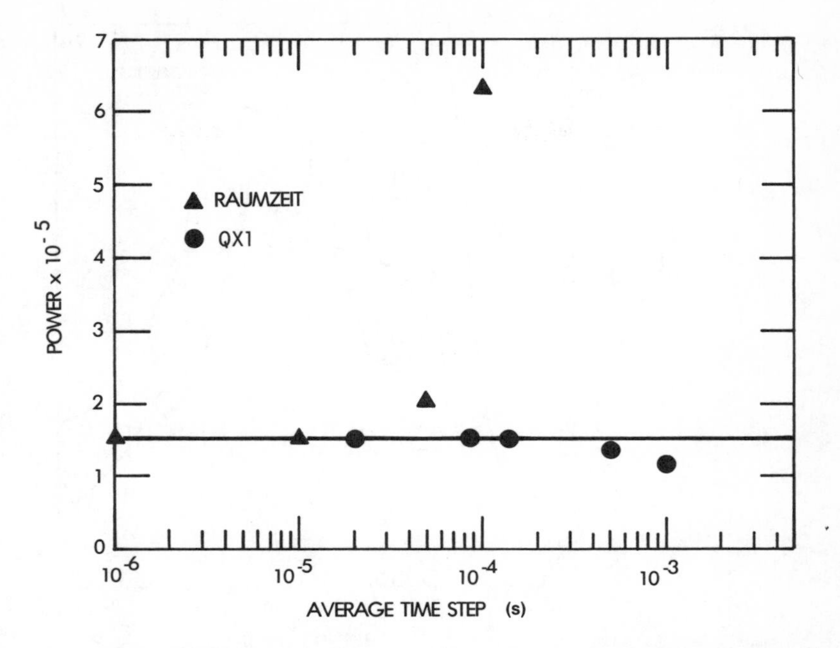

Fig. 11-11. Flux amplitude at 5 ms from a superprompt-critical asymmetric transient.[54]

If the point kinetics results were plotted onto Fig. 11-11, a single point would be plotted at the full length of the interval (i.e., at 5×10^{-3} s in Fig. 11-11). The results of the factorized approach would end up at this point if the interval between shape calculations were to be successively increased. The high degree of computational stability, i.e., movement from "exact results" to point reactor model results as the interval between shape calculations is increased, is unique to the flux factorization approach.

The benchmark problem does not allow for feedback. The error of the point kinetics treatment of reactor transients without feedback may be dramatic, possibly several orders of magnitude. If, however, nonlinear feedback is included in a reactor transient, the error of the point kinetics treatment is generally substantially reduced; for example, from several orders of magnitude without feedback to something like a factor of 2 or even less with feedback (compare the examples of Figs. 11-12 and 11-13).

The complicated curves shown in Fig. 11-12 represent the flux amplitude during a rapid flow coastdown accident simulation in a model of a large fast reactor.[45,46] The complicated structure of this transient is caused by the superposition of the reactivity that results from coolant

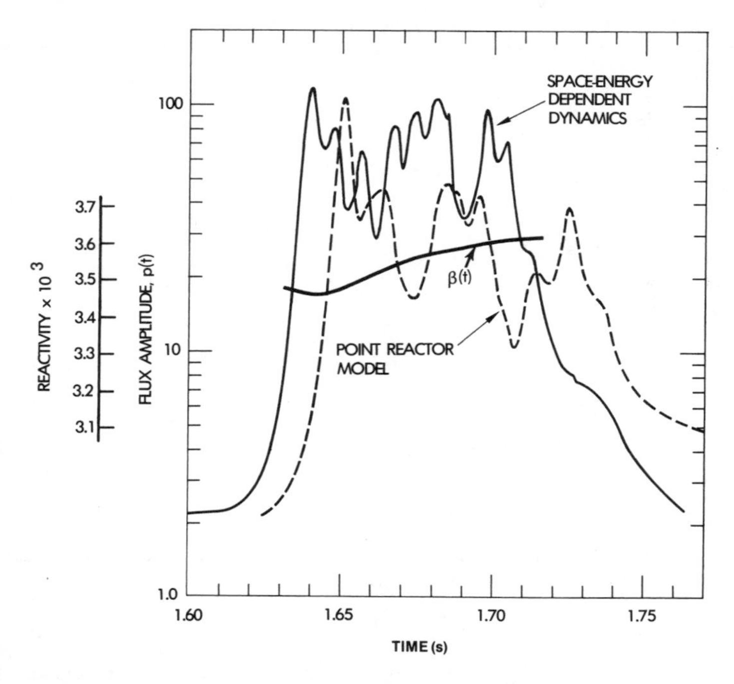

Fig. 11-12. Flux amplitude during a flow coastdown accident in an idealized large fast reactor.

boiling in various cooling channels. The solid curve is calculated with the QX2S space-energy dependent dynamics (SED) program that was developed at Purdue University by Smith, Bezella, and Ott.[44–46] The PRM result is calculated with the same program by eliminating the shape recalculation.

The SED flux in the pre-boiling phase is somewhat larger than the PRM flux. Therefore, boiling of the hottest channel starts somewhat earlier in the SED treatment, and the positive reactivity associated with sodium boiling in this large fast reactor model drives the power up at an earlier time than in the PRM treatment. At 1.64 s, the SED flux is ~25 times larger than the PRM flux. This large deviation is similar to that discussed earlier for kinetics without feedback. However, negative Doppler feedback starts turning the flux down at about the same maximum value, i.e., after a comparable temperature rise as indicated by comparable integrals over the first burst.

The subsequent progression of the transient is substantially different in the SED and PRM treatments. The second burst is more energetic in the SED than in the PRM. As a consequence of this, boiling in the

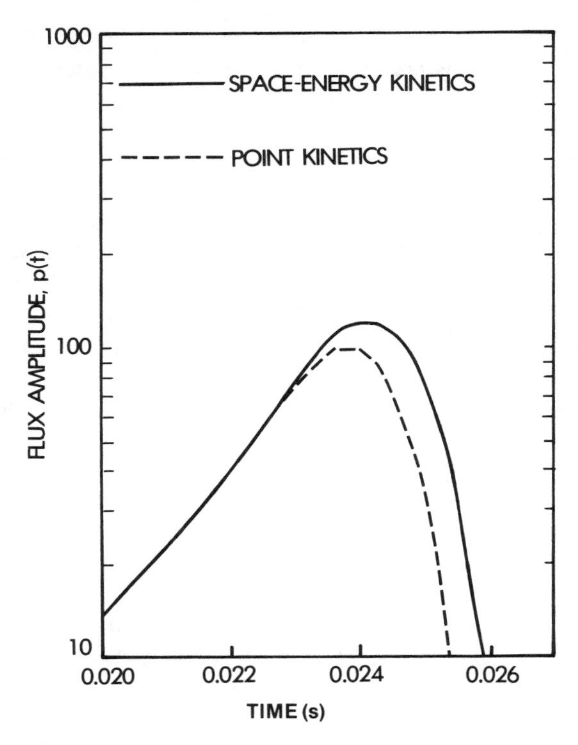

Fig. 11-13. Comparison of flux amplitudes during an idealized disassembly transient; $\rho_{max} = 0.93\$$.

next channel starts earlier in SED and the entire further superposition of reactivity insertions is different. The PRM treatment stretches this transient out farther in time. The overall energy release in this example is ~38% larger utilizing SED than in the PRM model approximation.

Figure 11-12 also shows the time variation of the effective delayed neutron fraction, $\beta(t)$: β changes by ~5% during this transient.

In Fig. 11-13, SED and PRM treatments of a core-disruptive accident are compared. The results of Fig. 11-13 are calculated with QX2S (Refs. 44, 47, and 55). Since core disassembly transients are shut down by displacement of fuel, significant differences between the SED and PRM results can be expected. To isolate these differences during the actual disassembly, the "initial" temperature was assumed to be at the onset point of disassembly, and the reactivity was inserted by increasing ν. This eliminates the differences between PRM and SED results during the rising phase of the transient. Thus, the only differences that occur in this artificial example show up during the decreasing phase.

The SED results are higher than the PRM results due to a higher reactivity resulting from flux deformation. The moving fuel takes part of the flux with it, which results in a higher flux weight factor for the moved fuel than in the PRM. The SED flux during this phase is a factor of 2 or more larger than the PRM result. The major part of the energy is released, however, during the rising phase of the transient and around the maximum where the deviation between the two treatments is small in this special example. A further contributing factor to the reduction of the deviation between PRM and SED results is the asymmetry of the transient; the transient decreases much faster than it increases due to the rapid disassembly acting in addition to the Doppler feedback effect. (Doppler feedback alone leads to a symmetrical transient; see Sec. 10-3C.) A direct consequence of the asymmetry of the transient is that the energy production during the decreasing phase is smaller than during the power increase. Therefore, the inaccuracy of the treatment during the disassembly does not have a very large effect on the energy release since this inaccuracy affects only the smaller part of the *overall* energy production. However, the energy above a certain threshold, e.g., the fuel boiling point, can be more significantly affected.

In realistic modeling of core-disruptive accidents as contrasted to the idealized case examined above, the reactivity insertion rate obtained from an SED treatment would generally be larger than in the PRM due to an increased prediction of the initiating reactivity. This in turn may lead to larger deviations of SED-type and PRM-type predictions for the overall energy release and even more so for energy increments.

In comparing the SED treatment of core disassembly to the PRM treatment, the total energy release is by far not as important as the part of the energy release available for mechanical work. Since the latter part results from the material of highest temperature, a comparatively small increase in the total energy could strongly increase the energy above a certain temperature limit.[47,55] Due to such consideration, the SED treatment can, for some accidents, lead to substantial differences in the prediction of the available work compared to the PRM.

11-5 Dynamic Reactivity Coefficients[c]

The reactivity coefficients are the classical point kinetics concepts that relate changes in a reactor to the corresponding variations of the reactivity. With the more sophisticated SED dynamics, the automatic extensions of these concepts are "dynamic reactivity coefficients."

---

[c]This section should not be included in an introductory course.

SED dynamics using the factorizing approach involves the dynamic reactivity, $\rho(t)$, directly. Other methods that aim at the calculation of the time-dependent flux, $\phi_g(\mathbf{r},t)$, can calculate the dynamic reactivity *a posteriori* by inserting $\phi_g(\mathbf{r},t)$ and the time-dependent macroscopic group cross sections in the definition of $\rho(t)$ by the scalar product, Eq. (5.26b).

Various feedback phenomena are normally affecting $\rho(t)$. Although none of the solutions of the dynamics problem in the approaches presented in this chapter makes direct use of reactivity coefficients, they are important for an evaluation of the integral feedback effects. Applications of such reactivity coefficients are for reactor operation; preoperational demonstration of safety characteristics; *a posteriori* separation of reactivity effects to better understand the transients involved in safety analysis, as the basis for design decisions on safety "parameters" (though they may not be constant); and for devising and analyzing experiments for an evaluation of the accuracy of computational predictions of the involved temperature and reactivity changes.

In general, reactivity coefficients are defined by partial derivatives:

$$\gamma_\xi = \frac{\partial \rho}{\partial \xi} \quad , \tag{11.28}$$

with ξ being the changing quantity. If ξ is an integral quantity, as is ρ, the result of the partial derivative is an integral quantity for the reactor. It may depend on ξ itself and on other parameters as well.

The typical example is the power coefficient, γ_p:

$$\gamma_p = \frac{\partial \rho}{\partial P} \simeq \frac{\Delta \rho}{\Delta P} \quad . \tag{11.29}$$

This is the major steady-state reactivity coefficient in which the various phenomena such as temperature effects, expansions throughout the system, as well as xenon and samarium poisoning (see App. A) appear in steady-state combinations. The power coefficient is not unique, as interim power levels may involve different coolant flow and thus different temperature conditions.

For other changes, the choice of ξ in Eq. (11.28) is not as obvious as for the power. An example is the temperature coefficient. Aside from the fact that prompt and delayed effects need to be distinguished, as discussed in Sec. 10-1, there is no obvious temperature "value" that could be used in the differential quotient, Eq. (11.28). This is because $\delta\rho$ is determined by a change in the temperature field, $\delta T(\mathbf{r})$, which is also different for fuel, cladding, coolant, and moderator. However, with a temperature field, one can uniquely associate a suitable integral concept; e.g., $\delta\overline{T}$, the corresponding change of the average temperature, $\overline{T}$, or as an alternate, a change in the average coolant outlet temperature. This gives then the *integral temperature coefficient,*

$$\gamma_T = \frac{\partial \rho}{\partial T} \simeq \frac{\Delta \rho}{\Delta T} \quad , \tag{11.30}$$

with a suitably defined temperature T. Even more so than in the case of the power coefficient, the resulting γ_T is not unique. It depends on the combination of the temperature field changes in the various reactor constituents, since fuel, coolant, and moderator reactivity contributions normally appear in different combinations, depending on the type of transient.

The combination of local contributions to reactivity changes and the space-dependent temperature field can be expressed in the following way. If the energy integration of the scalar product Eq. (5.26b) is carried out first, one obtains the "local reactivity contribution (per volume element)," $\delta\rho(\mathbf{r})$, appearing as an integrand of the remaining spatial integration:

$$\Delta \rho = \int_V \delta\rho_V(\mathbf{r})\, dV \quad . \tag{11.31}$$

Expanding the integrand with the local temperature change, $\delta T(\mathbf{r})$, gives:

$$\Delta \rho = \int_V \frac{\delta\rho_V(\mathbf{r})}{\delta T(\mathbf{r})}\, \delta T(\mathbf{r})\, dV = \int_V \gamma_{TV}(\mathbf{r})\delta T(\mathbf{r})\, dV \quad , \tag{11.32}$$

where the first factor in the integrand is the local temperature coefficient (again per volume element). The integral quantity $\Delta\rho$ primarily affects the transient of the amplitude function; the local contributions, $\Delta\rho_V(\mathbf{r})$, may distort the space distribution of the flux, the shape function.

A special computational application of local reactivity coefficients is in the development of *intermediate approaches* to spatial dynamics.

The comparison of results in the previous section revealed that the PRM may significantly underpredict power transients and the resulting energy release. On the other hand, the full SED treatment, even using the improved quasistatic approach, may be too lengthy and time consuming for routine application in some cases. A few flux shape recalculations (for example, for ten energy groups in three space dimensions) may require a substantial amount of computer time if the evaluation of detailed compositional changes requires a three-dimensional treatment. Furthermore, in two- and three-dimensional problems, the calculation of a large number of inner products—involving the adjoint flux, time-dependent group constants and scattering matrices, and the time-dependent flux to find ρ, β_k, and Λ—can be more expensive than the calculation of a shape function.

This suggests the desirability of further developments in the spatial dynamics area. These developments could consist of refinements of the quasistatic computation scheme, of augmentation of point reactor dynamics, or of "intermediate" approaches (between the PRM and complete SED dynamics).[55,56]

The computation time of the improved quasistatic method as applied to three-dimensional problems can probably be reduced by significant amounts. This requires elaborate extrapolation schemes for the flux amplitude and the reactivity in order to stretch the reactivity and the shape function time steps. Error-estimating procedures could be employed to avoid unnecessary flux shape and reactivity calculations. The time-dependent β_k and Λ calculations by inner products might be replaced by simple empirical or theoretical formulas or at least performed much less frequently than the reactivity calculation, e.g., only at shape steps.

An alternative route for improving the computation time in complicated dynamics problems could possibly be achieved by use of reactivity coefficients in the calculation of the reactivity for a given change in the configuration.

The safety analysis system, SAS, developed at the Argonne National Laboratory[57,58] employs a set of precalculated regional reactivity coefficients:

$$\gamma_{nR} = \left(\frac{\partial \rho}{\partial \xi_n}\right)_R , \qquad (11.33)$$

with $\partial \xi_n$ describing a change in a region R. A region represents an axial segment of a "channel." The table of reactivity coefficients for SAS is precalculated, using the initial flux as weighting, i.e., applying the PRM. The coefficient table is used to account for the reactivity effect of regional changes in composition and temperature of the various core constituents—fuel, cladding, coolant, and structural material.

A specific inaccuracy of the reactivity coefficients calculated with the initial flux shapes consists of the neglect of the coupling of the various effects. For example, voiding part of the core of sodium leads to a spectral hardening, which in turn reduces the Doppler coefficient. Thus, a change in the energy-dependent flux can appear as *coupling* between reactivity coefficients. This particular coupling, i.e., the reduction of the Doppler coefficient during sodium voiding, is approximately accounted for in SAS, as a modification of the uncoupled coefficients, Eq. (11.33).

It appears feasible to develop a scheme of reactivity coefficients that accounts for the major coupling effect. The approach could essentially consist of considering the uncoupled coefficients as first-order terms in a Taylor expansion *and* extending the expansion to more terms. This automatically provides terms that involve simultaneous changes of two or more kinds. The basic features of such an approach have been developed by Malloy and Ott.[59] For *two* changes—say, fuel temperature, T, and coolant void, v—the Taylor expansion for the feedback reactivity, $\Delta\rho$, up to the third-order terms is given by

$$\begin{aligned}\Delta\rho = &\left(\frac{\partial\rho}{\partial T}\Delta T + \frac{\partial\rho}{\partial v}\Delta v\right)\\ &+ \frac{1}{2}\left(\frac{\partial^2\rho}{\partial T^2}\Delta T^2 + 2\,\frac{\partial^2\rho}{\partial T\partial v}\Delta v\Delta T + \frac{\partial^2\rho}{\partial v^2}\Delta v^2\right)\\ &+ \frac{1}{6}\left(\frac{\partial^3\rho}{\partial T^3}\Delta T^3 + 3\,\frac{\partial^3\rho}{\partial T^2\partial v}\Delta v\Delta T^2 + 3\,\frac{\partial^3\rho}{\partial T\partial v^2}\Delta v^2\Delta T + \frac{\partial^3\rho}{\partial v^3}\Delta v^3\right)\\ &+ \ldots\ ,\end{aligned} \tag{11.34a}$$

with

$$\Delta T = T_1 - T_0,\ \Delta v = v_1 - v_0,\ \Delta T^2 = (T_1 - T_0)^2,\ \text{etc.} \tag{11.34b}$$

As in all Taylor expansions, the partial derivatives in Eqs. (11.34) are taken at the initial conditions ($T = T_0$ and $v = v_0$). The terms in Eqs. (11.34) can be rearranged so that one obtains three groups of terms, describing the pure T and v dependencies and the coupling:

T dependence:

$$\gamma_T(T) = \left(\frac{\partial\rho}{\partial T} + \frac{1}{2}\frac{\partial^2\rho}{\partial T^2}\Delta T + \frac{1}{6}\frac{\partial^3\rho}{\partial T^3}\Delta T^2 + \ldots\right)\ , \tag{11.35}$$

v dependence:

$$\gamma_v(v) = \left(\frac{\partial\rho}{\partial v} + \frac{1}{2}\frac{\partial^2\rho}{\partial v^2}\Delta v + \frac{1}{6}\frac{\partial^3\rho}{\partial v^3}\Delta v^2 + \ldots\right)\ , \tag{11.36}$$

coupling of T and v dependence:

$$\gamma_{Tv}(T,v) = \left(\frac{\partial^2\rho}{\partial T\partial v} + \frac{1}{2}\frac{\partial^3\rho}{\partial T^2\partial v}\Delta T + \frac{1}{2}\frac{\partial^3\rho}{\partial T\partial v^2}\Delta v + \ldots\right)\ . \tag{11.37}$$

Equation (11.35), completed by the higher terms,[d] describes the temperature dependence of the Doppler coefficient, $\gamma_T(T)$; Eq. (11.36) gives the void dependence of the void coefficient, $\gamma_v(v)$; and Eq. (11.37)

---

[d]Since $\gamma_T(T)$ depends inversely on T, the convergence radius of Eq. (11.30) is limited to $T < 2T_0$. This should not present a practical problem, especially if Eq. (11.30) is summed up and the T dependence (e.g., $1/T$) is used explicitly.

represents a coupling term, including its T and v dependence. Although γ_{Tv} is symmetrical in T and v, it is common to interpret the coupling term as a modification of the Doppler coefficient:

$$\gamma_T(T,v) = \gamma_T(T) + \gamma_{Tv}(T,v)\Delta v \quad . \tag{11.38}$$

To find the generalization of the regional first-order coefficients, Eq. (11.32), one breaks up the spatial integral of the reactivity calculations for Eq. (11.38) into the regional contributions. This gives for the void-fraction-dependent Doppler coefficient:

$$\gamma_{TR}(T_R,v_R) = \gamma_{TR}(T_R)\left[1 + \frac{\gamma_{TvR}(T_R,v_R)}{\gamma_{TR}(T_R)}\,\Delta v_R\right] \quad . \tag{11.39}$$

The term in brackets describes the regional reduction of the Doppler coefficient due to coolant voiding.

Reactivity coefficients with appropriate coupling can also be employed to stretch the intervals between the inner product and the shape calculations in the factorized spatial dynamics. Then, both routes for developing intermediate approaches—speeding up the SED dynamics and augmenting the PRM—might eventually merge into a single method for the treatment of dynamics problems with any desired degree of accuracy.

Review Questions

1. What is space-energy dependent dynamics?
2. Give the two reasons that make space-energy dependent dynamics necessary.
3. Explain why the point reactor model is not conservative (for fuel and absorber insertions).
4. Which phenomena, in addition to the ones described by the kinetics equations, have to be included in a more complete description of the dynamics problem?
5. Which delayed neutron precursors are volatile at temperatures that could lead to fuel damage?
6. Discuss the effect of the loss of volatile precursors (a) on β-effective, (b) on the flux level in a delayed supercritical transient, (c) on the flux decrease after achieving subcriticality, and (d) on k_{eff}.
7. Name the four basic approaches of space-energy dependent dynamics.

8. Describe and discuss advantages and disadvantages of the following four approaches to space-energy dependent dynamics: (a) the finite difference solution approach, (b) the nodal approach, (c) the modal approach (space-time synthesis), and (d) the quasistatic approach.
9. What is the general idea of the flux factorization approaches?
10. Describe a sequence of three flux factorization approaches that are between point kinetics and the finite difference method for the time-dependent flux.
11. Discuss the relation of the quasistatic and the adiabatic methods: (a) Which approximation is common to both, and (b) what are the key differences?
12. Describe the typical time-step hierarchy that is used in the quasistatic method.

REFERENCES

1. T. J. Thompson and J. G. Beckerley, Eds., *The Technology of Nuclear Reactor Safety,* The MIT Press, Cambridge, Massachusetts (1973).
2. J. H. Rust and L. E. Weaver, Eds., *Nuclear Power Safety,* Pergamon Press, New York (1976).
3. H. H. Hummel and D. Okrent, *Reactivity Coefficients in Large Fast Power Reactors,* American Nuclear Society, La Grange Park, Illinois (1970).
4. J. Graham, *Fast Reactor Safety,* Vol. 8 of Nuclear Science and Technology Series, Academic Press, New York (1971).
5. E. E. Lewis, *Nuclear Power Reactor Safety,* John Wiley & Sons, New York (1977).
6. W. M. Stacey, Jr., *Space-Time Nuclear Reactor Kinetics,* Nuclear Science and Technology, Academic Press, New York (1969).
7. A. F. Henry, "Review of Computational Methods for Space-Dependent Kinetics," p. 9 in *Dynamics of Nuclear Systems,* D. L. Hetrick, Ed., The University of Arizona Press, Tucson (1972).
8. A. F. Henry, "Status Report on Several Methods for Predicting the Space-Time Dependence of Neutrons in Large Power Reactors," *Proc. Specialist Mtg. Reactivity Effects in Large Power Reactors,* Ispra Italy, October 28–31, 1970.
9. D. A. Meneley, G. K. Leaf, A. J. Lindeman, T. A. Daly, and W. T. Sha, "A Kinetics Model for Fast Reactor Analysis in Two Dimensions," p. 483 in *Dynamics of Nuclear Systems,* D. L. Hetrick, Ed., The University of Arizona Press, Tucson (1972).
10. K. F. Hansen and S. R. Johnson, "GAKIN—A One-Dimensional Multigroup Kinetics Code," GA-7543, GA Technologies (Aug. 1967).
11. J. Barclay Andrews, II and K. F. Hansen, "Numerical Solution of the Time-Dependent Multigroup Diffusion Equations," *Nucl. Sci. Eng.,* **31,** 304 (1968).

12. K. F. Hansen, B. V. Koen, and W. W. Little, "Stable Numerical Solutions of the Reactor Kinetics Equations," *Nucl. Sci. Eng.*, **22,** 51 (1965).
13. W. R. Cadwell, A. F. Henry, and A. J. Vigilotti, "WIGLE—A Program for the Solution of the Two-Group Space-Time Diffusion Equations in Slab Geometry," WAPD-TM-416, Westinghouse Electric Co. (Jan. 1964).
14. A. F. Henry and A. V. Vota, "WIGL2—A Program for the Solution of the One-Dimensional Two-Group, Space-Time Diffusion Equations Accounting for Temperature, Xenon and Control Feedback," WAPD-TM-532, Westinghouse Electric Co. (Oct. 1965).
15. A. V. Vota, N. J. Curlee, Jr., and A. F. Henry, "WIGL3—A Program for the Steady-State and Transient Solution of the One-Dimensional, Two-Group, Space-Time Diffusion Equations Accounting for Temperature, Xenon, and Control Feedback," WAPD-TM-788, Westinghouse Electric Co. (Feb. 1969).
16. J. B. Yasinsky, M. Natelson, and L. A. Hageman, "TWIGL—A Program to Solve the Two-Dimensional, Two-Group, Space-Time Neutron Diffusion Equations with Temperature Feedback," WAPD-TM-743, Westinghouse Electric Co. (Feb. 1968).
17. Wm. H. Reed and K. F. Hansen, "Alternating Direction Methods for the Reactor Kinetics Equations," *Nucl. Sci. Eng.*, **41,** 431 (1970).
18. A. L. Wight, K. F. Hansen, and D. R. Ferguson, "Application of Alternating-Direction Implicit Methods to the Space-Dependent Kinetics Equations," *Nucl. Sci. Eng.*, **44,** 239 (1971).
19. D. R. Ferguson and K. F. Hansen, "Solution of the Space-Dependent Reactor Kinetics Equations in Three Dimensions," *Nucl. Sci. Eng.*, **51,** 189 (1973).
20. R. S. Denning, "ADEP, One- and Two-Dimensional Few-Group Kinetics Code," Topical Report, Task 18, Battelle Columbus Laboratories (July 1971).
21. S. Kaplan, O. J. Marlowe, and J. Bewick, "Application of Synthesis Techniques to Problems Involving Time Dependence," *Nucl. Sci. Eng.*, **18,** 163 (1964).
22. S. Kaplan, "Synthesis Methods in Reactor Analysis," Vol. 3, p. 233 in *Advances in Nuclear Science and Technology,* Paul Greebler and Ernest J. Henley, Eds., Academic Press, New York (1966).
23. J. B. Yasinsky, "The Solution of the Space-Time Neutron Group Diffusion Equations by a Time-Discontinuous Synthesis Method," *Nucl. Sci. Eng.*, **29,** 381 (1967).
24. G. Kessler, "Space-Dependent Dynamic Behavior of Fast Reactors Using the Time-Discontinuous Synthesis Method," *Nucl. Sci. Eng.*, **41,** 115 (1970).
25. R. Avery, "Theory of Coupled Reactors," *Proc. Second Int. Conf. Peaceful Uses of Atomic Energy,* Vol. 12, p. 182, International Atomic Energy Agency, Vienna (1958).
26. C. G. Chezem, G. H. Hansen, H. H. Helmick, and R. L. Seale, "Los Alamos Coupled Reactor Experience," LA-3494, Los Alamos National Laboratory (Mar. 1966).
27. H. Plaza and W. H. Köhler, "Coupled-Reactors Kinetics Equations," *Nucl. Sci. Eng.*, **26,** 419 (1966).
28. R. G. Cockrell and R. B. Perez, "On the Kinetic Theory of Spatial and

Spectral Coupling of the Reactor Neutron Field," *Proc. Symp. Neutron Dynamics and Control,* CONF-650413, U.S. Atomic Energy Commission (1966).
29. F. T. Adler, S. J. Gage, and G. C. Hopkins, "Spatial and Spectral Coupling Effects in Multicore Reactor Systems," *Proc. Conf. Coupled Reactor Kinetics,* C. C. Chezem and W. H. Köhler, Eds., Texas A&M Press, College Station, Texas (1967).
30. M. R. Wagner, "Synthese von Mehrdimensionalen Grob-und Feinmaschenrechnungen," KT6, p. 35, Tagungsbericht der ReactorLagung, Deutchen Atomsforums, Karlsruhe, FRG (Apr. 1973).
31. M. R. Wagner, "Nodal Synthesis Method and Imbedded Flux Calculations," *Trans. Am. Nucl. Soc.,* **18,** 152 (1974).
32. A. Birkhofer, S. Langenbuch, and W. Werner, "Coarse-Mesh Method for Space-Time Kinetics," *Trans. Am. Nucl. Soc.,* **18,** 153 (1974).
33. E. L. Fuller, "Formulation of Space-Energy-Time Iterative Synthesis," *Nucl. Sci. Eng.,* **47,** 483 (1972).
34. K. O. Ott and D. A. Meneley, "Accuracy of the Quasistatic Treatment of Spatial Reactor Kinetics," *Nucl. Sci. Eng.,* **36,** 402 (1969).
35. D. A. Meneley, K. O. Ott, and E. S. Wiener, "Influence of the Shape-Function Time Derivative on Spatial Kinetics Calculations in Fast Reactors," *Trans. Am. Nucl. Soc.,* **11,** 225 (1968).
36. K. Ott, "Quasistatic Treatment of Spatial Phenomena in Reactor Dynamics," *Nucl. Sci. Eng.,* **26,** 563 (1966).
37. A. F. Henry, "The Application of Reactor Kinetics to the Analysis of Experiments," *Nucl. Sci. Eng.,* **3,** 52 (1958).
38. J. B. Yasinsky and A. F. Henry, "Some Numerical Experiments Concerning Space-Time Reactor Kinetics Behavior," *Nucl. Sci. Eng.,* **22,** 171 (1965).
39. J. T. Madell, Appendix to "Quasistatic Treatment of Spatial Phenomena in Reactor Dynamics," *Nucl. Sci. Eng.,* **26,** 564 (1966).
40. K. O. Ott, D. A. Meneley, and E. S. Wiener, "Quasistatic Treatment of Fast Reactor Excursions," ANL-7210, p. 370, Argonne National Laboratory (1966).
41. D. A. Meneley, K. O. Ott, and E. S. Wiener, "Space-Time Kinetics, The QX1 Code," ANL-7310, p. 471, Argonne National Laboratory (1967).
42. D. A. Meneley, K. O. Ott, and E. S. Wiener, "Fast-Reactor Kinetics—The QX1 Code," ANL-7769, Argonne National Laboratory (1971).
43. D. A. Meneley to K. O. Ott, Private Communication (1968).
44. L. L. Smith, "The Synthesis Quasistatic Model and Spatial Effects in Two Dimensional Neutron Dynamics Models for Fast Reactor Accidents," PhD Thesis, Purdue University (June 1971).
45. W. A. Bezella, "A Two Dimensional Fast Reactor Accident Analysis Model with Sodium Voiding Feedback Using the Synthesized Quasistatic Approach," PhD Thesis, Purdue University (June 1972).
46. W. A. Bezella and K. O. Ott, "Two-Dimensional Spatial Effects in Fast Reactors During Sodium Voiding," *Trans. Am. Nucl. Soc.,* **14,** 737 (1971).
47. L. L. Smith and K. O. Ott, "Space-Energy Dependent Fast Reactor Disassembly Analysis," *Trans. Am. Nucl. Soc.,* **15,** 821 (1972).
48. W. T. Sha, T. A. Daly, A. J. Lindeman, E. L. Fuller, D. A. Meneley, and

G. K. Leaf, "Two-Dimensional Fast-Reactor Disassembly Analysis with Space-Time Kinetics," *Proc. Conf. New Developments in Reactor Mathematics and Applications,* CONF-710302, Idaho Falls, Idaho, March 29–31, 1971.
49. D. R. Ferguson, T. A. Daly, and E. L. Fuller, "Improvement of and Calculations with the Two-Dimensional Space-Time Kinetics Code FX2," *Proc. Conf. Mathematical Models and Computational Techniques for Analysis of Nuclear Systems,* CONF-730414, Ann Arbor, Michigan, April 9–11, 1973.
50. L. Meyer and H. Bachmann, "KINTIC: A Program for the Calculation of Two-Dimensional Reactor Dynamics of Fast Reactors with the Quasistatic Method," KFK-1627, EURFNR-1083, Kernforschungszentrum Karlsruhe, FRG (1972).
51. L. L. Smith, J. E. Boudreau, C. R. Bell, P. B. Bleiweis, J. F. Barnes, and J. R. Travis, "SIMMER-I, An LMFBR Disrupted Core Analysis Code," *Proc. Int. Mtg. on Fast Reactor Safety and Related Physics,* Chicago, Illinois, October 5–8, 1976, CONF-761001, Vol. III, P. 1195, National Technical Information Service (1977).
52. H. L. Dodds, Jr., J. W. Stewart II, and C. E. Bailey, "Comparison of the Quasistatic and Direct Methods for Two-Dimensional Thermal Reactor Dynamics," *Trans. Am. Nucl. Soc.,* **15,** 786 (1972).
53. C. H. Adams and W. M. Stacey Jr., "RAUMZEIT—A Program to Solve Coupled Time-Dependent Neutron Diffusion Equations in One Space Dimension," KAPL-M-6728, Knolls Atomic Power Laboratory (1967).
54. E. L. Fuller, "One-Dimensional Space-Time Kinetics Benchmark Calculations," ANL 7910, Argonne National Laboratory (1972).
55. K. O. Ott, "Advances and Current Problems in Reactor Kinetics," *Proc. Natl. Topl. Mtg. New Developments in Reactor Physics and Shielding,* Kiamesha Lake, New York, September 12–15, 1972, CONF-720901, p. 168, Technical Information Center, U.S Atomic Energy Commission.
56. D. R. Ferguson, T. A. Daly, and E. L. Fuller, "Improvement of and Calculations with the Two-Dimensional Space-Time Kinetics Code FX2," *Proc. Conf. Mathematical Models and Computational Techniques for Analysis of Nuclear Systems,* CONF-730414, Ann Arbor, Michigan, April 9–11, 1973.
57. J. C. Carter, G. J. Fischer, T. J. Heames, D. R. MacFarlane, N. A. McNeal, W. T. Sha, C. K. Sanathanan, and C. K. Youngdahl, "SASIA, A Computer Code for the Analysis of Fast-Reactor Power and Flow Transients," ANL-7607, Argonne National Laboratory (1970).
58. F. E. Dunn, W. R. Bohl, T. J. Heames, G. Hoppner, M. G. Stevenson, L. L. Smith, J. R. Travis, W. Woodruff, L. W. Person, J. M. Kyser, and D. R. Ferguson, "The SAS3A LMFBR Accident Analysis Computer Code," ANL/RAS 75-17, Argonne National Laboratory (1975).
59. D. J. Malloy and K. O. Ott, "Neutronic Coupling of Feedback Effects," *Proc. Int. Mtg. on Fast Reactor Safety and Related Physics,* Chicago, Illinois, October 5–8, 1976, CONF-761001, Vol. II, p. 482, National Technical Information Service (1977).

APPENDIX A

Reactivity Effects of Fission Products

A-1 Introduction

The fission product (FP) pair that is generated in fissioning a heavy nucleus in a power reactor largely stays in place until the fuel element is discharged. During operation with constant power, i.e., constant fission rate, the number of FP nuclei in the reactor then increases linearly. However, as most fission products are radioactive they are converted through beta decay into neighboring nuclides. Their beta decay is a β^- emission, converting a neutron into a proton, since FP nuclei are neutron-rich. Also, neutron capture contributes to the conversion of FPs.

The conversion of fission products through neutron capture ceases after the shutdown of a reactor; the beta decay, however, still continues, though at a rapidly decreasing rate. The energy liberated in these post-shutdown beta decays appears as "decay heat" or "afterheat," which immediately after shutdown typically amounts to several percent of the original power.

The neutron capture reactions of the fission products constitutes a negative reactivity effect, which needs to be considered in the design and operation of reactors. Historically, the neutron capturing FPs are called "poison." The evaluation of the "poisoning" of a reactor by FPs would have to be based in general on the treatment of the concentration of the *individual* FP nuclides and their capture cross sections. In most practical applications, however, a more cursory description is fully adequate, employing the classification of "saturating" and "nonsaturating" fission products, but singling out, for special treatment, ^{135}Xe and ^{149}Sm.

As *delayed neutron precursors* are also fission products, the balance equations are the same as for other FPs. The notations are commonly different. The categorization of FPs formally corresponds to the classification of precursors in delay groups.

The buildup of *higher actinide nuclei* is described by balance equations similar to the ones for fission products, though the source of the higher actinides is neutron capture rather than fission. The buildup and burnup of these actinide nuclides are the subject of fuel cycle analyses, where the reactivity effects of fission products are also considered.

The buildup and burnup of FPs and higher actinides are not kinetics problems, as the flux can be calculated as a sequence of steady states. Neither the flux derivative nor precursors or delayed neutrons play a role in these problems. The treatment of these problems is therefore not covered in this text. Only xenon and samarium are discussed in this appendix because of their effects on shutdown and restart. The categorization of other fission products is only briefly addressed.

A-2 Saturating and Nonsaturating Fission Products

Fission products appear in the atomic weight range from ~75 to 160, i.e., over an A span of ~85. For each A, several Z values are possible. Some nuclides appear as direct FPs, others are generated only through β^- decays; sometimes these beta decay chains contain up to six successive decays. All in all, there are >300 different nuclides in the FP population.

At the startup of a reactor with fresh fuel, there are no FPs. After the flux is raised to the full-power condition, the direct FPs are generated at a rate, say S_i, with

$$S_i = a_i R_f \quad , \tag{A.1}$$

where R_f is the fission rate (density) and a_i the "abundance" of nuclide i, i.e., their number per fission (the space dependence is not explicitly indicated).

Neutron capture, with a microscopic rate of $\sigma_i\phi$, and beta decay, with λ_i as the decay constant, together with S_i make up the balance equation:

$$\dot{N}_i = -\lambda_{ri}N_i + S_i \quad , \tag{A.2}$$

where

$$\lambda_{ri} = \lambda_i + \sigma_i\phi \tag{A.3}$$

denotes the total (microscopic) removal rate. For a time-independent flux level ϕ and source rate S_i, the solution of Eq. (A.2) is given by

$$N_i(t) = N_i^0 \exp(-\lambda_{ri}t) + N_i^{as}[1 - \exp(-\lambda_{ri}t)] \quad ; \tag{A.4}$$

again, a decay of the initial concentrations results and a buildup of an asymptotic value,

$$N_i^{as} = \frac{S_i}{\lambda_{ri}} = \frac{a_i}{\lambda_{ri}} R_f \quad . \tag{A.5}$$

All FPs with a short lifetime—seconds, minutes, hours, or even some days—saturate quickly, assuming their asymptotic concentrations, N_i^{as}, in a period of several lifetimes. One can then form the macroscopic cross section of the saturating category of FPs:

$$\Sigma^{sat} = \sum_i \sigma_i \frac{a_i}{\lambda_{ri}} R_f \quad . \tag{A.6}$$

As the concentration of the short-lived nuclides remains small because of the large value of λ_{ri}, the total reactivity effect is also small. About 1.0% is a typical value for LWRs. A slight increase of the original enrichment can provide this reactivity.

The above evaluation of $N_i(t)$ pertained to direct FPs. Most FP nuclides are formed through beta decay. Then, the balance equation, Eq. (A.2), has another source term, the decay rate of the parent nuclide, either in addition to or instead of the direct source, S_i. However, the parent nuclides are generally farther away from the stable "valley" of nuclides and therefore most have a shorter lifetime than the daughter nuclides. Since they then achieve their saturation even earlier than the daughter nuclides, they can be combined with S_i, using the respective asymptotic number:

$$S_{i+1}^t = S_{i+1} + \lambda_i N_i^{as} \quad , \tag{A.7}$$

where the daughter nuclide is labeled "$i+1$" and the superscript t (for "total") indicates the combined effect.

On the other extreme, one has the long-lived and stable FPs. If their capture cross sections are also small, the corresponding λ_{ri} is so small that $\lambda_{ri}t$ in the exponential function of Eq. (A.4) remains small compared to one for the entire residence time of a fuel element. Then, the exponential function in Eq. (A.3) can be expanded and one obtains for nonsaturating FPs:

$$N_i(t) \simeq S_i^t t = a_i^t R_f t \quad , \tag{A.8}$$

i.e., a linear buildup (a_i^t combines the direct and the indirect production).

Again, the effect of all stable or long-lived nuclides is combined to find the total buildup of poison. As the abundances and cross sections of the stable FPs and long-lived FPs are well known and since their reactivity effect is important, the combined effects of these FPs are expressed in terms of microscopic group constant sets as the basis for a detailed space- and time-dependent evaluation of the poisoning in the framework of fuel cycle analysis.

A-3 Xenon and Samarium Concentrations in Steady-State Operation

As indicated above, there are two FP nuclides that are subjected to special treatment, ^{135}Xe and ^{149}Sm. Both have extremely large capture cross sections for thermal neutrons.

The xenon cross section has a resonance near 0.2 eV where σ rises above 3×10^6 barns. It decreases on both sides of the resonance, but not by very much on the low energy wing before it resumes its $1/v$ rise as $E \rightarrow 0$. The spectrum averaged thermal σ_c is $\sim$1.5 to 2×10^6 barns.

A similar situation occurs with ^{149}Sm. The capture cross section for thermal neutrons is much smaller than for xenon; $\sim 5 \times 10^4$ barns. But since ^{149}Sm is stable, there is no competition of neutron capture and beta decay. Therefore, practically all ^{149}Sm nuclei formed will eventually capture a neutron (except for the nuclei formed shortly before a shutdown with subsequent fuel removal).

Xenon-135 is a direct fission product, but most of it is generated by beta decay of ^{135}I (the half-lives are given above the arrows; the abundances for thermal-neutron fission of ^{235}U, below the nuclides):

$$^{135}\text{Sb} \xrightarrow{1.6\text{ s}} {}^{135}\text{Te} \xrightarrow{29\text{ s}} {}^{135}\text{I} \xrightarrow{6.7\text{ h}} {}^{135}\text{Xe} \xrightarrow{9.15\text{ h}} {}^{135}\text{Cs} \xrightarrow{2\times 10^6\text{ yr}} {}^{135}\text{Ba (stable)}$$

a_i = ···> ···> 0.061 0.003

The abundances for ^{135}Sb and ^{135}Te need not be considered explicitly; both nuclei are converted so quickly into iodine that their abundances can just be combined with that for ^{135}I. Let N_I and N_X be the concentrations of these two poison nuclei, λ_I and λ_X the respective decay constants, and σ_I and σ_X the corresponding one-group cross sections. The steady-state balance equations are then given by:

$$\frac{dN_X}{dt} = 0 = -\lambda_{rX} N_X^{as} + \lambda_I N_I^{as} + a_X R_f \tag{A.9}$$

and

$$\frac{dN_I}{dt} = 0 = -\lambda_{rI} N_I^{as} + a_I R_f \quad . \tag{A.10}$$

From Eqs. (A.9) and (A.10), it follows at first:

$$N_X^{as} = \frac{\lambda_I N_I^{as} + a_X R_f}{\lambda_{rX}} \quad , \tag{A.11}$$

with

$$N_I^{as} = \frac{a_I R_f}{\lambda_{rI}} \quad . \tag{A.12}$$

Neglecting the very small capture rate of iodine and inserting Eqs. (A.3) and (A.12) into Eq. (A.9) gives:

$$N_X^{as} = \frac{(a_I + a_X)}{\lambda_X + \sigma_X \phi} \Sigma_f \phi = \frac{0.064}{\lambda_X + \sigma_X \phi} \Sigma_f \phi \quad . \tag{A.13}$$

Thus, N_X^{as} is proportional to the combined abundances in the chain; through competition of conversions through beta decay and neutron capture, it has a nonlinear dependence on the flux level ϕ.

Figure A-1 shows the flux level dependence of the asymptotic ^{135}Xe concentration, which is reached several xenon lifetimes after startup, say, after approximately one day. As long as $\sigma_X \phi$ is small compared to $\lambda_X = 2.10 \times 10^{-5}$/s, $N_X^{as}(\phi)$ increases linearly with ϕ; at

$$\phi_{1/2} = \frac{\lambda_X}{\sigma_X} = \frac{2.10 \times 10^{-5}}{2 \times 10^6 \times 10^{-24} \text{cm}^2 \text{s}} \simeq 1.05 \times 10^{13} \text{ neutron/cm}^2\text{s} \quad ,$$

$N_X^{as}(\phi)$ equals one half of its maximum value at very high fluxes. Flux levels of several times 10^{15} neutron/cm^{2}s are typical for the so-called

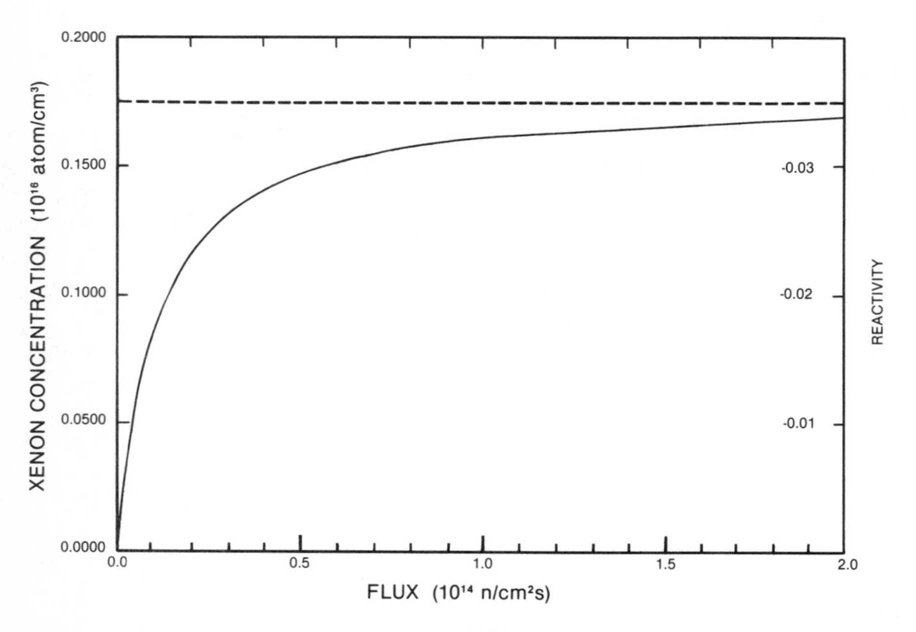

Fig. A-1. Concentration and reactivity effect of ^{135}Xe for a thermal reactor in a steady state as a function of the flux level ϕ.

"high-flux reactors" that are used for research at various laboratories in the United States and in other countries. At these flux levels, $N_X^{as}(\phi)$ reaches its maximum value. As beta decay is then negligible compared to neutron capture, every ^{135}Xe nucleus eventually captures a neutron. If $\sigma_X\phi$ is about ten times λ_X, the lifetime of a ^{135}Xe nucleus in such a high flux is ~1 h.

The stable ^{149}Sm nuclide results from the following beta decay chain:

$$^{149}\text{Pr} \xrightarrow{2.3\text{ min}} {}^{149}\text{Nd} \xrightarrow{1.7\text{ h}} {}^{149}\text{Pm} \xrightarrow{53\text{ h}} {}^{149}\text{Sm (stable)}$$

$$a_i = \cdots> \qquad \cdots> \qquad 0.011$$

Again, the first two decays occur so quickly that the sum of the respective abundances (in thermal neutron fission of ^{235}U) can be attributed to ^{149}Pm. The balance equations are then given by:

$$\frac{dN_P}{dt} = 0 = -\lambda_P N_P + a_P R_f \tag{A.14a}$$

and

$$\frac{dN_S}{dt} = 0 = -\sigma_S\phi N_S + \lambda_P N_P \quad , \tag{A.14b}$$

where the samarium and promethium quantities are distinguished by the S and P indices. The neglect of the very small capture rate of promethium during its small mean lifetime in Eqs. (A.14) allows us to combine the two equations, Eqs. (A.14), and solve for N_S^{as}. The result is:

$$N_S^{as} = \frac{a_P R_f}{\sigma_S\,\phi} = a_P\,\frac{\Sigma_f}{\sigma_S} \quad . \tag{A.15}$$

The asymptotic samarium concentration is independent of the flux level.

A-4 The Reactivity Effects of Xenon and Samarium in Steady-State Operation

Both asymptotic concentrations of ^{135}Xe and ^{149}Sm depend on space through $\Sigma_f(\mathbf{r})$; N_X^{as} has the additional space dependence through $\phi(\mathbf{r})$. Thus, accurate reactivity calculations have to be based on a solution of the diffusion equation for the poisoned core. Perturbation calculation can be employed to obtain an estimate, if the flux shape deformation due to the space-dependent xenon concentration is not too strong.

Using average values of Σ_f and of k_∞ allows an estimate of the reactivity effect of N_S^{as} and $N_X^{as}(\infty)$ by assessing the influence of the poison on the f factor. As the operating reactor is critical, the reactivity ρ is the proper measure for the description of the effects of the poisons:

$$\rho_p = 1 - \frac{1}{k_p} \simeq 1 - \frac{1}{k_0}\frac{f_0}{f_p} = 1 - \frac{f_0}{f_p} \quad , \tag{A.16}$$

where the index p refers to the system with the poison, and 0 to the unpoisoned reactor for which $k_0 = 1$. The f factor is approximately given by the ratio of fuel and total absorption cross sections. The two f factors,

$$f_0 = \frac{\Sigma_a^{\text{fuel}}}{\Sigma_{a0}} \text{ and } f_p = \frac{\Sigma_a^{\text{fuel}}}{\Sigma_{a0} + \Sigma_{ap}} \quad , \tag{A.17}$$

inserted in Eq. (A.16) give the estimate for ρ_p:

$$\rho_p = 1 - \frac{\Sigma_{a0} + \Sigma_{ap}}{\Sigma_{a0}} = -\frac{\Sigma_{ap}}{\Sigma_{a0}} = -\frac{N_p\sigma_{ap}}{\Sigma_{a0}} \quad . \tag{A.18}$$

The macroscopic cross sections for Σ_{ap} are found from Eqs. (A.13) and (A.15) and Σ_{a0} is expressed by ν and k_∞ (Σ_f in R_f is denoted here by Σ_{f0}):

$$\rho_X(\phi \to \infty) = -\frac{(a_I + a_X)\Sigma_{f0}}{\Sigma_{a0}} = -\frac{0.064}{\nu}k_\infty \tag{A.19}$$

and

$$\rho_S(\text{all } \phi) = -\frac{a_S\Sigma_{f0}}{\Sigma_{a0}} = -\frac{0.011}{\nu}k_\infty \quad . \tag{A.20}$$

With $\nu = 2.43$ and $k_\infty \simeq 1.22$, one obtains the following estimates:

$$\rho_X^\infty = \rho_X(\phi \to \infty) = -0.032 \tag{A.21}$$

and

$$\rho_S(\text{all } \phi) = -0.0055 \quad . \tag{A.22}$$

The ρ_X^∞ value can be reduced by f_ϕ:

$$\rho_X(\phi) \simeq \rho_X(\phi = \infty)f_\phi = -\frac{0.064}{\nu}k_\infty \frac{\sigma_X\phi}{\lambda_X + \sigma_X\phi} \tag{A.23}$$

as suggested by Eq. (A.13) to account approximately for a finite flux level. Of course, spatial flux distortions and adjoint weighting are then disregarded.

A-5 Xenon and Samarium Poisoning After Shutdown

After shutdown of the reactor power, the removal of ^{135}Xe and ^{149}Sm through neutron capture ceases. However, the production of both nuclides through beta decay of the respective parent nuclides still continues.

In the case of samarium, this simply means that the ^{149}Sm content at shutdown is eventually increased by the decay of all available ^{149}Pm. Taking both asymptotic concentrations from Eqs. (A.14) gives

$$N_S^{\text{after shutdown}} = N_S^{as} + N_P^{as} = a_P\left(\frac{\Sigma_f}{\sigma_S} + \Sigma_f \frac{\phi}{\lambda_P}\right) \quad . \qquad \text{(A.24)}$$

Multiplying this equation with σ_S and inserting it in the reactivity formula, Eq. (A.18), yields for the samarium reactivity after shutdown ($\Sigma_f = \Sigma_{f0}$):

$$\begin{aligned} \rho_S^{\text{shutdown}} &= -\frac{\Sigma_{f0}}{\Sigma_{a0}} a_P\left(1 + \frac{\sigma_S\phi}{\lambda_P}\right) \\ &= -\frac{0.011}{\nu} k_\infty\left(1 + \frac{5 \times 10^4 \times 10^{-24} \times \phi}{3.62 \times 10^{-6}}\right) \\ &= \rho_S^{\text{operation}} (1 + 1.38\phi/10^{14}/\text{cm}^2\text{s}) \quad . \end{aligned} \qquad \text{(A.25)}$$

Thus, ρ_S is increased after shutdown by ~140% for an average flux of 10^{14}/cm^{2}s, i.e., from ρ = 0.0055 to 0.0132. In a high-flux reactor, with about ten times that flux level, ρ_S is increased by a factor of 14.

The case of ^{135}Xe after shutdown is more complicated as the residing xenon decays with a half-life of ~9 h, but new xenon is produced at a larger rate by the faster decay of ^{135}I. The iodine decay is simply exponential:

$$N_I(t) = N_I^{as} \exp(-\lambda_I t) \quad , \qquad \text{(A.26)}$$

where t now denotes the time after shutdown.

The decay rate of iodine is the source of xenon:

$$\frac{dN_X}{dt} = -\lambda_X N_X + \lambda_I N_I^{as} \exp(-\lambda_I t) \quad . \qquad \text{(A.27)}$$

Applying Eq. (A.6) gives at first for the solution:

$$\begin{aligned} N_X(t) &= N_X^{as} \exp(-\lambda_X t) \\ &+ \lambda_I N_I^{as} \int_0^t \exp(-\lambda_I t') \exp[-\lambda_X(t - t')]\, dt' \quad ; \end{aligned} \qquad \text{(A.28)}$$

thus

$$N_X(t) = N_X^{as} \exp(-\lambda_X t) + \frac{\lambda_I}{\lambda_I - \lambda_X} \times N_I^{as}[\exp(-\lambda_X t) - \exp(-\lambda_I t)] \quad . \tag{A.29}$$

Inserting $N_X(t)$ in the simple reactivity estimate formula, Eq. (A.18), gives a corresponding $\rho_X(t)$. The same reservations with respect to the neglect of the flux distribution that were expressed along with Eq. (A.23) also apply here. In addition, the reactivity effect of xenon after shutdown would have to be expressed as an increment below the shutdown reactivity and not around critical, as does Eq. (A.18).

Figure A-2 shows the results. The reactivity increment of the changing xenon concentration is plotted as a function of the time after shutdown for several operating flux levels. As the original xenon reactivity

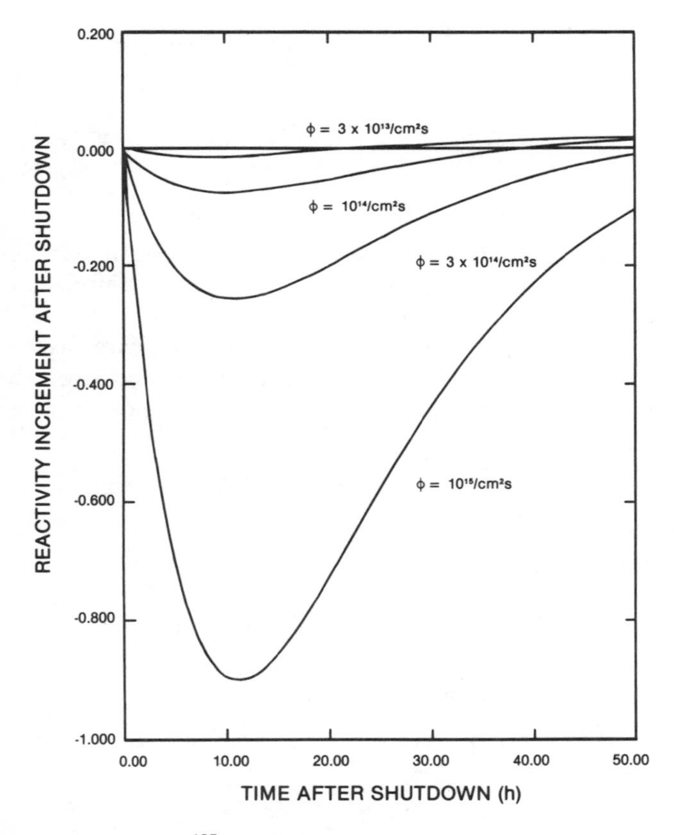

Fig. A-2. Reactivity effect of ^{135}Xe after shutdown as a function of time with the prior steady-state flux level as a parameter.

is compensated during operation, all reactivity increment trajectories start at $\rho = 0$, decrease then toward a minimum and eventually increase again. For a restart after a few hours, one has to "override" this negative xenon reactivity. For $\phi_0 = 10^{15}/cm^2s$, the minimum of ρ_x as obtained from this simple estimate is at -0.94. Therefore, it is just not practical in a high-flux reactor to provide the huge excess reactivity that would be needed to override the xenon reactivity at all times. It is better to settle for a day or two of waiting time before the reactor can be started up again.

All reactivity curves in Fig. A-2 will asymptotically approach the positive reactivity values that had been added in operation to compensate for the steady-state xenon. Naturally, additional control reactivity must be inserted to prevent the reactor from becoming critical if it is not desired.

A-6 Load Changes and Xenon Oscillations

In Sec. A-5, it was shown that the xenon concentration after a reactor shutdown will increase for several hours (provided that the initial flux is sufficiently large). A similar situation will arise in the case of a load reduction. If—through proper control—the flux is kept constant at the new reduced level, xenon will increase at first, because of the reduced removal through neutron capture, while the production through iodine decay still continues at the previous level. The xenon concentration will move over a maximum and then settle exponentially at its new lower asymptotic level.

Spatial effects occur when in a large, weakly coupled reactor, *local* offsetting load *changes* are initiated while the reactor is kept critical *at a constant total* power. Then, the xenon concentration on the reduced power side of the core will at first increase, and it will decrease on the side with an increased power. If only the overall power is managed, the increasing xenon reduces the power even further as the decreasing xenon will let it increase on the other side. This process will continue until it is reversed by the modified iodine production. However, unlike the case of a controlled load change discussed above, the flux and with it the xenon and iodine concentrations start to oscillate as the flux, in a sense, is pushed down by the increasing xenon, and comes up when xenon decreases. The overswinging of xenon is governed by the combined lifetimes of the iodine and xenon decays, which becomes apparent from Fig. A-2. Therefore, the time constant of this oscillation is of a similar magnitude. These slow changes can be controlled readily and represent no hazard.

APPENDIX B

Operators in Reactor Applications

Although conceptually quite simple, the detailed neutron balance equations, such as the energy-dependent Boltzmann equation and even the diffusion equations, are so lengthy that one can easily lose track of their meaning in the flood of all the symbols if they are used explicitly in detailed derivations. Furthermore, many derivations and relations that involve neutron balance equations are independent of the special approximation. There is, therefore, an incentive to express these equations in an appropriate shorthand that is independent of the special approximation. Such a shorthand is available. It is the concept of *operators,* which can be applied directly to neutron balance equations. The definition of operators is briefly reviewed in this appendix.

In addition to using operators as a convenient shorthand *and* to apply them in general approximation-independent derivations, there is a third application, the adjoint neutronics problem, that gains greatly in clarity if operators are employed rather than explicit formulations. The operator concept is extensively applied in reactor statics. The adjoint operator concept (see Sec. B-3) is primarily useful in dynamics as well as in static perturbation theory.

B-1 Definition of Operators

Operators express certain mathematical operations or prescriptions to be carried out with a function or a vector. Applying an operator to a function yields another function. Four types of operators are used in the following: differential, matrix, integral, and multiplication operators. Examples of differential operators are:

$$\frac{d}{dx}, \frac{d}{dt}, \text{ and } \frac{d^2}{dx^2} \quad .$$

Their application produces the derivative function; for example,

$$\frac{d}{dx} f(x) = f'(x) \quad . \tag{B.1}$$

The simplest matrix operator is:

$$\begin{pmatrix} a_{11} & a_{12} \\ a_{21} & a_{22} \end{pmatrix} \quad . \tag{B.2}$$

Its application to a vector yields another vector:

$$\begin{pmatrix} a_{11} & a_{12} \\ a_{21} & a_{22} \end{pmatrix} \begin{pmatrix} x_1 \\ x_2 \end{pmatrix} = \begin{pmatrix} y_1 \\ y_2 \end{pmatrix} \quad . \tag{B.3}$$

The continuous analogue to a matrix is an integral operator, $\int K(E' \to E) \ldots dE'$. Its application to an appropriate function yields another function:

$$\int_{E'} K_s(E' \to E)\phi(E')\, dE' = S_s(E) \quad . \tag{B.4}$$

For example, $K_s(E' \to E)$ may be the scattering kernel; $S_s(E)$ is then the energy distribution of the scattered neutrons.

Also, fission can be expressed by a kernel:

$$K_f(E' \to E) = \nu\Sigma_f(E')\chi(E) \quad ;$$

its application to the flux yields the fission source:

$$\chi(E) \int_{E'} \nu\Sigma_f(E')\phi(\mathbf{r},E')\, dE' = S_f(\mathbf{r},E) \quad . \tag{B.5}$$

For completeness, one also introduces a "multiplication operator," such as $\Sigma_a(\mathbf{r},E)$.

The application of operators to functions or vectors that yield other functions or vectors of the same basic type can be called a "transformation." One can also consider operators as quantities that perform a mapping of one space on another. It is not necessary here to make use of these mathematical properties of operators.

Some other basic properties of operators that are needed are briefly reviewed here.

An operator can only be applied to an appropriate class of functions, characterized by several properties concerning structure, smoothness, singularities and asymptotic behavior, and boundary conditions:

1. The function must have the appropriate structure and dependence; for example, d/dt is to be applied to a function of time such as $p(t)$ or $\phi(\mathbf{r},t)$. The matrix operator $\begin{pmatrix} a_{11} & a_{12} \\ a_{21} & a_{22} \end{pmatrix}$ is to be applied to a vector of two components.
2. The result of the application of the operator must exist; for

example, $p(t)$ must have a first derivative. Or, in order for $\frac{d^2}{dx^2}\phi(x)$ to be formed, the second derivative of $\phi(x)$ must exist, etc.

3. There is a third and very important property that these functions must have: Their scalar product must exist, i.e., the respective integrals must converge.

For vectors, the scalar product is often called the "dot product":

$$\mathbf{y} \cdot \mathbf{x} = (\mathbf{y},\mathbf{x}) = (y_1 y_2)\begin{pmatrix} x_1 \\ x_2 \end{pmatrix} = y_1 x_1 + y_2 x_2 \quad , \tag{B.6}$$

with $(y_1 y_2)$ being a row vector, and $\begin{pmatrix} x_1 \\ x_2 \end{pmatrix}$ a column vector. One of the vectors may already be transformed by a matrix. The corresponding scalar product is then given by:

$$(\mathbf{y},\mathbf{Ax}) = (y_1 y_2)\begin{pmatrix} a_{11} & a_{12} \\ a_{21} & a_{22} \end{pmatrix}\begin{pmatrix} x_1 \\ x_2 \end{pmatrix} \quad . \tag{B.7}$$

Operator symbols are identified here in a block-style bold-face type.

If continuous functions are defined in a certain domain $-a \leq x \leq a$, then the scalar product is the integral of the product of the functions, integrated over the domain. For example,

$$(\psi,\phi) = \int_{-a}^{a} \psi(x)\phi(x)\, dx \quad , \tag{B.8}$$

or with a differential operator included:

$$(\psi,\nabla_x^2\phi) = \int_{-a}^{a} \psi(x)\, \frac{d^2}{dx^2}\phi(x)\, dx \quad , \tag{B.9}$$

or with an integral operator included:

$$(\psi,\mathbf{K}\phi) = \int_E \psi(E) \int_{E'} K(E' \rightarrow E)\phi(E')\, dE'\, dE \quad . \tag{B.10}$$

The entirety or class of functions to which an operator can be applied forms a *functional space.* In the case of a matrix, this is simply a finite dimensional *vector space.* For differential and integral operators, this space normally has an infinite number of dimensions; it is then a "Hilbert space."

The two functions or vectors in the scalar product may be interchanged:

$$(\psi,\phi) = (\phi,\psi) \text{ or } (\mathbf{y},\mathbf{x}) = (\mathbf{x},\mathbf{y}) \quad . \tag{B.11}$$

In the following, these mathematical concepts are applied to neutron balance equations. The operators will find extensive application in an area called "perturbation theory," which is presented in Chapter 4, and in derivations of the kinetics equations (Chapters 3, 5, and 11).

B-2 Neutron Balance Equations in Operator Form

In the operator formulation of the energy-dependent diffusion equation or the Boltzmann equation, one generally identifies three terms to describe different physical effects. The first two are:

$$S(\mathbf{r},E) = \text{independent source}$$

and

$$\mathbf{F}\Phi = \chi(E)\int_0^\infty \nu\Sigma_f(\mathbf{r},E')\phi(\mathbf{r},E')\,dE' = \text{fission source} \quad , \tag{B.12}$$

where $\mathbf{F}$ is the fission operator. The remainder of the neutron balance equation is combined into one term, which is abbreviated as $\mathbf{M}\Phi$ with $\mathbf{M}$ as the "migration and loss operator," which represents the third term in the operator form of the balance equation. In diffusion approximation, $\mathbf{M}$ is given by:

$$\begin{aligned}\mathbf{M}\Phi = &-\nabla\cdot D(\mathbf{r},E)\nabla\phi(\mathbf{r},E) + \Sigma_t(\mathbf{r},E)\phi(\mathbf{r},E)\\ &-\int_{E'}\Sigma_S(\mathbf{r},E'\rightarrow E)\phi(\mathbf{r},E')\,dE' \quad .\end{aligned} \tag{B.13}$$

This brings the energy-dependent diffusion equation to the form:

$$\mathbf{M}\Phi = \mathbf{F}\Phi + S \quad . \tag{B.14}$$

The general flux that appears in the operator equation is denoted by Φ without an argument, since the operator formulation should be applicable to any approximation.

Equation (B.13) can also be written with the corresponding terms of the Boltzmann equation, which gives a different definition of $\mathbf{M}$. Neutron migration and loss can also be expressed in one of the simpler models used above. In one-group diffusion theory, the operators and S for all neutrons are given by:

$$\mathbf{M} = -\nabla\cdot D(\mathbf{r})\nabla + \Sigma_a(\mathbf{r}) \quad ,$$

$$\mathbf{F} = \nu\Sigma_f(\mathbf{r}) \quad ,$$

and

$$S = S(\mathbf{r}) \quad . \tag{B.15}$$

In the simplest neutronics model, in which all space and energy dependencies are eliminated, all operators are reduced to numbers (multiplication operators):

$$\mathbf{M} = DB^2 + \Sigma_a \quad ,$$

$$\mathbf{F} = \nu\Sigma_f \quad ,$$

and

$$S = S \quad . \tag{B.16}$$

The formal identity of all neutron balance equtions shows the advantage of the use of operator notation. By means of operators, the neutron balance equation can be expressed in a general form that is independent of the specific approximation employed. For example, Φ then denotes:

the angular flux, $\phi(\mathbf{r},E,\mathbf{\Omega})$,
the flux, $\phi(\mathbf{r},E)$,
the total flux, $\phi(\mathbf{r})$, or
the integrated flux, $\hat{\phi}$,

depending on whether **M** and **F** are the operators of

the Boltzmann equation,
the E-dependent diffusion equation,
the one-group diffusion equation, or
the one-group, space-integrated diffusion equation.

B-3 Adjoint Operators[a]

An important concept for the treatment of "perturbations" of eigenvalue problems is that of "adjoint operators" that operate on "adjoint functions." Their application allows the estimation of an eigenvalue perturbation without actually solving the often very complicated "perturbed" problem (see Chapter 4).

The adjoint operator is defined by the scalar product equation

$$(\Psi,\mathbf{H}\Phi) = (\mathbf{H}^*\Psi,\Phi) = (\Phi,\mathbf{H}^*\Psi) \quad , \tag{B.17}$$

to hold for *all* allowed Ψ and Φ, i.e., for all adjoint functions and "real" functions, Ψ and Φ, out of the respective functional space. In Eq. (B.17), $\mathbf{H}^*$ is the adjoint operator.

---

[a]For a more detailed presentation, see advanced textbooks on nuclear engineering, for example, Refs. 1 through 3.

The way the adjoint operator is defined by the scalar products in Eq. (B.17) is demonstrated in several examples. The first example is a 2×2 matrix. Note that the real and adjoint "functions" are column and row vectors, respectively:

$$(x_1 x_2)\begin{pmatrix} a_{11} & a_{12} \\ a_{21} & a_{22} \end{pmatrix}\begin{pmatrix} y_1 \\ y_2 \end{pmatrix} = (y_1 y_2)\begin{pmatrix} a_{11}^* & a_{12}^* \\ a_{21}^* & a_{22}^* \end{pmatrix}\begin{pmatrix} x_1 \\ x_2 \end{pmatrix} \quad . \tag{B.18}$$

Carrying out the matrix multiplication and rearranging the terms yields:

$$x_1 y_2 (a_{11} - a_{11}^*) + x_1 y_2 \, (a_{12} - a_{21}^*)$$
$$+ x_2 y_1 (a_{21} - a_{12}^*) + x_2 y_2 (a_{22} - a_{22}^*) = 0 \quad . \tag{B.19}$$

Since Eq. (B.19) is to be satisfied for *any* $x_i y_k$ values, the values of all parentheses must be zero. This then yields the components of the adjoint matrix:

$$\left.\begin{array}{l} a_{11}^* = a_{11} \\ a_{12}^* = a_{21} \\ a_{21}^* = a_{12} \\ a_{22}^* = a_{22} \end{array}\right\} \text{ i.e., } \mathbf{A}^* = \mathbf{A}^T \quad . \tag{B.20}$$

In other words, for a "real" matrix, the adjoint matrix equals the transposed one, $\mathbf{A}^T$.

The definitions of adjoint differential operators are obtained through integration by parts, which requires the consideration of boundary conditions. The main results are given first.

Let $\mathbf{L}$ be the linear differential operator defined by

$$\mathbf{L}\Phi = a_0(x) \frac{d^n}{dx^n}\phi(x) + a_1(x) \frac{d^{n-1}}{dx^{n-1}}\phi(x) + \ldots$$
$$+ a_n(x)\phi(x) \quad . \tag{B.21}$$

The corresponding adjoint operator as obtained from the general definition is given by

$$\mathbf{L}^*\Phi = (-1)^n \left\{ \frac{d^n}{dx^n} [a_0(x)\phi(x)] - \frac{d^{n-1}}{dx^{n-1}} [a_1(x)\phi(x)] + - \ldots \right.$$
$$\left. + (-1)^n \, a_n(x)\phi(x) \right\} \quad . \tag{B.22}$$

Examples are:

$$\left(\frac{d}{dt}\right)^* = -\frac{d}{dt} \quad , \tag{B.23}$$

and

$$\left(\frac{d^2}{dx^2}\right)^* = \frac{d^2}{dx^2} \quad ; \tag{B.24}$$

the same also holds for the three-dimensional Laplace operator:

$$(\nabla^2)^* = \nabla^2 \quad . \tag{B.25}$$

Equations (B.24) and (B.25) indicate that the Laplace operator is "self-adjoint." In general, "self-adjoint" means that the adjoint operator equals the operator itself. If

$$\mathbf{H}^* = \mathbf{H} \quad , \tag{B.26}$$

then $\mathbf{H}$ is self-adjoint.

From Eq. (B.17) one can deduce readily that the adjoint of an adjoint operator again equals the original operator:

$$(\Psi,\mathbf{H}\Phi) = (\Phi,\mathbf{H}^*\Psi) = (\Psi,\mathbf{H}^{**}\Phi) \quad ; \tag{B.27}$$

thus

$$\mathbf{H}^{**} = \mathbf{H} \quad . \tag{B.28}$$

The fact that certain boundary conditions need to be imposed on adjoint functions is demonstrated for the simplest operator, Eq. (B.23); for the corresponding discussion of the neutronics problem, see Sec. B-4. From the definition of the adjoint operator follows:

$$\int_0^\infty \psi(t) \frac{d}{dt} \phi(t)\, dt = \int_0^\infty \phi(t) \left(\frac{d}{dt}\right)^* \psi(t)\, dt \quad . \tag{B.29}$$

Integrating the left side of Eq. (B.29) by parts yields:

$$\int_0^\infty \psi(t) \frac{d}{dt}\phi(t)\, dt = [\psi(t)\phi(t)]_0^\infty - \int_0^\infty \phi(t) \frac{d}{dt}\psi(t)\, dt \quad . \tag{B.30}$$

Whereas the boundary conditions for the real problems are dictated by the corresponding physics, the boundary conditions for the corresponding adjoint problem are derived by implementing the definition, Eq. (B.17), or here Eq. (B.29). This requires the bracket in Eq. (B.30) to vanish. Suppose the physical boundary condition for $\phi(t)$ is $\phi(t) \rightarrow 0$ for $t \rightarrow \infty$, then $\psi(0) = 0$ must be the boundary condition for the adjoint function to obtain Eq. (B.23).

After the adjoint operator has been defined by requiring Eq. (B.17) to hold for *all* allowed pairs of ϕ and ψ, it can now be applied for any specific pair to obtain:

$$(\Psi,\mathbf{H}\Phi) = (\Phi,\mathbf{H}^*\Psi) \quad ; \tag{B.31}$$

in words, revolving the functions in a scalar product turns the operator **H** into its adjoint $\mathbf{H}^*$.

B-4 Adjoint Neutronics Problems[b]

The neutronics balance equation in operator notation was given in Sec. B-2. In a source-free problem, one has to include an eigenvalue in front of the fission source to obtain a nontrivial solution; thus

$$\mathbf{M}\Phi = \lambda\mathbf{F}\Phi \quad . \tag{B.32}$$

The corresponding adjoint problem is written as

$$\mathbf{M}^*\Phi^* = \lambda\mathbf{F}^*\Phi^* \quad . \tag{B.33}$$

Before giving the explicit form of the adjoint equation, some fundamental aspects are discussed that are independent of the particular approximation:

1. The fundamental eigenvalue in the adjoint equation, Eq. (B.33), is the same as in the original equation. The proof is simple: multiplying Eqs. (B.32) and (B.33) scalarly with Φ^* and Φ, respectively, gives:

$$(\Phi^*,\mathbf{M}\Phi) = \lambda(\Phi^*,\mathbf{F}\Phi)$$

and

$$(\Phi,\mathbf{M}^*\Phi^*) = \lambda^*(\Phi,\mathbf{F}^*\Phi^*) \quad , \tag{B.34}$$

where the eigenvalue in the adjoint equation has been temporarily denoted by λ^*. Revolving the functions in the second equation and subtracting the result from the first equation gives:

$$(\Phi^*,\mathbf{M}\Phi) - (\Phi^*,\mathbf{M}\Phi) = 0 = (\lambda - \lambda^*)(\Phi^*,\mathbf{F}\Phi) \quad . \tag{B.35}$$

Since the scalar product on the far right side of Eq. (B.35) is positive if Φ^* and Φ are the respective fundamental modes, it follows that $\lambda^* = \lambda$.

2. Since the Laplace operator is self-adjoint, and so are multiplication operators, the one-group diffusion equation is self-adjoint.

3. The basic non-self-adjoint problem in neutronics is the slowing down problem, or any problem involving predominant downscattering.

4. The adjoint flux can only be obtained as a solution of the cor-

---

[b]For a more detailed presentation, see advanced textbooks on nuclear engineering; for example, Refs. 1 through 3.

responding adjoint problem, e.g., Eq. (B.33); it cannot be constructed in any way out of the flux itself.

5. The adjoint problem normally needs to be defined for each particular approximation of the real problem, applying the definition, Eq. (B.17). An independent derivation from the adjoint of a more general equation, e.g., the Boltzmann equation, may yield a result that is not adjoint to the real problem on the intended level of approximation. The definition of the particular "approximation" may also have to include the specification of the finite differencing scheme applied in the numerical solution procedure (see Ref. 1, p. 271). That the adjoint problem is to be defined for a particular approximation of the real problem becomes obvious for the multigroup approximation derived from a continuous energy-dependent formulation. The group constants of the real problem appear flux weighted; they are also employed by the adjoint problem, for which a direct derivation could only yield group constants with adjoint flux weighting.

The adjoint neutronics operators corresponding to Eq. (B.13) and (B.12) are given by:

$$\mathbf{M}^*\Phi^* = -\nabla\cdot D(\mathbf{r},E)\nabla\phi^*(\mathbf{r},E) + \Sigma_t(\mathbf{r},E)\phi^*(\mathbf{r},E) - \int_{E'}\Sigma_s(\mathbf{r},E\rightarrow E')\phi^*(\mathbf{r},E')\,dE'$$

and

$$\mathbf{F}^*\Phi^* = \nu\Sigma_f(\mathbf{r},E)\int_{E'}\chi(E')\phi^*(\mathbf{r},E')\,dE' \quad . \tag{B.36}$$

The derivative term in $\mathbf{M}^*$ is second order and thus has, according to Eq. (B.22), the same sign as $\mathbf{M}$. The adjoint flux applies to the end energy in the scattering as well as in the fission integrals; these integrals are therefore carried out over the energies of the emerging neutrons.

In the proof that the adjoint operators, Eqs. (B.36), satisfy the definition, Eq. (B.17), one considers the individual terms. It follows readily, with Ψ in Eq. (B.17) replaced by Φ^*:

$$\int_V\int_E \phi^*(\mathbf{r},E)\Sigma_t(\mathbf{r},E)\phi(\mathbf{r},E)\,dE\,dV = \int_V\int_E \phi(\mathbf{r},E)\Sigma_t(\mathbf{r},E)\,\phi^*(\mathbf{r},E)\,dE\,dV$$

and

$$\int_V\int_E \phi^*(\mathbf{r},E)\int_{E'}\Sigma_s(E'\rightarrow E)\phi(\mathbf{r},E')\,dE'\,dE\,dV$$

$$= \int_V\int_E \phi(\mathbf{r},E)\int_{E'}\Sigma_s(E\rightarrow E')\phi^*(\mathbf{r},E')\,dE'\,dE\,dV \quad . \tag{B.37}$$

In the latter integrals, the order of integration in the definite integrals over the same $E' \rightarrow E$ domain can be interchanged, which gives the desired result.

The treatment of the leakage term is more complicated. It also yields the boundary condition for the adjoint flux. In this derivation, the following vector relation is employed:

$$\phi \nabla \cdot D \nabla \phi^* = \nabla \cdot \phi D \nabla \phi^* - D \nabla \phi \cdot \nabla \phi^* \quad . \tag{B.38}$$

This relation can be easily verified by carrying out the differentiation of the product ϕ and $D\nabla\phi^*$ (the first term of the right side). This gives the two other terms. The dot product notation is needed here to indicate the extent of the differentiation; i.e., the last term is a dot product of two gradients, the other two are vector gradients. The definition of the adjoint operator, Eq. (B.17), applied to the leakage terms of Eqs. (B.13) and (B.36) gives:

$$(\Phi^*, \nabla \cdot D \nabla \Phi) = (\Phi, \nabla \cdot D \nabla \Phi^*) \quad . \tag{B.39}$$

Using Eq. (B.38) on both sides of Eq. (B.39), and canceling the terms containing the product of the two gradients leads to

$$(\nabla, \Phi^* D \nabla \Phi) = (\nabla, \Phi D \nabla \Phi^*) \quad ,$$

or explicitly

$$\int_E \int_V \nabla \cdot \phi^* D \nabla \phi \, dV \, dE = \int_E \int_V \nabla \cdot \phi D \nabla \phi^* \, dV \, dE \quad . \tag{B.40}$$

The volume integration can be transformed by applying Gauss' theorem (see Sec. C-3 in App. C) giving:

$$\int_E \int_A \mathbf{n} \cdot \phi^* D \nabla \phi \, dA \, dE = \int_E \int_A \mathbf{n} \cdot \phi D \nabla \phi^* \, dA \, dE \quad . \tag{B.41}$$

The diffusion theory boundary conditions relate fluxes and currents linearly:

$$b\phi = -\mathbf{n} \cdot D \nabla \phi \tag{B.42a}$$

and

$$b\phi^* = -\mathbf{n} \cdot D \nabla \phi^* \quad , \tag{B.42b}$$

with the same b. Inserting the boundary conditions in Eq. (B.41) shows that both sides are equal. Thus, Eqs. (B.36) represent the desired adjoint operators of the neutronics problem and Eq. (B.42b) gives the boundary condition for the adjoint problem.

The physical interpretation of the adjoint angular flux is derived from placing a neutron at **r** of energy E and direction of motion $\mathbf{\Omega}$ in a critical reactor, and observing the resulting asymptotic rise in the flux level. This flux level rise is proportional to the adjoint angular flux, $\phi^*(\mathbf{r},E,\mathbf{\Omega})$. Similarly, the interpretation of the adjoint flux is obtained when neutrons with the angular distribution at the particular point **r** are inserted in a critical reactor. The flux rise is proportional to $\phi^*(\mathbf{r},E)$. See Ref. 1 for a detailed discussion of the interpretation of adjoint fluxes and Sec. 7-6 for a derivation of this interpretation from microkinetics.

B-5 The Leakage Term in Perturbation Theory

The derivation of the perturbation theory formulas in Chapter 4 requires the consideration of the differences in the neutronics operators of *two* problems: perturbed and the unperturbed. These differences can be formed readily for the terms of the neutronics operators that are proportional to macroscopic cross sections. The leakage term, however, requires special considerations. Let **L** denote the leakage part of **M**, i.e.,

$$\mathbf{L}\Phi = -\nabla \cdot D\nabla\Phi \tag{B.43a}$$

and

$$\mathbf{L}_0\Phi = -\nabla \cdot D_0\nabla\Phi \quad . \tag{B.43b}$$

The difference of the leakage contribution appears in the form [see Eqs. (4.36) and (4.39)]:

$$(\Phi_0^*,\Delta\mathbf{M}\Phi) = (\Phi_0^*,[\mathbf{M} - \mathbf{M}_0]\Phi) \quad , \tag{B.44}$$

which is obtained by a formal derivation from which the explicit formulas can be obtained for the cross-section terms. However, special considerations are required for the leakage terms presented here.

The difference contained in Eq. (B.44) appears originally in Eqs. (4.38) as

$$\Delta_M = (\Phi_0^*,\mathbf{M}\Phi) - (\Phi,\mathbf{M}_0^*\Phi_0^*) \quad . \tag{B.45}$$

With **L** being the leakage part of **M**, the respective component of Δ_M is given by:

$$\begin{aligned}\Delta_L &= (\Phi_0^*,\mathbf{L}\Phi) - (\Phi,\mathbf{L}_0^*\Phi_0^*) \\ &= -(\Phi_0^*,\nabla \cdot D\nabla\Phi) + (\Phi,\nabla \cdot D_0\nabla\Phi_0^*) \quad .\end{aligned} \tag{B.46}$$

Applying the vector relation, Eq. (B.38), as above gives

$$\Delta_L = \int_E\int_V(-\nabla\cdot\phi_0^*\,D\nabla\phi + D\nabla\phi_0^*\cdot\nabla\phi$$

$$+\nabla\cdot\phi D_0\nabla\phi_0^* - D_0\nabla\phi\cdot\nabla\phi_0^*)\,dV\,dE$$

$$= \int_E\int_V(-\nabla\cdot\phi_0^*D\nabla\phi + \nabla\cdot\phi D_0\nabla\phi_0^*)\,dV\,dE$$

$$+ \int_E\int_V(D - D_0)\nabla\phi_0^*\cdot\nabla\phi\,dV\,dE \quad . \tag{B.47}$$

The first integral on the right side of Eq. (B.47) can be converted as in Eq. (B.41) to an integral over the system boundary, where the boundary conditions, Eqs. (B.42), can be inserted, giving δ_L:

$$\delta_L = -\int_E\int_V \mathbf{n}\cdot(\phi_0^*D\nabla\phi - \phi D_0\nabla\phi_0^*)\,dA\,dE$$

$$= \int_E\int_A(b\phi_0^*\phi - b_0\phi_0^*\phi)\,dA\,dE \quad , \tag{B.48}$$

with b values that may be different for the perturbed and unperturbed problem. Thus, the two terms in Eq. (B.48) are not necessarily equal as in Eq. (B.41) with Eqs. (B.42) as boundary conditions. However, b may be equal to b_0 if the diffusion constant near the outer boundary and thus the extrapolation distance is unperturbed. Often, the boundary condition with the extrapolation distance is replaced by a zero flux boundary condition at an extrapolated boundary. Then $\phi = 0$ at A, and δ_L disappears.

Even if δ_L is not equal to zero for the stated reasons, it is normally negligibly small because the product $\phi_0^*\phi$ is very small at the outer boundary.

With δ_L being zero or negligibly small, Δ_L is given by

$$\Delta_L = \int_E\int_V(D - D_0)\nabla\phi_0^*\cdot\nabla\phi\,dV\,dE \quad . \tag{B.49}$$

Thus, the leakage component of the scalar product, Eq. (B.44), is explicitly given by Eq. (B.49):

$$(\Phi_0^*,\Delta\mathbf{L}\Phi) = \int_E\int_V(D - D_0)\nabla\phi_0^*\cdot\nabla\phi\,dV\,dE \quad . \tag{B.50}$$

REFERENCES

1. G. I. Bell and S. Glasstone, *Nuclear Reactor Theory,* Van Nostrand Reinhold Co., New York (1970).
2. A. F. Henry, *Nuclear Reactor Analysis,* The MIT Press, Cambridge, Massachusetts (1975).
3. H. Greenspan, C. N. Kelber, and D. Okrent, Eds., *Computing Methods in Reactor Physics,* Gordon and Breach Science Publishers, New York (1968).

APPENDIX C

Mathematical Formulas

C-1 Solutions to First-Order Linear Differential Equations

Let

$$\frac{dp(t)}{dt} + a_1(t)p(t) = a_2(t) \tag{C.1}$$

be a linear first-order differential equation. Its general solution (for an unspecified initial condition) is the following:

$$p(t) = \exp\left[-\int^t a_1(t')\,dt'\right] \times \left\{C + \int^t a_2(t')\exp\left[\int^{t'} a_1(t'')\,dt''\right] dt'\right\} \quad . \tag{C.2}$$

Specifically, if $p(t)$ is given at $t = 0$,

$$p(0) = p_0 \quad , \tag{C.3}$$

then Eq. (C.2) yields:

$$p(t) = \exp\left[-\int_0^t a_1(t')\,dt'\right] \times \left\{p_0 + \int_0^t a_2(t')\exp\left[\int_0^{t'} a_1(t'')\,dt''\right] dt'\right\} \tag{C.4}$$

or

$$p(t) = p_0 \exp\left[-\int_0^t a_1(t')\,dt'\right] + \int_0^t a_2(t')\exp\left[\int_t^{t'} a_1(t'')\,dt''\right] dt' \quad . \tag{C.5}$$

If a_1 is constant, one has simply:

$$p(t) = p_0 \exp(-a_1 t) + \int_0^t a_2(t') \exp[a_1(t' - t)]\, dt' \quad . \tag{C.6}$$

C-2 The Factorial and the Γ Function

An important definite integral in nuclear reactor statics and kinetics is

$$\int_0^\infty x^a e^{-x}\, dx = a! = \Gamma(a + 1) \quad , \tag{C.7}$$

where the Γ function is the continuous "interpolation" of the factorial, $n!$. In most applications, a is an integer, n. Then the well-known factorial $n!$ gives the values for the integrals of Eq. (C.7). In addition, integrals for half-integer a values are also used in this text, for which the factorial is less well known:

$$\int_0^\infty \sqrt{x}\, e^{-x}\, dx = \left(\frac{1}{2}\right)! = \frac{1}{2}\sqrt{\pi} \tag{C.8}$$

and

$$\int_0^\infty \frac{1}{\sqrt{x}} e^{-x}\, dx = \left(-\frac{1}{2}\right)! = \sqrt{\pi} \quad . \tag{C.9}$$

The recursion formula that defines the factorial for all a, gives the factorial between $a = 0$ and $a = 1$, that is,

$$a! = a(a - 1)! \quad , \tag{C.10}$$

allows us to obtain the other factorials for half-integer values $a = n + 1/2$ with n being an integer; for example, for $a = 1/2$ and $3/2$, one obtains

$$\left(\frac{1}{2}\right)! = \frac{1}{2}\left(-\frac{1}{2}\right)! \quad , \text{ thus } \left(-\frac{1}{2}\right)! = 2\left(\frac{1}{2}\right)! \quad , \tag{C.11}$$

$$\left(\frac{3}{2}\right)! = \frac{3}{2}\left(\frac{1}{2}\right)! \quad , \tag{C.12}$$

etc.

C-3 Gauss' Theorem

Gauss' theorem states that the volume integral of the divergence of a vector equals the surface integral of the outward pointing component of the vector itself. Let $\mathbf{w}(\mathbf{r})$ be a vector field defined in the volume V, with a surface A. In the notation normally applied in nuclear engineering, "div" is written for the vector gradient, using either dot product or scalar product notation:

$$\operatorname{div} \mathbf{w} = \nabla \cdot \mathbf{w} = (\nabla, \mathbf{w}) \quad . \tag{C.13}$$

Gauss' theorem is written as:

$$\int_V \nabla \cdot \mathbf{w} \, dV = \int_A \mathbf{n} \cdot \mathbf{w} \, dA \quad , \tag{C.14}$$

where $\mathbf{n}$ is a unit vector normal to A, pointing outward. Its dot product with $\mathbf{w}$ gives a component of $\mathbf{w}$ in the direction of $\mathbf{n}$.

Gauss' theorem is applied here for various vectors $\mathbf{w}$; e.g., $\mathbf{w} = D\nabla\phi$, giving

$$\int_V \nabla \cdot D\nabla\phi \, dV = \int_A \mathbf{n} \cdot D\nabla\phi \, dA \quad . \tag{C.15}$$

APPENDIX D

The δ Function

Dirac introduced the δ function in the mid 1920s for application in quantum mechanics. It turned out to be a very powerful concept, now widely applied in theoretical and applied science fields. The δ function is not a function in the mathematical definition of this concept, it is merely defined by the properties of its integral, i.e., it is a "generalized function," a "distribution." Nevertheless, the historical name, δ function, is preferred to the correct name "δ distribution."

The standard definition of Dirac's δ function as it appears in mathematical textbooks is:

$$\delta(x - x') = 0 \text{ for } x' \neq x \tag{D.1}$$

and

$$\int_a^b f(x')\delta(x - x')\,dx' = \begin{cases} f(x) & \text{for } x \in (a,b) \\ 0 & \text{for } x \notin [a,b] \end{cases} , \tag{D.2}$$

where (a,b) is the open interval

$$a < x < b$$

and the closed interval $[a,b]$ includes the boundaries

$$a \leq x \leq b \quad .$$

The function $f(x)$ must be continuous in (a,b).

For $f(x) = 1$ in particular, the following relationship holds:

$$\int_a^b \delta(x - x')\,dx' = \begin{cases} 1 & \text{for } x \in (a,b) \\ 0 & \text{for } x \notin [a,b] \end{cases} . \tag{D.3}$$

The δ function singles out the value of a function upon which it acts at any particular point x. This is sometimes called the "sifting property" of the δ function.

The δ function is an *even* "generalized function," i.e., a change of the sign in its argument does not change the results; thus

$$\int_a^b f(x')\delta(x' - x)\,dx' = \int_a^b f(x')\delta(x - x')\,dx' \quad . \tag{D.4}$$

In the same fashion, derivatives of the δ function are defined; e.g., $\delta'(x - x')$:

$$\int_a^b f(x')\delta'(x - x')\, dx' = f'(x) \quad , \tag{D.5}$$

with $f'(x)$ being the derivative of f at a point x within the same interval as above. Then $f(x)$ must be differentiable in the interval.

INDEX